ON THE TOWN IN NEW YORK

ON THE TOWN IN NEW YORK

The Landmark History of Eating, Drinking, and Entertainments
from the American Revolution to the Food Revolution

Michael and Ariane
BATTERBERRY

ROUTLEDGE

New York and London

Published in 1999 by
Routledge
29 West 35th Street
New York, NY 10001

Published in Great Britain in 1999 by
Routledge
11 New Fetter Lane
London EC4P 4EE

Printed in the United States of America on acid-free paper
Typography updates: Jack Donner

10 9 8 7 6 5 4 3 2 1

Library of Congress Cataloging-in-Publication Data

Batterberry, Michael and Ariane
On the town in New York : the landmark history of eating, drinking,
and entertainments from the American Revolution to the food revolution /
by Michael and Ariane Batterberry.
p.   cm.

Originally published: New York: Scribner, 1973.
Includes bibliographical references and index.

ISBN 0–415–92020–5

1. Restaurants—New York (State)—New York—History. 2. Hotels-New York
(State)—New York—History. 3. Amusements—New York (State)—New York—History.
4. New York (N.Y.)—Social life and customs. I. Batterberry, Ariane Ruskin. II. Title.
TX909.B32   1998
647.95747'1'09—dc21   98–23443
CIP

For our parents,
who first taught us to
love New York

Table of Contents

———❦———

Acknowledgments

For the 1998 anniversary edition, our thanks go to Melissa Rosati, Routledge's unflaggingly helpful and attentive Publishing Director, Food and Culture; her tireless assistant Eric Nelson; and the members of the Routledge staff who have worked so effectively to make this updated version a seamless reality. For their hospitality and support, we wish to express gratitude to Director Betsy Gottbaum and the New York Historical Society. We also thank *Food Arts* magazine Publisher M. Shanken Communications, and most especially the photographers Courtney Grant Winston, Brooks Walsh, Paul Warchol; and thanks to Laurence D. Harvey, Executive Director of Catering, The Plaza Hotel, for his photographs. Needless to say, no amount of applause would repay our debt to the many talented and resilient friends who have made New York's restaurant and hotel industries the gold standard of the world.

For the first edition of *On The Town in New York* published in 1973, we again wish to thank Mr. Albert K. Baragwanath, Senior Curator of the Museum of the City of New York, for his invaluable advice regarding points of historical accuracy in our manuscript. We also wish to thank the following for their generous aid: Miss Grace Meyer of the Museum of Modern Art, and the staffs of the New York Historical Society Library and Print Room, the New York Public Library Print Room and Picture Collection, the Sons of the Revolution, and the Sleepy Hollow Restorations, as well as Mr. James E. Brodhead III, Mr. Jordan Mayro of the Burlington Book Shop, The Picture Decorator, Inc., and Mrs. James Thurber.

ON THE TOWN IN NEW YORK

Chapter I

1776-1800

Little Birds are choking   Taught, I say, to splinter
Baronets with bun,   Salmon in the winter——
Taught to fire a gun: Merely for the fun.

Lewis Carroll

There is nothing which has yet been contrived by man, by which so much happiness is produced, as by a good tavern or inn.
SAMUEL JOHNSON

American history books do little to describe the plight of New York City during the Revolution for the simple reason that it was an enemy base. Washington prepared to defend Manhattan early in January of 1776, and within a month fully a quarter of the panicked residents had fled. By August, Staten Island and New York harbor were in the clutches of a monster army of thirty-two thousand redcoats under the command of two brothers, Admiral the Right Honorable Richard, Lord Vincent Howe, of the Kingdom of Ireland, more commonly known as "Black Dick," and General William Howe, who, but for his dark eyes, so closely resembled Washington that he could be mistaken for him at a distance. Washington's eighteen thousand inadequately trained troops were no match for this, the largest expeditionary force in British history, and in September the British were able to land unimpeded at Kip's Bay as the last of the revolutionaries retreated to Harlem and then to Westchester. For the duration of the war such patriots as there were in the vicinity lurked in the north, descending into New York for an occasional raid.

The British marched into a ghost town of about four thousand buildings, a thousand of which were swept away within a month by a mysterious fire. But soon the city took on the character of a wartime capital as the deserted houses became crammed with soldiers, officers and their families, and refugee Tories of all classes from all thirteen colonies.

Meanwhile captured revolutionaries, the only men of patriotic persuasion on Manhattan, were crushed into squalid jails or rotting prison ships anchored in the East River. New York settled down to seven years of something like the life of an eighteenth-century siege town, with the horrors of crowded quarters, black markets, disease, and cold, offset by the glamor of military assemblies, dress uniforms, and the general suspense of war.

The Tory tavernkeepers, those who had remained behind or emigrated to the city, made a fortune serving the heavy-drinking English. And it is important to remember that taverns were the local hotels, restaurants, and meeting houses of the eighteenth century. As competition became heated, a flurry of freshly painted shingles went up throughout the city, bearing such staunchly partisan legends as The Sign of Lord Cornwallis, The King's Arms, and The Prince of Wales. More impressed by the boom than by protestations of loyalty, the authorities proclaimed in Rivington's *Royal Gazette* on January 8, 1780, that "many evils daily arise from unlimited numbers of Taverns and Publick Houses within the City and its Precincts" and restricted licenses to two hundred, any of which would be immediately withdrawn "from such as shall be known to harbor any riotous or disorderly companies."

The more astute tavernkeepers, far from soliciting the company of the riotous and disorderly, posted advertisements for such entertainments as would theoretically appeal to tasteful young bloods straight from London.

John Kenzie, proprietor of Ranelagh Gardens, formerly of the Mason's Arms, offers dinners on the shortest notice, breakfast, relishes, etc. all the forenoon; tea, coffee, etc., all the afternoon; a band of music to attend every Saturday evening . . . the best wines that can possibly be had in this city . . . and being a veteran in this Most Gracious Majesty's service, he shall hope for the Smiles, Protection, and Encouragement of the gentlemen of the army and navy as well as the Respectable Public. N.B. In the superb Garden there is the most elegant boxes [sic] prepared for the reception of the Ladies, and the most perfect enjoyment of the evening Air.

Precisely how superb Mr. Kenzie's Garden was is a matter for conjecture—almost every tree on Manhattan Island had long since been chopped down for firewood.

Despite the detached frivolity of New York wartime society, privation in the occupied city was appalling. For two winters the Hudson actually froze over. Shantytowns sprang up in the burnt-out sections, but even they could not house the destitute. Many of the Royalist refugees escaped to New York virtually penniless; the tavernkeepers rallied to their aid by organizing lotteries and playing host to theatrical benefits performed by British officers. No aid, however, was given to the American prisoners of war, who were freed in 1777 for lack of rations. Many were so debilitated by malnutrition that they fell dead in the streets of New York before they could reach the vessels that were to transport them to New Jersey. At one point potatoes sold for a guinea a bushel, and biscuits of oatmeal containing as little nourishment as ground straw were served to the British troops. Only the very rich could afford the black-market prices of produce brought in from outlying farms.

Taverns had food to offer at a healthy price, and their supply of drink was frequently and lavishly replenished by privateers and Tory traders. But what there was could not have been prepared in the style most British demanded, for Mr. Ephraim Smith "of London" announced that he would reopen his tavern in Water Street "at the desire of many gentlemen of the Royal Army and Navy to be a Steak and Chop House in the London Stile, much wanted in the City. The best of wines, punch, and draft porter, with steaks, chops and cutlets will be served every day from one o'clock til four." For some enigmatic reason Smith's good intentions mush have backfired. His landlord advertised within the year: "House to rent—*No Tavern keeper need apply.*"

The senior sycophant of the Tory tavernkeepers, a Mr. Looseley, was proprietor of the King's Head tavern in Brooklyn Heights. A fatuous, fawning egomaniac, with unquenchable aplomb, he published his own *Royalist Gazette,* a conglomeration of loyalist editorials and advertisements for his tavern, in which he characteristically referred to himself in the third person. Looseley would tolerate no nonsense on the part of Washington and his men. On September 20, 1780, he proclaimed in his paper: "The anniversary of the Coronation of our ever good and gracious King will be celebrated at Looseley's, 22nd instant. It is expected that no rebel will approach nearer than Flatbush Wood." When prose could no longer contain his fervor, he expressed himself in verse:

On June 20 a dinner exactly British, after which there is no doubt but that the song of "Oh! the Roast Beef of Olde England" will be sung with harmony and glee.

> *This notice gives to all who covet*
> *Baiting the bull and dearly love it,*
> *Tomorrow's very afternoon*
> *At three or rather not so soon*
> *A bull of magnitude and spirit*
> *Will dare the dog's presuming merit.*
>
> *Taurus is steel to the backbone*
> *And canine cunning does disown*
> *True British blood runs through his veins*
> *And barking number he disdains*
> *Sooner than knavish dogs shall rule*
> *He'll prove himself a true John Bull.*

On May 29, 1782, he wrote: "On this Day firm reestablishment being given to Monarchy—Tyranny and its Republic are no more. Looseley the Boniface of Brooklyn Hall, will produce a dinner at three o'clock fit for a conquering Sovereign." An exceptional invitation considering the fact that Cornwallis had surrendered at Yorktown seven months before. Reality caught up with Looseley soon enough. On November 23, 1782, the following advertisement appeared, not in the *Royalist Gazette:* "Auction Sale at Looseley's Inn, Brooklyn Ferry. Paintings, Pictures, Pier Glasses, an organ, billiard table, twenty globe lamps, flag staff, etc. The Landlord intends for Nova Scotia immediately."

As the Peace of Paris was negotiated, many a Tory tavernkeeper joined Looseley in his flight north, while still others were "intending for England" or "intending for the Indies." The Boniface of Brooklyn Hall, in fact, dragged out his last days as the proprietor of an unsuccessful hotel at Port Roseway.

In May of 1783 Washington and Sir Guy Carlton made the final arrangements for the evacuation of the British Army and safe conduct for any remaining Tories. On November 25 the triumphant Americans marched down Broadway.

New Yorkers, who have always hated missing a party, rushed back to their ravaged city for a week-long victory celebration. Sam

Sam Fraunces. *Courtesy of the Sons of the Revolution.*

Fraunces was asked by Governor Clinton to organize the first public dinner for Washington immediately following the march.

Fraunces, as Sam Francis styled himself after the Revolution, was a mulatto born in the French West Indies, and had served as proprietor of a New York tavern, the Mason's Arms, before 1763. In that year he became master of the establishment that made him famous, known before the war as the Queen's Head and situated in what had been the DeLancey mansion on the corner of Broad and Pearl streets. A dapper figure, he was, like most noted professional hosts, a connoisseur, an extrovert, and an autocrat. Above all, he was a sworn revolutionary. The Revolution itself was fomented in the urban "public houses" of the colonies, and it was at the Queen's Head that the Sons of Liberty and the Vigilance Committee met in 1774 to protest the landing of British tea and to lay plans for dumping it.

According to a popular Revolutionary poem by Philip Freneau, new York's foremost tavernkeeper had been among the first in the city to taste the wrath of the British:

> *Scarce a broadside was ended*
> *'til another began again—*
> *By Jove! It was nothing but*
> *Fire away Flanagan!*
> *At first we supposed it*
> *was only a sham,*
> *'Til he drove a round-shot*
> *thru the roof of Black Sam.*
> *The town by their flashes*
> *was fairly enlightened*
> *The women miscarry'd, the beaus*
> *were all frightened. . . .*

Fraunces was kept a virtual prisoner on his own premises by the British during their occupation of New York, but managed to present a façade of such unruffled geniality that he was able to continue his anti-Royalist activities. For secretly aiding prisoners of war he was later awarded two thousand dollars by Congress "in consideration for singular services" and was appointed Steward to the President by Washington himself.

There is a tale that Frances' daughter, Phoebe, had afforded her country services no less singular than her father's. Unlucky Phoebe was serving as housekeeper at Washington's Richmond

Hill headquarters when she fell in love with a young soldier named Thomas Hickey, a member of the commander's Personal Guard. Just before lunch one day she came upon her lover sprinkling poison over a dish of spring peas, a particular favorite of Washington's. Sworn to silence, Phoebe did as she was ordered and bore the dish to the dining table. But at the crucial moment honor prevailed and she snatched the lethal peas from beneath the general's nose and pitched them out the window. There, in the garden, they were pecked at by a neighbor's chickens, every one of which promptly fell dead, and the notorious "Hickey Plot" was exposed. Sam Fraunces' pride in his daughter's heroism must have been cold comfort to Phoebe when on June 28, 1776, Thomas Hickey was hanged at the corner of Bowery Lane and Grand streets, the first man to be executed for treason by the United States Army.

It could be said that in many ways the triumphal dinner organized by Fraunces in honor of General Washington was a climactic one. Paneled walls and pewter gleamed in the firelight and waiters in long white aprons scurried between the trestled tables refilling glasses and tankards as the company rose to the toast, "May a close Union of the States guard the Temple they have erected to Liberty." But the truly death-defying revel of the week took place on December 2, and was given, not surprisingly, by the French ambassador. His one hundred and twenty guests succeeded in dispatching one hundred and thirty-five bottles of Madeira, thirty-six bottles of port, sixty bottles of English beer, thirty bowls of punch, sixty wineglasses, and eight glass decanters. No further festivities were recorded for the next forty-eight hours.

On the morning of December 4, Washington's officers were told that their commander would leave the city for Mount Vernon that same day. At noon they hurried through the rain to Fraunces' Tavern, where Washington had arranged to meet them in the Long Room. After proposing a toast to his men, Washington embraced each of them in turn. Among those present were Generals Knox, Steuben, Schuyler, Gates, Putnam, Kosciuszko, and Hamilton, Governor Clinton, Major Fish, and Colonel Tallmadge, in whose *Memories* is an eye-witness report of the meeting:

Such a scene of sorrow and weeping I have never before witnessed, and I hope may never be called upon to witness again. . . .

But the time of separation had come, and waving his hand to his griev-

ing children around him, he left the room and passing through a corps of
light infantry, who were paraded to receive him, he walked silently on
to Whitehall, where a barge was in waiting.

We all followed in mournful silence to the wharf, where a prodigious
crowd had assembled to witness the departure of the man who, under
God, had been the great agent in establishing the glory and indepen-
dence of these United States.

As soon as he was seated, the barge put off into the river, and when out
in the stream, our great and beloved General waved his hat, and bid us
a silent adieu.

In 1784, New York City became the temporary capital of the
country. Its ten thousand residents were apparently not discouraged by the shambles that lay about them and immediately
launched an energetic program for general reconstruction.
Within the next four years the population doubled and the rush
was on. Boom towns always attract a surrealistic swarm of immigrants from within their own country and abroad. Western
Europe was politically restless, and the ranks of local merchants,
financiers, and speculators were promptly invaded by their foreign counterparts. The Chamber of Commerce and the New York
Insurance Company reopened their doors; the first independent
bank was founded, and customs were drastically reorganized. A
temporary crisis arose when Caribbean trade was cut off not only
by Britain but by France and Spain as well. Undaunted, New
Yorkers dispatched the *Empress of China* to Canton and welcomed her back a year later laden with tea and silk, signaling the
start of the great China Trade.

The entire population of New York in 1784 could have gone to
work daily in today's Pan-American Building. The city boasted
four thousand houses set on about one square mile of land below
Chambers Street, and business was conducted in a leisurely fashion at home or in a local tavern. But by 1800 the population had
taken a leap to sixty thousand, and the city's boundaries expanded
to Worth Street on Broadway, Harrison Street on the North River,
and Rutgers Street on the East River. Artists and engravers of the
period depicted New York as a sort of citified New England port
town, serene and immaculate—a wonderland of green gardens,
schooner masts, and graceful colonial architecture. But diarists
saw it for what it was, a confusing paradox of luxury and squalor.

In writing his memoirs in the nineteenth century, one local
resident recalled: "N. Y. was a dull and dirty town in 1789. It was

a city without a bath, without a furnace, with bedrooms which in winter lay within the Arctic zone, with no ice during the torrid summers, without an omnibus, without a moustache, without a match, without a latch key." In May 1788, the Grand Jury had reported the streets to be dirty and many of them impassable. Pigs were the only scavengers. All wood delivered to stores or houses (there was no coal) was sawed and split on the street after delivery. Street lamps had been introduced in 1762; but they were few and poor, apt to go out, and often left unlighted. The city water works consisted chiefly of the Tea Water Pump on Chatham Street. The water came from the Collect Pond, where the public washed its dirty linen. Highway robbery was common. This last bit of nostalgia is corroborated by a 1797 newspaper in which it was noted that the city watchmen had in September made a "previous haul" of twenty-one vagrants, and the public was cautioned to keep off the streets after nine o'clock as the city had become infested with robbers and pickpockets.

Fires were frequent and yellow fever epidemics an annual menace. The state of public works was such as to arouse this typical complaint: "The streets from between the Friends' Meeting and the head of Queen's Street, particularly Rutgers St., descend into the swamp on both sides and are very dangerous; the waters now lying in the center of the street have actually swallowed several children, and a number of grown persons have also been swamped in this pond of a dark night, as to be obliged to call for aid." The victim who survived a plunge into the depths could presumably restore himself at an establishment on Crown Street which advertised warm baths on a half hour's notice.

Despite its infinite variety of discomforts, New York was remarkable for its love of luxury, even in the eyes of a Frenchman. Brissot de Warville, journalist and traveler later guillotined in The Terror, reported, in his *New Travels in the U. S. A.*:

The presence of Congress with the diplomatic body and the concourse of strangers, contribute much to extend here the ravages of luxury. The inhabitants are far from complaining of it. They prefer the splendor of wealth and the show of enjoyment to the simplicity of manners and the pure pleasures resulting from it. If there is a town on the American continent where the English luxury displays its fallacies, it is New York. You will find here the English fashions. In the dress of the women you will see the most brilliant silks, gauzes, hats, and borrowed hair. Equipages are rare; but they are elegant. The men have more simplicity in their dress, they disdain gew-gaws, but they take their revenge in the *luxury of the table*.

Americans were excellent farmers and hunters, their soil was rich, and game prolific. Small wonder Continental visitors confused "quantity" with "luxury." New Yorkers thrived on a high protein diet. Beef and mutton were available the year round, and wild fowl and venison in season. Spring yielded a glut of veal and shad, followed in summer by herring, bass, weakfish, and turtle. In 1724 seven hundred thousand bass were taken from a single pond in New Jersey, loading fifty carts, a thousand horses, and several boats. Salt fish, principally cod, augmented winter menus. Oysters were so popular that even before 1776 it was a common sight to see at least two hundred heavily laden canoes returning to port each evening from the Bergen shore. By 1792 these beds had grown poor, and oystermen had to explore the more distant waters of Long Island and New Jersey. An exotic touch was frequently provided by citrus fruits and pineapples brought by schooner from the Indies.

Although their economy was shaky, New Yorkers persisted in overeating to such an extent that John Thurman, a local businessman, was moved to lament, "Many of our new merchants and shopkeepers set up since the war have failed. We have nothing but complaints and hard times. Labor, prices, and rents are all high. Feasting and every kind of extravagance go on—reconcile these things if you can. Gloomy joys."

By contemporary culinary standards, it would be easy enough to equate "feasting and every kind of extravagance" with "gloomy joys." Variety and imagination were foreign to post-Revolutionary tavern kitchens. Any romantic notions one might have are rapidly dispelled when confronted with an authentic breakfast menu of the period: "Cold meat with a pint of good ale or cyder." But for the early American traveler, anything would have been an enticement to sprint out of bed and down the stairs with all possible speed. In the first place, he could rarely find such a thing as a private bedroom or even a private bed (although the authorities did frown upon excesses—three to a berth was the legal limit). Men and women slept in segregated dormitories, as sparely furnished as an army barracks, with several beds (occasionally collapsible), a chest of drawers, and a washbowl and pitcher which in winter months were perpetually ice-bound. Women with children fared somewhat better, and if there was a bedroom fireplace, a looking glass, a bedwarmer, or an upholstered chair, it fell to them. For younger children, two large slatted chairs, ingeniously designed, could be interlocked to form a crib. Cradles were supplied for babies.

Shortly after midday, overnight guests would be joined by the solid businessmen of the city, who filed into their favorite taverns for dinner. This was the heartiest meal of the day, the last being an early evening supper much like an English "high tea." There were no night dives, late dinners, or sandwich lunches except for travelers actually en route. At dinnertime (the hour grew later as foreign ideas became popular), the public rooms filled with pipe smoke, and sizzling roasts were removed from the spit and served with several kinds of relishes, cheese, bread, butter and preserves, as well as a variety of pies. The menu for supper was essentially the same, except that the meat was cold.

Most taverns provided diners with a commons room for private parties, usually at one or two tables, where women could be served, and a taproom with a proper bar and communal tables, where women could absolutely not be served. Salvers, chargers, plates, jugs, and tankards were of indestructible pewter or heavy pottery; utensils included bone-handled knives, two-tined or three-tined forks, and spoons. Gadgets and gimmicks appeared: a hungry late arrival could sup in his room off a hollow pewter hot plate, its cavity filled with boiling water, and the Puzzle Jug was a forefather of the modern gag shop's Dribble Glass. Uninhibited Hogarth and Rowlandson prints peppered the taproom walls and in the fireplace there frequently stood a wrought-iron contraption on which long-stemmed clay pipes, provided by the management, were regularly sterilized. Leather fire buckets, filled with water or sand, hung by the chimneys and in the hall over the registry desk.

The actual kitchen or cookhouse was a purist's joy. Under a hooded chimney flickered a huge wood fire, its heat reflected by a sheet of metal shaped like a fanciful tombstone; spits, pots on chains, and a coffee roaster resembling a cylindrical corn popper were adjusted over the flames. Embers were shoveled into perforated copper pot warmers on which kettles of water were kept hot. A beehive oven, built into the wall, served for baking.

Not all tavern cooking was so rudimentary. There were a few tavernkeepers who, like Fraunces, offered an escape from the general monotony. Old Tom's specialized in venison, bear meat, and homemade cherry bounce. In 1787, Nicholas LaFargue opened a "Boarding and Lodging House" where he cooked in both French and English modes. "Dishes sent on three hours notice anywhere in the city. Alamode beef cooked and dressed to keep good for four months even in the West Indies." The idea of preparing dishes for either home consumption or travel caught on quickly. In the October 21, 1788 *Daily Advertiser*, Thomas Rattone announced "oysters cooked in various manners, and served up with the greatest expedition. Oyster suppers to send to gentlemen's houses when called upon. Also Oysters and Lobsters pickled in the best manner for exportation."

Many taverns made up johnnycake for transient guests; it was durable and could be conveniently carried in their saddlebags (hence its original name, "journey cake"). That no cook was particularly renowned for its preparation may be explained by the following standard recipe:

To make a pleasing Johnny-Cake

Take yr. 4 parts corn meal well ground, 4 parts wheat flour, 4 parts cow's milk, four eggs, 1 part hot lard and stir until well mixed. Set in oven to Bake, Cool, Cut, and Reserve.

A relish, on the other hand, frequently made the reputation of a kitchen. All sorts were popular and most were concocted from secret and comparatively complicated recipes. In the first of this country's published cookbooks, *American Cookery, by Amelia Simmons, an American Orphan,* the preparation of a favorite relish is given:

To pickle or make Mangoes of Melons

Take green melons, as many as you please and make a brine strong enough to bear an egg; then pour it boiling hot on the melons, keeping them down under the brine; let them stand 5 or 6 days; then take them out, slit them down on one side, take out all the seeds, scrape them well in the inside, and wash them clean with cold water; then take a clove of garlick, a little ginger and nutmeg sliced, and a little whole pepper; put all these proportionably into the melons, filling them up with mustard seeds; then lay them in an earthen pot with the slit upwards, and take one part of mustard and two parts of vinegar, enough to cover them, pouring upon them scalding hot, and keep them close stopped.

On the whole, a Yankee was more fastidious about what he drank than what he ate. Americans, and New Yorkers in particular, drank heavily from earliest times, and before the Revolution there was a tavern for every forty-five residents (man, woman, and child) in the city. The British tradition, whereby Queen Elizabeth's Maids of Honor were issued two gallons of ale each day for breakfast, seems to have been maintained, and there are frequent accounts of women, and even children, drinking themselves to death, although not at taverns—in 1764 Margaret Jones of Manhattan, for example, "drank too freely of spirituous liquors, fell from the Main Deck into the hold of the Coventry man of war . . . and was killed."

The New Yorker preferred his wine fortified, and consumed port, Madeira, and sherry in unstinting quantities, regardless of price. His respect for good brandy was such that he willingly parted with ten shillings for a bottle as opposed to one shilling for a night's lodging. Measures of English gin were poured for him from dark brown ceramic bottles, Indies rum from pale two-gallon jugs in the shape of funerary urns. Spirits were often served in "firing glasses," so called because the racket produced by pounding their heavy bases on a table following a toast could be mistaken for an artillery barrage. What light wines were consumed appeared in squat dark bottles similar to the ones used today for Chilean Riesling. Beer, ale, and cider were served in pewter measuring pitchers and jugs. Drinking was permitted only at the tables, never at the bar, which for security purposes was inevitably fitted out with a latticed grill that could be pulled down to prevent damage to the stock during a brawl or pilferage after closing hours.

While the local beer, ale, and cider were considered excellent, imported porter was so widely appreciated that John Morgan opened a large porter room at the Sign of the Grand Master that sold only "the best London porter brought by Captain Corpar on *The Favorite.*" Prepared drinks, with the exception of mint juleps and cherry bounce concoctions, ran to the rich and heavy. Martin Doyle opened a "Purl House at the Sign of the Faithful Irishman and Jolly Sailors where may be had at any minute in the morning, if you please, a glass of excellent Purl piping hot." Not surprisingly, purl, a medicated malt liquor in which wormwood and aromatics were combined, is no longer in demand; nor its cousin, posset, a similar emetic to which milk was added. Mixed punches and "sangaree" were available the year round; grogs, eggnogs, and syllabubs were cold-weather specialties. One recipe for the last is straightforward:

To make a Fine Syllabub from the Cow

Sweeten a quart of cyder with double-refined sugar, grate nutmeg into it, then milk your cow into your liquor, when you have thus added what quantity of milk you think proper, pour half a pint or more in proportion to the quantity of syllabub you make, of the sweetest cream you can get all over it.

An account of an evening's tippling in New York is unexpectedly included in the bible of French gastronomy, Brillat-Savarin's *Physiologie du Goût.* Many eminent Frenchmen who fled the guillotine arrived penniless in the New World, and were forced to live by whatsoever wits or talents they possessed. Anthelme Brillat-Savarin, gastronomer, political philosopher, one-time Mayor of Belley, and cousin of Madame Recamier, could be heard fiddling away every night in the pit of a New York theater. After the performance he frequently spent the evening at a "sort of cafe-tavern of a Mr. Little, where you could get a bowl of turtle soup or a supper of welsh rarebit washed down with ale or cider."

It was there, at Michael Little's Coffee House, that Brillat-Savarin and two fellow *émigrés,* the Vicomte de la Massue and Jean Rodolphe Fehr, became acquainted with a Mr. Wilkinson, an Englishman from Jamaica, and his friend, whose name Brillat-Savarin never discovered, but whom he described as being "a very taciturn man, with a square face, keen eyes, and features as expressionless as those of a blind man, who appeared to notice everything but never spoke; only, when he heard a witty remark or merry joke, his face would expand, his eyes close, and opening a mouth as large as the bell of a trumpet, he would send forth a sound between a laugh and a howl, called by the English 'horselaugh.'"

One evening these two sauntered over to the Frenchmen's table and invited them to a dinner party. The invitation was proffered with many pained smirks and sidelong glances. Having accepted, Brillat-Savarin and company were not surprised to learn from a sympathetic waiter that the Britishers intended to turn the dinner into a serious drinking contest.

The appointed day found Brillat-Savarin and his friends armed with a strategem for staying sober based on two maxims: (1) conceal bitter almonds about your person and chew on them when-

Brillat-Savarin, who lived to tell the tale of a ferocious drinking bout at Mr. Little's Tavern. *Courtesy of the New York Public Library.*

ever the enemy's attention is distracted; (2) try not to drink too much. They were cheered by the amount of food that had been ordered. To a corner table for five were brought an enormous roast of beef, turkey cooked in its own gravy, vegetables, raw cabbage salad, and a jam tart. The entree was accompanied by more claret than the hosts could keep track of, but their guests managed to dump every second or third glass in a handy beerpot. When the table was finally cleared, port followed by Madeira was poured lavishly and to these they "stuck for a long time. 'Desert' came on—butter, cheese, coconuts, and hickory nuts. Toasts were proposed. After the wine came rum, brandy, raspberry brandy, and songs neck and neck." To avoid hard liquor, Brillat-Savarin asked if he might have punch. The hosts had foreseen this contingency and a brimming bowl was brought out already prepared—enough for forty. The toasting gained momentum. Suddenly Wilkinson rose to his feet, bellowed "Rule, Britannia," collapsed backward into his chair, and rolled under the table. "Horselaugh" succumbed to a final fit of whinnies, and stooping to assist his friend, fell by his side.

Highly pleased, Brillat-Savarin rang for Little to come up, and "after addressing him in the conventional phrase 'see to it that these gentlemen are properly cared for' drank with him their health in a parting glass of punch." The waiter, with his assistants, soon came in and bore away the vanquished, whom they carried out, "according to the rule, *feet foremost,* which expression is used in English to designate those *dead* or *drunk,* Mr. Wilkinson still trying to sing *Rule Britannia,* his friend remaining absolutely motionless."

An account of the disaster appeared in the newspapers the following day. The Englishmen were still quite ill. Insult was doubtless added to injury when they received Little's bill. New York tavern dining prices were, in fact, notoriously high and complaints were frequent. One traveler bought space in a local newspaper to denounce the cost of board as being at least twice as much as in Philadelphia. Shortly after, a city editor took up the cry, accusing boarding-house owners and tavernkeepers of injuring not only themselves but the city by overcharging and causing congressmen to feel imposed upon.

Something of the scale can be seen from the fact that at a tavern in which a luxurious single bed (all to oneself) cost one shilling and the stabling of a horse, hay included, for one night cost two shillings, grog could be had at one shilling a serving, sangaree for four shillings a bowl, and oysters for up to two shillings a portion. Meat, too, seemed proportionately expensive, with ham at one shilling twopence a portion, and beef at one shilling a cut, but it was chosen with great care. This was necessary, as meat was always served *au natural* and its quality alone maintained a tavern-keeper's reputation. If he did not trust his own judgment in selection, he could turn to a pamphlet on the subject published in 1786 and entitled *Directions for Catering or the procuring of the best Viands, Fish, etc.*

How to Choofe Flesh

Beef. The large stall fed ox beef is the beft, it has a coarse open grain, and oily smoothnefs, dent it with your finger and it will immediately rife again; if old, it will be rough and spongy, and the dent remain.

Veal is soon lost—great care therefore is necessary in purchasing. Veal brought to market in panniers, or in carriages, is to be preferred to that brought in bags, and flouncing on a fweaty horse. . . .

The champion flesh chooser of the period was John Jacob Astor's brother Henry. He first came to America during the Revolution with the British troops, managed to escape their service, and entered into that of the "Art and Mystery of Butchering." When his brother arrived steerage in 1786, Henry was well enough established to offer him a position as clerk, but John Jacob chose rather to become a journeyman baker and achieved instant success with the "Astor Roll," a creation of his own still popular in New York. But his sights were higher. After a brief, brain-picking apprenticeship to a fur trader, John Jacob borrowed five hundred dollars from Henry, and set out doggedly into the wilds.

Henry's fellow butchers were soon to discover that he was no less enterprising than John Jacob. The Bull's Head in Bowery Lane regularly served Astor's celebrated cuts of meat—a signal honor, as the tavern was known to be the unofficial clubhouse of drovers and butchers. Whenever news reached the city that a drove of cattle had been sighted approaching, the local butchers would collect in the taproom to await its stately arrival. They were a convivial and patient crew, unperturbed at the prospect of standing each other drinks all day if necessary. But Astor was never one to lallygag. When no one was looking, he would nip out the back door, leap into the saddle, and gallop up the Boston Post Road to intercept the oncoming drove and buy up the best of the lot.

No amount of reproof could deter him from this devious habit, and in self-defense the other butchers, joints cracking and paunches pounding, were obliged at regular intervals to join in motley pursuit. As this grotesque steeplechase became a common event, the Bull's Head found itself deserted more often than not, and in desperation, the owner, Richard Varian, pulled up stakes and moved several miles up the Boston Post Road. There, in the Manhattan countryside, he built the New Bull's Head. It became even more popular than its predecessor, particularly with the drovers for whose convenience pens were set up to accommodate fifteen hundred head of cattle. To attract the sporting element from a nearby race track Varian organized frequent turkey shoots and turtle feasts.

It was at the Old Bull's Head that Washington Irving in his youth "used to hear the tales of travelers, watch the coaches and envy the more pretentious country gentlemen in Castor hat, cherry-derry jacket and doeskin breeches." Years later, Irving modeled his "Hogg's Porter House" in Salmagundi upon fond recollections of the Bull's Head: "Hogg's is a capital place for hearing the same stories, the same jokes, and the same songs every night in the year except Sunday nights . . . a fair school for young politicians too—some of the longest and thickest heads in the city come there to settle the nation."

In fact, many matters were settled in taverns and coffeehouses. The larger establishments provided upper-floor "long rooms" where the beaux and belles could catch a well-chaperoned glimpse of each other at formal "Assemblies." These long rooms were also convenient for meetings of groups and societies of every description, and, in fact, they were the very heart and soul of city life.

The list seems endless. At the Merchant's and Tontine coffeehouses alone (facing each other at the corner of Wall and Water streets) the New York Chamber of Commerce was founded, as were the New York Hospital, the New York Insurance Company, and the Bank of New York; there too met the Society of Cincinnati, the Marine Society, the General Society of Merchants and Tradesmen, the Manufacturing Society, and the Society for Promoting the Manumission of Slaves. Cornelius Bradford, proprietor of the Merchant's Coffee House, supplied a variety of public services; he kept a log in which sea captains could enter accounts of their voyages, as well as a registry of refugees, a public auction room, and a general postal delivery. Outside, a group of merchants met one day under a buttonwood tree and signed the agreement which formed the New York Stock Exchange, while across the street in front of Gaines' Publication Office, a paper vendor regularly bawled, "News, News, Bloody News, Great News!"

A grab-bag of other political and social societies met at taverns throughout the city. Martling's on Nassau Street served as the Freemason's Hall as well as the Wigwam for the sons of "Immortal Tammany of Indian Race, Great in the field and foremost in the Chase!" That is, until 1811, when the cornerstone was finally laid for a permanent Tammany Hall. Moreover, apart from Federal Hall, there were few public buildings, and tavern owners were only too happy to volunteer their premises for public ceremonies.

In February 1784, James Duane, the first American mayor of New York City, was sworn into office—at a tavern chosen for no particular distinction other than the popularity of its owner, John Simmons. Old John Simmons was a fixture on Wall Street and the fattest man in New York. "In the summer he filled the whole bench on the stoop, and in the winter the whole of one front window." No worthy cause escaped his paternal bear hug. While he occasionally played host to the Society for the Relief of Distressed Debtors, his pet projects were the Society for the Manumission of Slaves and the Committee for the Establishment of a School for Children of Free Negroes. When he died in 1795, it was with grief that his patrons hacked down the pier between his front door and window so that his stupendous coffin could be crammed through, although his last will and testament raised a few eyebrows: to his daughter he left "my two small silver salts, also my Negro boy slave named Phil."

Peter Bayard's, a "country" restaurant on Grand Street, served as the scene for Alexander Hamilton's public triumph over his political enemies, Governor George Clinton and the Anti-Federalists. By the summer of 1788 the Constitution, favoring a strong federal government, had been ratified by nine states. New York remained divided. To enlist popular support for immediate ratification, Hamilton and the so-called "Feds" pulled out all the stops and treated the city to a slambang spectacle.

First came a parade. Under gaudily painted banners, guildsmen stomped down Broadway brandishing their wares. The bakers balanced cakes in either hand, the coopers beat out march time on their kegs, and the butchers, menacing the air with gleaming knives and cleavers, were followed by pewterers, goldsmiths, silversmiths, wigmakers, hatters, and tailors. The floats were ambitious. The printers' bore a giant press on which they turned out copies of an "Ode to the Occasion" and tossed them to the crowd. Not to be outdone, the blacksmiths fearlessly mounted a fire and anvil on theirs and forged an anchor while en route. Their thunder was stolen, in turn, by the ship carpenters' float, "the Hamilton," a twenty-seven-foot model of a frigate at full sail. Fitted out with thirty-two guns and a wooden figurehead of Alexander Hamilton, she was manned by upwards of thirty seamen under Commodore James Nicholson and drawn by ten horses. She got under way "with her topsails a-trip and coursers in the brails," and at Beaver Street she hoisted a jack and fired a gun, at which signal a "pilot boat" appeared on her weather quarter, the pilot was duly received, and his boat dismissed. When she reached the capital she fired a thirteen-gun salute, which was followed by three cheers and politely answered by the gentlemen of Congress. After this she was hauled all the way to Bayard's tavern, taking the hills like waves and fighting off squalls in the fields. The "Hamilton" eventually hove to abreast of the dining tables, where the officers disembarked for dinner.

Here the festivities reached their peak when five thousand celebrants descended to devour a free dinner. Bayard's seemed an unusual choice; a quiet place, known chiefly for its fine cellar and superb turtle soup, it was patronized by "portly up-river patrcons with low-crowned, broad-brimmed hats, massive fob chains and seals, and spacious coat-flaps, sedate, prim men from Pennsylvania, and snuff-colored denizens of Jersey." Still and all, Bayard saw the banquet through with unperturbable Yankee know-how, somehow managing to seat all five thousand guests at tables. The happily overstuffed voters shambled home at sunrise, and New York ratified within the week.

After all the states had signed the Constitution, an innovation in etiquette, stringent as a Japanese tea ceremony, was introduced and enthusiastically received, despite national disdain for excessive formality: dinner was to be followed by thirteen toasts. "Our American Fabius, the illustrious Washington" was toasted. "The Patriotic Hamilton" was toasted. So were "The Federal Edifice," "Bunker Hill," "Death Rather Than Submission to Foreign Controls of Influence," "The Day," and "The Fair Sex." In November of 1790, the Society of Black Friars raised their glasses to the hopes that "the Poor and Distressed throughout the World might find a Quick and Sure Asylum in America."

On April 30, 1789, Washington took the presidential oath at Federal Hall. He wore, for the occasion, a suit of Connecticut-made brown broadcloth as an advertisement of American products. And then, on May 5, the inaugural ball was held at the City Tavern (later known as the City Hotel). The guest list was dazzling enough. The New Society included Vice President John Adams, John Hancock, James Monroe, James Madison, Alexander Hamilton, John Jay, members of Congress, and an exotic parade of foreign emissaries.

Social life in the capital got off to a frenzied start. Cooks, dancing masters, and dressmakers were worked to the bone. There were teas, dinners, levees, assemblies, theater galas, fox hunts, and card and sleighing parties. Martha Washington, who was confusedly called "Lady Washington," was the first hostess of the land in name only. She had not bothered to attend the inauguration or the ball following it, nor did she arrive from Mount Vernon until a month later. Her attitude was hardly calculated to endear her to the local hostesses, all of whom were making supreme efforts. When she finally did get around to entertaining, some wished she had never left Virginia. Ladies' teas were her tour de force, and, stony-faced and close-lipped, she would preside over the table at 10 Cherry Street, the presidential residence, radiating gloom. She claimed that she felt "more like a state prisoner than anything else." Fortunately, the President was of a more gregarious nature and arranged occasional levees for gentlemen guests, presided over by Sam Fraunces, appointed first steward to the President.

"He may discharge me, he may kill me if he will, but while he is President of the United States, and I have the honor to be his Steward, his establishment shall be supplied with the very best of everything that the whole country can afford." These are strange words to come thundering down the annals of our history, and definitely an unusual twist to the theme of revolutionary defiance. In fact, General Washington, a careful man who had won his country's liberty on a shoe string, was now faced with the inflexible standards of Sam Fraunces, the finest tavern keeper in New York and the first in an unbroken chain of culinary perfectionists to terrorize its kitchens. This particular outburst of pride was occa-

Arrival of General George Washington in New York City, April 23, 1789, prior to his inauguration as first President of the United States, by A. Rivey. *Courtesy of the New-York Historical Society.*

sioned by an early shad that had caught Fraunces' eye while shopping for the presidential breakfast. When it was served, delicately broiled and perfectly boned, Washington examined it with practiced suspicion and inquired as to its price, which happened to be three dollars. The tirade that followed was not recorded verbatim, but it is known that Washington, despite his fondness for Fraunces, eventually discharged him for extravagance, only to reengage him when it was discovered that subsequent stewards were equally extravagant but far less capable. In this particular case the offending shad was banished from the house, but was "greatly enjoyed at the servants' table."

Fraunces' major competition as arbiter of the capital's table came from the chef of the Comte de Moustier, the French Ambassador. De Moustier had caused some consternation in American drawing rooms by appearing all tricked out in red heels and a single gold earring. A teetotaler and notoriously fussy about his food, the ambassador went so far as to take his chef along to a dinner at the home of John Adams, imagining the Vice President's menu would fall beneath his standards. The company was equally repelled by his presumption and by a dish his chef produced for their delectation—game birds perfumed with a dubious delicacy called truffles. The host himself disapproved of everybody. Adams wrote: "With all the opulence and splendor of this city, there is very little good breeding to be found. I have not seen one real gentleman, one well-bred man, since I came to town. At their entertainments there is no conversation that is agreeable; there is no modesty, no attention to one another. They talk very loud, very fast, and all together." The *Boston Gazette* piously concurred: "our beloved President stands unmoved in the vortex of folly and dissipation which New York presents."

The more puritanical New York City fathers also felt that the pursuit of happiness had got out of hand. They expressed their displeasure in a "memorial" addressed to the legislature in which the theater and tavern were referred to as "Evils which threaten our City and State." After this terse edict, vice was kept under closer wraps. A law had been passed in March of 1788 subjecting tavernkeepers to fine and imprisonment should they allow cockfighting, gaming, cardplaying, dice, billiards, or shuffleboard in their establishments. Drunkenness was punishable by three shillings, or two hours in the stocks, "such a conviction being without appeal."

The reformers were not completely successful in stamping out gambling; bets could still be made on the street in front of a tavern, as in the case of a countryman who, according to the *New York Gazette* of May 29, 1786, "for a trifling wager, et fifty boiled eggs, shell and all. He performed the task in about fifteen minutes being elevated on a butcher's block during the operation."

Such spontaneous performances could not be counted on with any regularity, however, and many tavern owners spent anxious nights dreaming up entertainments to lure patrons away from their competitors. Judging from contemporary advertisements, the results were frequently bizarre and sometimes staggering:

Jacob Jocalemon's Tavern—"The Exhibition of a Large Sea Dog, at a shilling a person."

Aorson's Tavern—"The Only Lecture of the Season . . . by a man more than thirty years an Atheist."

The Garden of the Academy—"An Exhibition of small Italian shades (a type of magic lantern slide) showing a Sea Battle . . . a large Chinese Shade in which some of the Actors will vault prodigious heights up and down. . . . There will also be a grand illumination upon Pyramids, to prevent any kind of accident by fire."

The Merchant's Coffee House (for obvious reasons Bradford sponsored the following in a house across the street, rather than on its premises)—

Dr. King, lately from South America, has arrived from Charleston with a collection of Natural Curiosities . . . a Male and Female of the surprising species of the Ourang Outang or the Man of the Woods . . . the Sloth, which from its sluggish disposition will grow poor from travelling from one tree to another . . . the Baboon, of different species and a Most Singular Nature . . . Monkey, Porcupine, Ant-Bear, Crocodile, Lizard, and Sword Fish, Snakes of various kinds and Very Extraordinary; Tayme Tiger and Buffalo. 10 A.M.—10 P.M.

But best of all, in the taproom of the New England and Jersey Hotel, on Courtland Street—

The Learned Pig . . . Every Evening! . . . his sagacity too well known to need a vain, puffing, elusive advertisement . . . the proprietor will state only what the Pig Actually Performs! . . . Reads! Spells! Tells the time of day, the date, day of the month! Distinguishes color! Tells how many people are present! . . . and to the astonishment of many of the spectators will Add, Subtract, Multiply and Divide . . . also will Divine Cards, besides a variety of entertaining matters on Politics, Love, and Matrimony!

What the Learned Pig failed to divine, though, was the swiftly approaching demise of the traditional tavern, and that its death knell would soon be tolled on the dinner gongs of New York's upstart hotels.

Chapter II

---•❧•---

1800-1830

Little Birds are hiding Blessed, I say, though beaten
 Crimes in Carpet bags   Since our friends are eaten
 Blessed by happy stags: When the memory flags.

"Balloon Ascensions. Valetudinarians may experience a restoration of Health, the motion being transported in a safe and easy carriage 1500 feet per minute, nearly 20 miles an hour, but slower if they chuse."

For the seasoned hypochondriac, such a therapeutic entertainment could be enjoyed simply by presenting oneself at the garden of Mr. Phineas Parker during the summer months of 1800. The management took over from there. Upon paying a minimal fare, an intrepid invalid would be hoisted into the basket with a great deal of hearty encouragement. Once unlashed from the moorings, it was up and away over the heads of the crowd, a fainthearted lot who, after mustering a round of nervous applause, tucked back into their melting ice cream and tepid lemonades. Soaring majestically aloft, the valetudinarian could, between deep inhalations, contemplate the rapidly receding scene—bushes in flower, winding gravel paths, bobbing ruffled bonnets, the glint of a brass band —in short, all the giddy "appointments" of the new rage, the Pleasure Garden.

Pleasure Gardens were an eighteenth-century import from England, where they enjoyed a great and scandalous success. But like so many European institutions, they suffered a moral purification upon arrival; a Tom Jones would have been sore put to arrange an assignation or instigate an intrigue in the staid and dreary retreats set up by early American imitators of the Vauxhall and the Ranelagh.

Sam Fraunces himself briefly owned a "Vauxhall" as early as 1765. His competition was insignificant, as New Yorkers had yet to feel the stifling constrictions of an expanding city and skyrocketing population. But they soon did. By the end of the century, sixty thousand residents were crowded together at the southernmost tip of Manhattan, and with the exception of a few promenades along the river and the Battery, there were no public parks in which to escape the sweltering summer heat. The opening of Joseph Corré's spectacular Mt. Vernon Garden, on the corner of Broadway and Leonard Street, marked a turning point. Another Frenchman, Joseph Delacroix, quickly followed suit, purchased the old Bayard homestead on Grand and Mulberry streets, and in 1798 launched the first American Vauxhall worthy of the name. These two compatriots embarked on a feud that was to last for years, and, in trying to outdo each other, plied the public with so many irresistible attractions that the tavern, at least during the summer months, went into eclipse.

From the outset Delacroix was the more flamboyant showman. He converted the entrance of his garden into an illuminated forest for the Fourth of July festivities in 1797; a band blared, and singers from the John Street Theater made one of their frequent appearances. For six shillings a celebrant was entitled to a glass of ice cream, punch, or lemonade. "To obviate difficulties, no other liquors will be served this evening." Two years later Delacroix moved to Broome Street, "well out in the country," where it became necessary to scatter lanterns around upper Broadway so that the guests would not stray into the fields. There he was able to present the sort of grandiose fireworks' display that would have proved hazardous downtown:

A Grand Firework, with music and illuminations . . . a grand balustrade, 25 feet long, a large brilliant fixed sun, with a transparent red face of Chinese fire, a grand pyramid of Roman candles, a large musical wheel, a horizontal wheel throwing out stars and serpents, a vertical sun with a variety of maroons [sic], the Windmill, a large flower pot, a roly-poly (a new piece, a dragon that will come from a distance to light the first piece), the fountain of St. Cloud, the magic box or devils let loose. To commence at 8:00 o'clock.

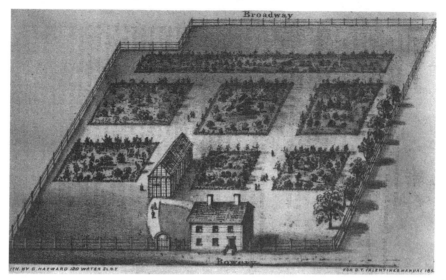

Delacroix's Vauxhall Garden in 1803. *Courtesy of the New York Public Library.*

Cato's Tavern on the Boston Post Road. *Courtesy of the New York Public Library.*

The previous year, at the Mount Vernon Garden, Corré had opened the city's first summer stock theater with a production of *Miss in Her Teens*. The company was recruited from the Park Theater upon conclusion of the winter season, and brought with it a faithful audience. Delacroix watched the stampede to his rival's with impotent fury. Pyrotechnic spectacles had, for the moment at least, lost their excitement. Back at the Vauxhall, he fulminated over plans for a Grand Amphitheater which was completed in time for a gala opening the next summer—"no gentleman admitted unless accompanied by a lady." At last the local ladies, denied the solace of their friendly neighborhood tavern and scarcely tolerated in the theater pit, were openly requested to present themselves publicly in the genteel air and ice cream atmosphere of the Gardens.

Decorum was shattered only on such occasions as the night Delacroix, armed with a horn blower, vilified Corré's patrons as they arrived for a performance of *The Comedy of the Child of Nature*. As the very same production had been rained out not two nights before at the Vauxhall, he denounced its presentation elsewhere as a gross breach of ethics. The next day the *Spectator* was crisp in its judgment of Delacroix's comportment. "We have no monopolies in this country" was the verdict.

Despite the owner's tantrums, the Vauxhall flourished to such a degree that larger quarters had to be found. With his keen nose for fast profit, John Jacob Astor paid the astronomic sum of nine thousand pounds for a complex of nurseries and greenhouses between the Bowery and Broadway, and leased them to Delacroix in 1803.

It took two years to convert the property into a maze of gravel walks, flowerbeds, rose arbors, trees, shrubs, fountains, and romantic statuary. A large equestrian figure of General Washington dominated the center of the scene. The greenhouses were turned into refreshment booths, and in one corner a theater was built in which plays, light musical pieces, and interludes were performed. The audience was seated in the open air while the orchestra played among the trees. A complicated apparatus was set up for fireworks and one could order special designs in advance. These included such specialties as "Emblems of Masonry, or the Portico of the Temple of Solomon in blue, white, and brilliant fires, which will be lighted by a flash of brilliant sun at the distance of 200 yards." Balloon ascensions, "prepared with neatness and certainty by Mr. Humbert," continued to draw large crowds for years after the Vauxhall's final move; increasingly fearsome feats of daring were expected of the Aeronauts, however, and a certain Mr. Guile was noted to have "successfully parachuted out, as planned."

As the public clamored for more and more outdoor distractions, it became clear that the Vauxhall and Mt. Vernon could not hold on to their monopoly forever, and in 1806 John H. Contoit opened his New York Garden at 355 Broadway. In the afternoons its long narrow plot of shade trees was a perfect aviary of chirruping ladies who applauded Contoit's introduction of the temperance plan; in the evenings, the gentlemen joined their wives for ice cream, pound cake, lemonade, and if they were known to the management, a surreptitious slug of claret or cognac.

As tavernkeepers had vied with each other for the most stunning divertissements, so did operators of pleasure gardens. Mr. Hogg of the Washington Garden solicited the concert-going crowd: "M. Cartier has kindly consented to add to our entertainment his astonishing powers on the clarinet." Art lovers could throng to the Columbia Garden to take in an exhibition of transparent paintings such as "Night flying before the God of Day, who opens his azure gate to distribute his favours on Earth."

Ice cream was a drawing card at all the gardens. In 1808 Charles Barnard opened the United States Garden, featuring ice cream prepared by Charles Collet, "well known as possessing ability in that line." Flavors were generally limited to vanilla and lemon until 1810, when Mr. Ensley announced that in the future pineapple, strawberry, and raspberry ice creams would be served at his "New and Elegant Columbian Garden." The texture was that of frozen custard.

Water ices were known to the Romans, but ice *cream,* a Chinese refinement, was first brought to Italy by Marco Polo and later to the court of Louis XIV by the Sicilian confectioner Procopio. By the turn of the nineteenth century, it had become a favorite on either side of the Atlantic. It has been said that George Washington made it, Mrs. Alexander Hamilton served it, and Dolly Madison popularized it. As the demand for ice cream grew, confectionaries multiplied all over New York. These frequently served food on the premises and sold excellent liqueurs and packaged sweets as well. It took death and the Delmonicos to topple the thrones of the two greatest confectioners, Delacroix and Corré, who maintained factories and distilleries outside their gardens.

Corré cut a memorable figure; many decades later one New Yorker, William Dunlop, recalled:

he stood knife in hand, in the full costume of his trade, looking as important as the mysteries of his craft entitle every cook to look, "with fair round belly, with good capon lined," covered with a fair white apron, and his powdered locks compressed by an equally white cap. His rotundity of face and rotundity of person—for he was not related to Hogarth's Cook at the gates of Calais—with this professional costume, made his figure, though by no means of gigantic height appear awfully grand, as well as outré, and it was stamped upon the young mind of his admirer in lights and shadows never to be erased.

This vivid impression of a *chef de cuisine* was made a decade before French tastes and fashions took New York by storm. An early nineteenth-century historian, John F. Watson, gives an account of the city's Francophilia in *Annals and Occurrences of New York City and State in the Olden Time;* Mr. Watson noted that, by 1794,

almost every vessel arriving, brought fugitives from the infuriated Negroes in Cape François, Port au Prince, etc.; or the sharp axe of the guillotine of France, dripping night and day with the blood of Frenchmen, shed in the name of liberty and the sacred rights of man. The city thronged with French people of all shades from the French colonies, and from old France, giving it the appearance of one great hotel or place of refuge for strangers hastily collected from a raging tempest . . . French boarding houses, marked "Pension Française," multiplied in every street. Before such houses, groups of both sexes were to be seen seated on chairs, embarrassing the streetwalkers, and the French in full converse; their tongues, shoulders, and hands in perpetual motion—"all talkers and no hearers." Mestizo ladies, with complexions of the palest marble, jet black hair, and eyes of the gazelle, with persons of exquisite symmetry, were to be seen escorted along the pavements by white French gentlemen, both dressed in the richest materials of West India cut and fashion; also coal black Negresses in flowing white dresses, and turbans of Mouchoire de Madras, exhibiting their ivory dominoes, in social walk with white or mixed creoles; altogether forming a lively contrast to our native Americans, and the émigrés from old France, most of whom still kept to the stately old Bourbon style of dress and manner; wearing the head full powdered à la Louis, golden headed cane, silver set buckles, and cocked hat, seemingly to express *en silence* their profound contempt for the pantaloons, silk shoestrings and "Brutus crop." While they remained, they gave an air of French to everything.

Prominent among these émigrés were two of the guests at Mrs. Loring's fashionable boarding house at number one Broadway. One was the brash and pint-sized Citizen Edmond-Charles Genet, who served as ambassador of the infant French Republic until he proposed the impeachment of Washington. Though instantly stripped of all official rank, he chose to linger in the States, and after some judicious shopping around, married Governor Clinton's daughter and lived happily ever after. The second and greatly more imposing figure, despite his unfortunate clubfoot, was Charles-Maurice de Talleyrand-Périgord, who spent the years 1794 to 1796 in America, biding his time between malleable regimes. Gulian Verplanck later recalled meeting him at the houses of Noah Webster and General Hamilton "with his passionless, immovable countenance, sarcastic and malicious even in his intercourse with children. He was disposed to amuse himself with gallantry too. But who does not know, or rather, who ever did know Talleyrand?"

The French introduced the cotillion, gold watches, gilded frames, mattresses, and high bedsteads. Happily their culinary talents were also appreciated, and soups, salads, sweet oil, tomatoes, ragouts, and fricassees began to brighten New York tables. They taught Americans the use of the double-barreled gun and, according to Mr. Watson, "shot and ate all manner of birds practically thinking that all depended upon the cooking." And needless to say, their predilection for perfumes, pastries, and bonbons spurred on the careers of such fellow countrymen as Collet, Corré, Delacroix, and others.

Further possible links with France were explored by one of New York's greatest publishers and printers, Louis Alexis Hocquet de Caritat, an energetic intellectual who, by the turn of the century, had assembled the largest lending library in the country. In a series of articles first published in 1797, and taken as gospel for the next twenty years, he expounded his theories and observations on Franco-American trade. They have been collected by George Gates Raddin in his study *The New York of Hocquet Caritat*.

It is through Caritat that we have some idea of the edibles shipped from abroad to the harbor of New York. He discouraged anyone with ideas of making a killing in the international pickled-goods trade unless he dealt in capers or anchovies, "which sell very well . . . tho' not in universal demand." Another bad bet was olive oil "or, as it is called in the United States, 'sweet oil,' an article of no very great consumption among them. In many houses that might even be termed genteel they have none at table; a great number among the Americans seeming not much to relish its taste, even in eating salad. Prudence would require not to send too much at once, especially as quantities are imported from Leghorn, and other foreign places up the Streights."

Caritat dwelt at length on American drinking habits. He found that among the laboring classes rum or brandy mixed with water were the favorites, that on occasion brandy was combined with several kinds of wine to imitate Madeira, Teneriffe, or sherry, and that Geneva, called "gin," was relished by all sorts of people. He deplored the fact that with the exception of Claret, Bordeaux, and a limited amount of Champagne, decent French wines were sorrily lacking, while Portuguese and Spanish Oporto, Madeira, sherry, and Teneriffe could be had by the boatload. As imported raisins and figs invariably sold out upon arrival in New York, Caritat suggested that exporters might do well to concentrate on soft-shelled almonds, grapes in jars, and olives, and thought it time to introduce the prune.

But on the whole, European food imports were inconsequential, for the American hinterland provided an abundance inconceivable in France. James Fenimore Cooper's fictitious "Travelling Bachelor," in one of his exhausting letters to the French Count de Bethizy, had this to say on the subject. "It is difficult to name fish, fowl, or beast that is not, either in its proper person, or in some species nearly allied to it, to be obtained in the markets of New York." He claimed that seventy to eighty varieties of fish, all edible and most excellent, could be sampled, along with a comparable catalogue of fowl. "It would do your digestive powers good to hear some of the semi-barbarous epicures of this provincial town expatiate on the merits of grouse, canvas-backs, brants, plover, wild turkeys and all the *et ceteras* of the collection." He amplified: "That delicious, wild, and peculiar flavour, that we learned to prize on the frontiers of Poland, and in the woods of Norway, exists in every thing that ranges the American forest." Of "the vulgar products of regular agriculture," all were cultivated, and there was "scarce a fruit that will endure the frost that is not found in a state nearly approaching to perfection." The apple and the peach grew side by side, and melons and cherries abounded in such richness that no effort was made to cultivate the rarer sort.

But these few were really remarkable. The French peach, The Bachelor pointed out, leaves a bitter aftertaste, and requires the addition of sugar; not so its sweet American counterpart, while as for melons, "these people would positively reject the best melon that ever appears on your table. America produces a little one that exceeds anything of its kind that I have ever admitted into the sanctuary of my mouth." With a final sweeping statement, calculated to enrage any Parisian, he concluded: "I know no spot of the habitable world to which the culinary scepter is so likely to be transferred, when the art shall begin to decline in your own renowned capital, as this city."

One thing that impressed Cooper's peripatetic hero was the extraordinary variety of tropical fruits for sale in New York. "Pineapples, large, rich, golden, and good, are sold for twelve to twenty-five sous; delicious oranges are hawked in the streets much cheaper than a tolerable apple can be bought in the shops of Paris, and bananas, yams and watermelons, etc. are as common as need be in the markets."

Of all the delicacies imported from the Carolinas and West Indies, it was not fruit, but the giant green turtle that was most highly prized. The turtle being too large and intractable a monster to be transformed into a simple supper by the average housewife, it could be served only by taverns and hotels whose patrons were able to finish one off in a day. This might be quite a project; heftier turtles weighed up to three hundred pounds.

The harassments involved in the preparation of turtle soup can be seen from a recipe that was published several decades later when smaller turtles became available:

A small turtle may be purchased at Fulton Market for from ten to twenty cents per pound and weighing from fifteen to forty pounds, the price varying according to the law of supply and demand. The only objection to small turtles is that they do not contain a very large percentage of the green fat so highly prized by epicures.

Procure a live turtle, cut off the head, and allow it to drain and cool over night; next morning place it on the working table, lay it on its back, and make an incision around the inner edge of the shell; then remove it. Now remove the intestines carefully, and be very careful that you do not break the gall; throw these away; cut off the fins and all fleshy particles and set them aside; trim out the fat, which has a blueish tint when raw; wash it well in several waters. Chop up the upper and under shells with a cleaver; put them with the fins into a large saucepan; cover them with boiling water; let stand ten minutes, drain and rub off the horny, scaly particles with a kitchen towel.

Scald a large saucepan, and put all the meat and shell into it (except the fat); cover with hot water; add a little salt, and boil four hours. Skim carefully and drain; put the meat into a large crock, remove the bones and boil the fat in the stock. This does not take very long if first scalded. When done, add it also to the crock, pour the stock into another crock; let it cool and remove all scum and oily particles; this is quite work enough for one day. Clean the saucepans used, and dry them thoroughly.

Next day fry out half a pound of fat ham; then add one chopped onion, one bay leaf, six cloves, one blade of mace, two tablespoons of salt, a teaspoonful of white pepper, and one quart of ordinary soup stock, now add a pint of dry sherry.

Do not let the soup come to a boil; taste for seasoning, and if herbs are needed tie a string to a bunch of mixed herbs, throw them into the soup, and tie the other end to the saucepan handle; taste often, and when palatable, remove the herbs. If the soup is not dark enough, brown a very little flour and add to it. Keep the soup quite hot until served; add quartered slices of lemon and the yolk of a hard boiled egg, quartered just before serving; send to the table with a decanter of sherry.

The yolks of the eggs may be worked to a paste and made into round balls to imitate turtle eggs if this is desired.

At the turn of the century the dressing of a turtle was an event and "turtle feasts" were announced in advance. In 1802, Hatfield's Tavern at John and Nassau streets advertised: "A mammoth turtle will be cooked in the best style on Wednesday next—served from one to ten o'clock in the evening. The lovers of good eating will have their taste gratified."

It was generally considered that the best turtle soup was to be had at Bayard's Tavern. "Turtle soup . . . worthy of the animal of which it was made; not a puree, which is served at some of our leading restaurants and clubs, not a thin consomme of that which might be calve's head or veal, but *bona fide* turtle, with callipash, callipee, and force meat balls." For the record, Webster defines calipash as "the carapace of a turtle . . . it contains a fatty, gelatinous, dull-greenish substance, esteemed as a delicacy" and calipee as "the plastion of a turtle . . . it contains a fatty, gelatinous, light-yellow substance, esteemed as a delicacy."

There seemed no limit to enthusiasm where turtles were concerned. A case in point was the ordeal of William Sykes. One brisk September day in 1822, this enterprising soul stepped down to the harbor to await the arrival of the schooner *Martha,* which was bound from Cuba and acrawl with turtles. She had no sooner docked than Sykes cornered the captain and purchased the entire cargo. Proceeding to the offices of the *New York Evening Post*, he took out an advertisement to announce his coup and spread the tidings that one or more of his menagerie would be served every day for the rest of the season at the Merchant Exchange and New York Coffee House. His custodial activities during the ensuing weeks may be guessed; one month later to the fateful day, Sykes, visibly rattled, reappeared at the *Post* and composed a second announcement: *"Lively Green Turtles for Sale!* The subscriber now offers for sale turtles of various sizes running around 100 pounds, which may be selected from about 40 at the Merchants Exchange."

In the inanimate line, all sorts of commodities could be found on the city's wharfs: boxes, hogsheads, barrels, bales, chests, and puncheons bursting with rice, flour, tea, rum, sugar, potash, cotton, wool, silk and linen, porcelain, wine, glass, salt provisions, and hardware from Holland, Ireland, Italy, China, England, Switzerland, and Spain. The port was constantly jammed with ships, and business in the city thrived correspondingly. New York was expanding at such an unbridled clip that real efforts were made— had to be made—to improve conditions. In 1795 a pesthouse was constructed at Bellevue, an estate located on the East River at what is now Twenty-third Street, and shortly after the turn of the century the Collect Pond was finally drained. Nevertheless, sewage was still dumped in the river and scavenger pigs were permitted to scrounge in the streets. The only pure drinking water was drawn from the Tea Water pump and sold by young boys from pushcarts. To remedy the last situation Aaron Burr organized the Manhattan Water Company in 1799; within a year, six miles of wooden pipes had been laid, and a reservoir dug at Reade and Center streets. Nevertheless, the supply of potable water could not meet the demands of the city, and various inventive citizens offered their own solutions. On April 13, 1802, Rembrandt Peale, a fashionable artist of varied talents, announced a demonstration of his newest brainstorm, the first recorded desalinization device in the country:

Pro Bono Publico. Putrid Water Made Sweet. Rembrandt Peale, having provided himself with a Apparatus to take to sea with him, with which water the most dirty and putrid may in a few minutes be made sweet and clear, will exhibit the manner of operation before merchants, masters of vessels, and such gentlemen as may feel interested with the subject, to attend at the Assembly Room, no. 68, William Street, precisely at 11 o'clock tomorrow morning.

There was a surge of cultural activity; the Park Theater and the Philharmonic Society were founded. Upwards of twenty newspapers were published, half of them dailies, as well as several magazines and "essays." (It was in one of these that Washington Irving observed an emerging New York mentality and dubbed the city "Gotham" after a proverbial English town of irrepressible wiseacres.)

With the War of 1812, the interruption of international trade had a dismal effect on New York, until February 11, 1815, when news of the Treaty of Ghent reached the city—two months late. It was then immediately resumed, and the boom was on. In 1818, a regular packet-boat service out of New York gave commerce a tremendous boost, and, to the merchants' jubilation, construction was started on the Erie Canal. By 1820 the population stood at 123,000. To add to the congestion, a constant flow of immigrants surged through the city, as well as swarms of businessmen from other states. Herein lay New York's most pressing problem: the housing and feeding of a growing migrant population.

Boardinghouses proliferated, and never before had such diverse people been thrown together over a breakfast table. As early as the turn of the century, one visitor, the English actor-manager John Bernard, recalled in his *Retrospections of America:*

The house I stopped at gave me a tolerable specimen of the varieties of society now converging at this great exchange . . . here was a French gentleman of *l'ancien régime* looking melancholy and mysterious in a bag-wig and point lace ruffles, who had two cards of address, the one styling him "Marquis," the other "Dancing-master." Here was an English agriculturist just arrived . . . here was a Kentucky land-owner proving London to be the Babylon of the Apocalypse, and predicting England's downfall . . . here were major-generals from Vermont, walking encyclopaedias of the war, and planters from "Caroliny" who were alternately explaining the free principles of their constitution and reading the description of runaway slaves. Here were Italians who had brought over Fantoncini to refine the taste of the infant country; Germans who had come to hunt out some distant relatives; lean and voracious Scotsmen looking as if they could swallow the continent, and Irish "jintlemen" of slender figures and fortunes. Each must be pleased, and we were accordingly provided with fish, ham, beef, boiled fowls, eggs, pigeons, pumpkin-pie, lobsters, vegetables, tea, cider, coffee, Sangaree, and cherry-brandy!

While boardinghouses were by no means an innovation, it was at the turn of the nineteenth century that what could honestly be called a "hotel" first came into existence. The only difference between the boardinghouse and the early hotel was that although both offered short-term and long-term rates for room and board, only the hotel retained the tavern's facilities for banquets, dances, and other public entertainments.

The first such institution in New York was the City Hotel, built in 1794. Ranking as one of the most impressive buildings in New York and the greatest hostelry in the States, it occupied a four-story red brick building which extended the entire length of one block on Broadway, between Cedar and Thames streets. The ceilings of the first two floors were unusually high and the dining room was considered exceptionally light and airy. The structure could claim three innovations: its slate roof was the first in the country, curtains were replaced by interior shutters, and its ground floor facing on Broadway was occupied by fashionable shops. All in all, it was a radical departure from the standard tavern.

Originally three meals were served daily: breakfast, noonday dinner, and tea. Many well-traveled patrons complained of the early dinner hour, however, and gradually the foreign custom of dining at three was adopted. This in turn caused provincial hackles to rise, and the harried management was obliged to serve the diehards at the original hour as well. At the sound of the dinner gong, a squadron of well-trained waiters brought out the plainly cooked food—twelve to sixteen meat dishes which might include venison, bear steak, wild turkey and duck, lobster, terrapin, crabs, oysters and pigeon, as well as local vegetables, fruit, and fish. The cuts of meat, with the exception of the mutton which was stringy, were of the highest quality. All was put on the table at once and consumed silently at a breakneck pace. Businessmen as well as resident guests made a habit of dining at the City Hotel, and there were often fifty to a hundred seated at one of its long communal tables. The guests helped themselves from large family-style serving dishes to as much or as little of the huge table d'hôte as they chose. The bar and wine cellars of the hotel became famous and a group of regulars, calling themselves "The Jolly Good Fellows," initiated vintage tastings of the finest Madeira, port, and sherry of the day. Champagne was available but unpopular; one reason for this may be that it was ceremoniously *warmed* before serving.

The City Hotel was unquestionably the most prestigious location for concerts and social functions. During the season of 1808, a sort of Jane Austen class war was waged in the heady climate of its ballroom. New York's social ramparts were zealously guarded by the members of the City Assembly, a series of subscription dances. For two dollars and fifty cents an evening, one was afforded tea, coffee, a cold collation, and the opportunity of engaging in the Louvre, Minuet, Dauphine, Bretagne, Allemand or English country dances with young ladies of the first rank of society, that is, if one were of the first rank of society oneself. (This class encompassed high government officers, divines, lawyers, doctors, the chief merchants, and property owners.) A major offensive campaign was launched against the City Assembly by insurgents from the upper reaches of the second class (lesser shopkeepers, professionals, clerks, etc.) who had the temerity to establish a new assembly. Theirs, too, was to be held at the City Hotel, and what is more, "in the spirit of jealousy and pride" they trounced their competition by charging three dollars admission and getting it. The new balls were so expertly conducted that many subscribers of the City Assemby bolted and joined the enemy camp, while others, racked by wavering loyalties, chose to patronize both.

Broadway's other physical glories made a deep impression on such diverse visitors as Rochefoucauld, the English Reverend Wintherbottom, and an anonymous "Resident of Philadelphia." All commented on its elegant brick houses, its elevated position, and the beauty of its proportions. The Philadelphian was dazzled by the splendor of the brilliantly lit shops, and the Reverend Mr. Wintherbottom, utterly carried away, declared that no city in the world possessed a handsomer street. Noisier would have been more like it. The air was raucous with vendors' cries and redolent of oysters, clams, fish, buns, yeast, and hot spiced gingerbread. Blind or lame cartmen could be heard bellowing:

Here's clams, here's clams, here's clams today
They lately came from Rockaway
They're good to roast, they're good to fry,
They're good to make a clam pot-pie
HERE THEY GO!

Baker's boys toting baskets of buns squawked "Tea, ruk, ruk, ruk, tea ruk" in jarring counterpoint to Negro women chorusing, "Hot corn, hot corn, here's your lily white hot corn, hot corn, all hot, just came out of the boiling pot. . . . Baked pears, baked pears, fresh baked, baked pears."

The hubbub on Broadway never subsided. By 1825, many of its residences had been converted into hotels and boardinghouses. Among them was the City Hotel's rival, the Washington Hotel, which occupied the large, Doric brick mansion of the former English governor, Sir Henry Clinton, at the corner of Marketfield Street.

What typified a completely fitted hotel of this period is outlined in an advertisement offering the Mechanic Hall for sale: "basement-kitchen and servants' quarters; first floor—coffee room, bar, dining, and Ladies' room; second floor—large hall with orchestra and drawing room; third floor—five reception rooms for clubs and parties; fourth floor—twelve bedrooms." As more and more such establishments flourished along Broadway, private organizations, the Washington Benevolent and Tammany Societies for example, invested in the building of halls to be operated as commercial hotels as well as meeting places.

On Monday, May 18, 1811, to the fascination of the hapless bystander, an apparition, a kind of aboriginal war dance, was seen snaking through the city in the general direction of Frankfort Street and Park Place. But upon close inspection, not even the ferocity of war paint and feathers could disguise the bland and businesslike features of the community's most eminent Democrats, bent on laying the cornerstone for a new and permanent Tammany Hall. The ceremony was marked by some ponderous oratory on the part of Abraham M. Valentine, Grand Marshal of the day, and Clarkson Crolius, the Grand Sachem, followed by a heavy dinner at Martling's Tavern, and capped off with yet another of Joseph Delacroix's commemorative *feux-de-joie* at the Vauxhall Garden. Once completed, Tammany Hall enjoyed considerable success as a hotel. From thirty to eighty people daily attended its substantial three o'clock dinners, the residents clustered at one end of the table and the dinner guests at the other. Game and oysters were the specialties of the house, eased down with tumblers of rum and water. After the meal those who were not rushing back to their offices could linger on to sing and smoke, or scan the news in the barroom.

For the distaff side, "Aunt Margaret" Mann maintained a Hotel for Women on Broadway. A forbidding-looking individual in cast-iron stays, her wispy hair imprisoned in a bun, she economized on decor to the exclusion of all but the most essential of bedroom furnishings, namely beds, which, with what she provided in the way of illumination, could barely be discerned. The kitchen was renowned,.however, and curiously enough she appealed to the theatrical crowd. Many leading actors, including Tyrone Power, frequently unstrung the inmates by dropping in for dinner.

Dyde's London Hotel initiated the theater-break supper served at nine thirty "between the farce and the tragedy," and in 1808, Thomas Hodgkinson, temperamental brother of the Park Theater's owner, took over the Punch and Porter House, an old-fashioned yellow brick structure at number seventeen Fair (Fulton) Street and renamed it the "Shakespeare Tavern." Through its green-baize-covered side door on Nassau Street the country's greatest actors, writers, and wits filed in steady procession for thirty years. Among them were Fitz-Greene Halleck, poet and confidential clerk to John Jacob Astor, the author Robert G. Sands, James K. Paulding, who was associated with Irving in the Salmagundi papers, and the poet Willis Gaylor Clark. Meeting constantly in the barroom with its old-fashioned circular bar, their clique helped to fill in the historical gap between the Turk's Head in London and the Hotel Algonquin's Round Table. Even though taverns were fast fading into obscurity (only Fraunces' and one or two others survived by 1830), nostalgia for "the tavern" lived on, and the Shakespeare did such a roaring trade that Hodgkinson added an extensive three-story wing in 1822.

Of the many social meetings that took place at the Shakespeare, it is doubtful that any could compete with the Kraut Club's annual festivities. An organization whose members were all purportedly descended from early Dutch settlers, the Kraut Club held a Grand Meeting once a year. By way of announcing the event, a cabbage was jammed onto the end of a long pole and stuck out of one of the Shakespeare's upper-story windows. Ceremonies commenced at nine in the morning. If any member had died during the year or was absent for any other reason, it was picturesquely stated that he had "wilted." After all the living had assembled, the morning was devoted to electing a Grand Kraut. The beer flowed. At one o'clock the ringlets (sausages) went the rounds. At three, his Magnificence the Grand Kraut, regally attired in a mantle of cabbage leaves and a hollowed-out cabbage crown (the workmanship of several ingenious ladies, and esteemed more splendid than that of George IV), took his seat between two smoked geese. Everyone then attacked a lavish banquet of more ringlets, geese, a variety of dishes in which either cabbage or saurkraut was the principle ingredient, and a tidal wave of beer. The revels continued well into the night or until such time as the management could stand the noise no longer and insisted on closing the bar.

In the case of George Frederick Cooke, an English actor, Hodgkinson refused service too late; Cooke died of acute alcoholism in the Shakespeare after two years in New York. His doctor said, "His case adds another lamented example to the long list of those who have prematurely fallen victims of intemperance." That this doctor was more candid than most is indicated by James Fenimore Cooper in *The Traveling Bachelor, or Notions on the Americans*: "Indeed, I am led to believe that New York, owing to its fine situation, is on the whole more healthy than most large towns. It has also been told me, that the deaths by consumption, as reported, are probably greatly magnified beyond the truth, since the family physician or friend of one who has died, for instance, by excessive use of ardent liquors, would not be apt to tell the disreputable truth, especially as it is not exacted under obligations of an oath."

There were an estimated fifteen hundred corner grocery stores, selling either wine or spirits or both, in New York by 1817; they came to be known as "grog-shops," and were a source of considerable profit to their owners, the majority of whom were Irish. Their fastest selling commodity was New England, or "Yankee," rum, which was consumed at any hour of the day, depending on when the thirst struck. An English visitor, Henry B. Fearon, in his *Sketches of America*, was pleasantly surprised to find no drunken brawlers in the streets, yet on second glance he realized that "though the beastly drunkard is a character unknown here . . . many are throughout the day under the influence of liquor, or what is not inappropriately termed 'half-and-half.' " Other newcomers were shocked to see young men in the streets smoking "segars," reserve supplies of which they sported in their hatbands. The best "segars" obtainable were to be found at Cato's Tavern on the Post Road at what is now Fifty-ninth Street and Second Avenue. Cato, a popular Negro barkeeper noted for his wit and

"The Actors' Monument—Edmund Kean, Esq. recontemplating the Tomb he caused to be erected to the Memory of George Frederick Cook," who fell, "premature victim of intemperance," at the Shakespeare Tavern. *Courtesy of the New York Public Library.*

unadulterated brandy, imported cigars made to his specifications directly from Havana.

While New Yorkers were not morally outraged by such vicious practices as smoking on the street, they did draw the line at the reported goings-on at an institution succinctly called The Road to Ruin. A self-appointed vice squad brought it to the attention of Mayor Clinton. It seemed that, among other things, after paying a fifty-cent entrance fee, one could dance with "lewd women" to the strains of a band of French Jacobins and English deserters, lose one's shirt in its five illegal gambling rooms, or obtain forged checks and counterfeit money. The owner had no fear of the law, as he admitted lawyers and justices gratis and paid constables and marshals two dollars a night—"while I have money I can clear myself." But most intolerable of all, read the complaint, were the iniquitous apartments where "many Married Women and many Daughters and many Servants are introduced and seduced because they will be in the Fashion." With the sudden notoriety of The Road to Ruin, other such lairs, dubbed "gin mills," tried to emulate its exhilarating ambiance, but none, not even a particularly lurid spot called The Finish, could come within a mile of it.

For the more finicky man-about-town, there were few evening diversions—only one theater, one circus, three billiard rooms, a bowling alley, and Scudder's lackluster museum. Horse, dog, and flower shows, pugilistic exhibitions, and anatomical and freak museums had yet to appear. Coffeehouses (the term had become a catchall phrase for any respectable establishment that wished to attract an orderly crowd during the evening as well as daytime hours) such as Pierre Combault's on Murray Street and the French Coffee House on Broadway, however, helped fill the gap. They served not only the traditional cider, Albany ale, port, and brown sherry, but some novel liqueurs—"Noyeau, Anisette and Kirsoraser."

Matters had improved for the footloose diner as well. The 1820s saw the evolution of what might be accurately termed "restaurants," as opposed to "bed and board" taverns or hotels with their unvarying fare. It was not yet the era in which a gentleman might be expected to take a respectable single woman out to dinner, but another phenomenon had arisen: as the city expanded, and the residential area crept further uptown, greater and greater distances separated the New Yorker's place of business from his home. Hence, a traditionally large midday dinner with his wife and family became unfeasible, and mealtime found hungry crowds milling the street of the business section. Joining hotel dining rooms in meeting the new demand were three categories of eating houses, two of which, the coffeehouse and the embryonic French restaurant, were already familiar to the public, while the third, the short-order house, a bumptious new arrival, pleased only those who were obliged to eat on the run.

Joseph Collet, the ice-cream wizard, and Francis Guerin operated restaurants that were more popular with women than men, who found French cooking an affront to the digestion of a plain-living man. At least at Collet's and Guerin's, relative leisure could be taken at mealtime, a digestive aid unknown to either hotels or short-order houses. Robert Burford, an English traveler, described hotel dining: "The meals are taken in the public room where fifty to a hundred persons sit down at the same time . . . a vast number of dishes cover the table, and the dispatch with which they are cleared is almost incredible; from five to ten minutes for breakfast, fifteen to twenty for dinner, and ten minutes for supper are fully sufficient; each person as soon as satisfied leaves the table without regard to his neighbors; no social conversation follows." Captain Basil Hall, a naval officer and author with the distinction of having interviewed Napoleon on St. Helena, reported in his *Travels in North America* that he, too, had been depressed by the sight of his fellow diners glumly chomping away. "It might have been thought we had assembled rather for the purpose of inhuming the body of some departed friend than of merrily keeping alive the existing generation."

If silence and speed were characteristic of the hotel, the short-order house was distinguished by total pandemonium. Nothing of the sort had existed before, or existed elsewhere. Again, as reported by Captain Hall:

On the 21st of May, 1827, I accompanied two gentlemen, about three o'clock, to a curious place called the Plate House, in the very center of the business part of the busy town of New York. We entered a long, narrow and rather dark room, or gallery, fitted up like a coffee house, with a row of boxes on each side made just large enough to hold four persons, and divided into that number by fixed arms limiting the seats. Along the passage, or avenue between the row of boxes, which was not above four feet wide, were stationed sundry little boys, and two waiters, with their jackets off—and good need, too, as will be seen. There was an amazing clatter of knives and forks, but not a word audible to us was spoken by any of the guests. The silence, however, on the part of the company, was amply made up for by the rapid vociferations of the attendants, especially of the boys, who were gliding up and down, inclining their heads for an instant, first to one box, then to another, and receiving the whispered wishes of the company, which they straight way bawled out in a loud voice, to give notice of what fare was wanted. It quite baffled my comprehension to imagine how the people at the upper end of the room, by whom a communication was kept up in some magical way with the kitchen, could contrive to distinguish betwen one order and the other. . . . The multiplicity and rapidity of these orders and movements made one giddy. Had there been one set to receive and forward the orders, and another to put them in execution, we might have seen better through the confusion; but all hands, little and big together, were screaming out with equal loudness and quickness—"Half plate beef, 4!"—"One potato, 5"— "Two apple pie, one plum pudding, 8" and so on.

There could not be, I should think, fewer than a dozen boxes, with four people in each; and as everyone seemed to be eating as fast as he could, the extraordinary bustle may be conceived. We were not in the house above twenty minutes, but we sat out two sets of company, at least.

Epicureans repelled by such hash houses were appeased by the proprietor of the Old Bank Coffee House, William Niblo, a restaurateur with considerably more dash than his competitors. Niblo was the first to take advantage of New Yorkers' awakening interest in dining well and to realize their tastes were maturing. As a self-proclaimed gourmet, he was able to accommodate them in a manner that no other caterer in the city could approximate. His agents combed the American hinterland for recherché game to extend his staggering menus, one of which included "bald eagle shot on the Grouse Plains of Long Island," a "remarkably fine Hawk and Owl, shot in Turtle-Grove, Hoboken," a raccoon, a six-foot wild swan, and "Buffalo tongues from Russia," as well as the usual roast beef, wild turkeys, canvas-backs and quails, and a great variety of wines.

While Niblo's efforts were certainly praiseworthy, this menu also indicates that New Yorkers, in their quest for unusual dishes, were allowing a false concept of epicureanism to be foisted on them. Ever since the days of the Roman Empire, when a plateful of parakeet tongues was considered more delicious if the birds had been previously taught to speak, frankly freak foods have revolted palates in the name of gastronomic refinement. To stun a wary diner into total submission, a spectacular "presentation" of a doubtful dish was in order. Niblo seems to have excelled in this, too. "Ten minutes before the appointed hour, the word was given. 'Clear the passage! Here comes the bear!' And sure enough a huge bear, smoking hot, was served up *whole and standing.*"

Niblo's brilliant career can be accredited not only to his showmanship, but to his unerring talent for turning any situation to his advantage, a talent that has always separated the merely diligent innkeeper from the born entrepreneur. The Bank Coffee House was his first venture. Located on Pine and William streets behind the Bank of New York, it opened in 1814. In 1822 a golden opportunity for Niblo's advance appeared in the form of a pestilence. Since 1790, New York had been periodically ravaged by yellow fever and cholera, and particularly heavy epidemics had struck in 1804 and 1819. When the fever raged, which was usually from August until the first frost, New Yorkers fled across the swamp that lay between the city and the pleasant farming village of Greenwich, three miles to the north. During these almost yearly migrations, Greenwich Village would be temporarily transformed into a jerry-built shantytown, with the better part of the New York business world encamped in the surrounding fields. In 1822, as one witness observed, "In this irregular and temporary city in the field, you might find in one group, banking houses, insurance offices, coffeehouses, auctioneers' salesrooms, dry goods, hardware and grocery stores, milliner's shops, barber's shops, and last, though not least, a suitable proportion of grog and soda-water shops." Another witness, Henry W. Lanier, scrawled dramatically in his journal: "saw corn growing on the corner of Hammond and 4th Streets on a Saturday morning, and on the following Monday, Niblo and Sykes had a house erected capable of accommodating three hundred boarders."

The Sykes mentioned, it may be recalled, was the same William Sykes of the turtle disaster. As original as Niblo in his own way, he invented the meal ticket, offering at his New York Coffee House "12 tickets $5.25—a Saving of 75¢—meals generally, 50¢." Sykes met a bloody end, accidentally shot by a young man examining a pistol. "He lived too fast to live long," sniffed one local puritan, still incensed by Sykes's decadent inclusion of strawberries on a breakfast menu.

Niblo learned from his Greenwich partnership with Sykes that it was possible not only to recoup one's initial outlay of capital but make a healthy profit by servicing a sufficiently large and famished crowd under unusual circumstances. The Great Horse Race of

May 1823, held at the Union Course on Long Island, provided him with another such chance to strike it rich. Fifty thousand people from all over the country descended on the Island for the well-publicized event; the excitement was generated by a rivalry between the North, represented by the prize-winning Eclipse, and the South, represented by its greatest champion, Sir Henry. Niblo rented a house by the track and did a landslide business in turtle soup, light refreshments, and hard liquor.

By 1828, the Bank Coffee House had become too confining for Niblo and was peremptorily sold. With the proceeds he bought a full square block at Broadway and Prince Street. The property had two brick houses on it, one of which had been occupied by James Fenimore Cooper, as well as a large structure called The Stadium, previously occupied by a circus. With the addition of shrubs, gravel, small tables, and caged birds, Niblo converted the entire area into a "rural resort," a restaurant-cum-pleasure garden which he called Sans-Souci. Shortly after the opening of Sans-Souci, the brand-new Bowery Theater burned to the ground, the last in a series of civic disasters tailored to Niblo's order; within fifteen days The Stadium was transformed into a new, if perilous, home for the Bowery Company.

Years later, when the last vestiges of the "rural resort" had disappeared, and the frequently rebuilt theater was literally encircled by the Metropolitan Hotel, it was still known as Niblo's Garden. There Adelina Patti sang at the age of eight and the notorious production of *The Black Crook*, featuring chorus girls in tights, held its inflammatory opening night.

The only other noteworthy resort to attract a great following during the eighteen-twenties was situated on a tiny island connected by a ninety-yard wooden bridge to the Battery. As the military fortress that stood upon it had fallen into disuse, Congress passed an act whereby the entire works would become city property. Controversy as to its future raged immediately. One faction demanded the speedy demolition of "that large red wart which bad military tactics—in an unpropitious hour at an enormous cost —fixed on the fair face of our city." But the local government paid no attention to the dissidents and rented it to a Mr. Marsh as a place of public recreation. Barely two weeks after the lease was signed the refurbished fort was opened.

To the amusement of James Fenimore Cooper, Marsh gave it the "mongrel appellation of Castle Garden." "There is no garden, unless the area of the work can be called one; but it seems that as the city abounds in small public gardens which are appropriated to the same uses as this rejected castle, it has been thought proper, in this instance, to supply the space which is found so agreeable, by a name at least." It would have been difficult not to have found it agreeable, particularly as a portion of its parapet, shaded by colorful awnings, commanded an unequaled panoramic view of the bay by day. By night the scene was one of "gaseous glory," thanks to the efforts of the newly founded New York Gas Light Company, which had laid underground pipes all the way from its works at Hester and Centre streets to Castle Garden.

It was here, on September 14, 1824, that an event took place which was, in the words of a contemporary, "beyond what the brightest visions of Romance or the most brilliant dreams of Poetry have suggested to the imagination."

On the previous August 16, at the express invitation of President Monroe, General Lafayette, accompanied by his son, M. Auguste LaVaseur Lafayette, had sailed into New York harbor aboard the *Cadmus,* an oceangoing packet-boat bound from Le Havre. A full half century of strife, prison, and disillusionment in Europe separated the legendary general from the heroic days of the American Revolution. He had at last returned for a firsthand look at the progress of the nation he helped to found, and to receive its thanks. New York had been preparing its accolades for months.

When the *Cadmus* passed Sandy Hook on a Sunday evening, Lafayette was taken to the home of Vice President Tompkins on Staten Island, so that his reception in the city would not break the Sabbath. It was not until Monday morning that the ship of honor, the *Chancellor Livingston,* arrived to bring him to New York harbor, where a crowd of a hundred thousand waited to roar a hero's welcome. But the wealthy nineteen-year-old hero had become a penniless civilian of sixty-seven, wearing a short-cropped brown wig, old-fashioned tailcoat, and baggy buff trousers. Still and all, his initial reception aboard the official ship, as related by James Fenimore Cooper's "Traveling Bachelor," was solemnly electric:

Lafayette entered the vessel amid a deep and respectful silence. A similar reception of a public man, in Europe, would have been ominous of a waning popularity. Not an exclamation, not even a greeting of any sort, was audible. A lane was opened through a mass of bodies that was nearly

General Lafayette at the time of the American Revolution, and upon his tumultuous return to New York, fifty painful years later. *Courtesy of the New York Public Library.*

Castle Garden on a bandbox. *Courtesy of the New York Public Library.*

solid, and the visitor advanced slowly along the deck towards the stern. The expression of his countenance, though gratified and affectionate, seemed bewildered. His eye, remarkable for its fire, even in the decline of life, appeared to seek in vain the features of his ancient friends. To most of those whom he passed, his form must have worn the air of some image drawn from the pages of history. Half a century had carried nearly all of his contemporary actors of the Revolution into the great abyss of time, and he now stood like an imposing column that had been reared to commemorate deeds and principles that a whole people had been taught to reverence.

Lafayette moved slowly through the multitude, walking with a little difficulty from a personal infirmity. On every side of him his anxious gaze still sought some remembered face; but though all bowed, and, with a deep sentiment of respect and affection, each seemed to watch his laboured footstep, no one advanced to greet him. The crowd opened in his front by a sort of secret impulse, until he had gained the extremity of the boat, where, lost in the throng, stood a grey-headed and tottering veteran. The honour of the first embrace was his. I should fail in power were I to attempt a description equal to the effect produced by this scene. The old man extended his arms and, as Lafayette heard his name, he flew into them like one who was glad to seek any relief from the feelings by which he was oppressed. They were long silently folded in each other's arms. I know not, nor do I care, whether there were any present more stoical than myself: to me, this sight, simple and devoid of pageantry, was touching and grand. . . . I do not envy the man who could have witnessed such a scene unmoved.

The harbor swelled with craft of every description; battleships fired one-hundred-gun salutes which were answered from land; sailboats, barges, fireboats, fishing smacks, and dinghies bobbed in the wake of the *Chancellor Livingston;* the press barge steered perilously close, trying to catch a glimpse of the General, and the white-caps were outshone by the dazzling spume of thousands of lace and linen handkerchiefs waving from the shore.

Once on land, Lafayette was drawn through the streets in a barouche under a rainfall of flowers. After a reception at City Hall, he proceeded to his lodgings at the City Hotel where further festivities had been prepared—a sumptuous dinner, a review of the Ninth Regiment, fireworks, and the ascension of a golden balloon named the Eclipse in honor of the race horse whose portrait appeared upon it.

Lafayette's presence became inescapable. The *Mirror* reported that jewelers were displaying Lafayette medals and scent bottles;

the dry goods stores, Lafayette gloves and ribands; the tailors, Lafayette stocks; the milliners, Lafayette stockings and petticoats; and the grocers, Lafayette punch, brandy, salt herring, and molasses.

At the end of a week of receptions, Lafayette continued on his way to Boston, but in early September he returned again to New York where the celebrations reached a climax. All previous efforts were put in the shade by the subscription ball for six thousand given at Castle Garden in the General's honor. The *Evening Post* claimed, "We hazard nothing in saying that it was the most magnificent fete, given under cover, in the world." Moreover, it was a happening "to transcend the conceptions and delineations of Fancy, dazzle and bewilder the mind, and utterly baffle the powers of description." Fortunately, a few observers were able to keep themselves sufficiently under control to record their impressions of the evening.

At the entrance to the bridge a stupendous pyramid of lamps stood fully fifty feet high, surmounted by a star representing Lafayette's glory. The bridge itself was carpeted and covered by a canopy, and three bedizened steamboats ferried over those guests who were saving their feet for the cotillions. The great archway at the entrance of the Castle was smothered in lights, wreaths, and festoons. Inside, a tall pole like that of a circus tent supported a white sailcloth awning that covered the entire amphitheater, the fresh tang of a small forest of evergreens hung on the air, and banners and arms were clustered in formal arrangements on the walls amid busts of Washington, Hamilton, Jefferson, and various other statesmen. Countless gaslights sparkling in colored glass shades drew attention to a succession of inspirational allegoric conceptions—thirteen transparent columns, representing the Union, and between the two main galleries, a "representation" of the Genius of America attended by an Eagle and bearing the inscription "Gratitude to the Faithful Patriot." A mammoth cake, presented by a Mr. Ferry, was trundled up to the refreshment area where it was set upon by ladies fetchingly done up for the most part in lace-covered black or white satin gowns, headdresses of flowers, combs, or ostrich plumes, and medallions bearing a likeness of Lafayette pinned at their throats, bosoms, or waists. As the orchestras struck up, eight sets and more took the floor. During the intervals the throng was serenaded by the West Point band with military marches and airs. The guest of honor found it impossible to make his departure until two o'clock in the morning, when the sidewheeler *James Kent* arrived to bear him off up the Hudson.

Throughout the marathon of public celebrations, balls, and testimonial dinners, Lafayette seemed gratified, yet curiously perplexed. The New York that he saw was so totally changed as to be another city. Not five hundred houses remained that were standing in 1776, and of these, few could be recognized. It had become a metropolis whose "turrets sparkle in the sky." The population had spiraled from ten thousand to a hundred and sixty thousand. Faced with this, the most cosmopolitan mass in the world, Lafayette searched in vain for the leather-aproned "hewers of wood and drawers of water" with whom he had waged a revolution some fifty years before. In his deep confusion, he turned to his escort and asked, "But where is *the people?*"

Chapter III

1830-1860

Little Birds are feeding   Rich, I say, in oysters
    Justices with jam,   Haunting shady cloisters
    Rich in frizzled ham: That is what I am.

The "people," upstanding, down to earth, forthright republicans, were being swept by a rampaging economy into increasingly polarized classes; within forty years their modest, worthy, plain-spoken houses were to be hemmed in and embarrassed on one hand by the opulent and overbearing mansions of Fifth Avenue and on the other by a mortifying skid row known as the "Five Points."

The Great Fire of 1835, the most destructive in local history, necessitated the total reconstruction of the business district. The catastrophic blaze broke out on December 16 at nine in the evening and raged uncontrollably despite all efforts of the fire fighters who, in the bitter cold, could do no more than futilely stand by watching the water from the hoses freeze as it fell. The losses were wrenching—seven hundred houses, the august marble-columned Merchants' Exchange, and all vestiges of the Old Dutch town had vanished by morning.

Disasters abroad had an even greater effect on New York. The potato famine in Ireland and political chaos on the European continent drove huge waves of desperate immigrants to America. While some simply passed through New York to settle elsewhere, many were forced by circumstances to remain, and by mid-century the population had topped the half-million mark. Shanty-towns spread for miles up the riverbanks, and at one point almost ten thousand homeless children were said to be wandering the streets. These barely survived on what pittance they could make as bootblacks, paper boys, cinder pickers, or thieves. With no battles to wage for God and country, the restless youth of New York turned to other activities offering military panache or a strong element of danger—the militia target companies, volunteer fire brigades, and finally, gang warfare.

Expanding trade was bringing vast wealth to the port of New York, which, linked as it was to the west by the Erie Canal, very soon outranked its competitors on the eastern seaboard. On Broadway the traffic was so thick one pedestrian complained that it took a full ten minutes to get across. The confusion was made complete by overcrowded omnibuses bearing stylish names such as Lady Clinton, Lady Washington, and Lady Van Rensselaer. In 1853, tracks were laid for horsedrawn trolleys, provoking further havoc among the crush of hotel stages, carts, hackneys, wagons, and private carriages. By mid-century, buildings in the area stood shoulder to shoulder and six stories high, and locations as far north as Murray Hill were considered within the city limits. Fifth Avenue had become the highway of the rich, but Broadway still blazed with the lights of New York's best hotels and theaters, to say nothing of the renowned Barnum's Museum, made noticeable by the huge and lurid paintings plastered to its façade, its flapping flags, and its deafening band. The newly coined word "millionaire" could be applied with accuracy to a lengthening list of local princes, snug and secure in their shiny new mansions staffed by battalions of flunkies.

Mrs. Frances Trollope, mother of Anthony and a bilious critic of the American scene, had surprisingly kind words for New York in her *Domestic Manners of the Americans:*

I must still declare that I think New York one of the finest cities I ever saw, and as much superior to every other in the Union (Philadelphia not excepted) as London to Liverpool, or Paris to Rouen. Its advantages of position are, perhaps, unequalled anywhere. Situated on an island, which I think one day it will cover, it rises, like Venice, from the sea, and like that fairest of cities in the days of her glory, receives into its lap tribute of all the riches of the earth. . . .

Both days of glory and the riches of the earth could be spent at any of Broadway's newest luxury hotels. Following the lead of the Tremont in Boston, the world's first hotel in the modern sense of

A panoramic view of Broadway, commencing at the Astor House, mid-
nineteenth century. *Courtesy of the New York Public Library.*

the word, they introduced to a receptive public all manner of ingenious innovations and refinements. The first, Holt's Hotel (later officially known as the U. S. Hotel, and unofficially as "Holt's Folly"), was as roughly comparable to the old City Hotel as Scheherazade to Harriet Beecher Stowe.

Abandoning his career as a cabinetmaker in Salem, Steven Holt had, in 1808, opened a "victualing house" in New York where over the next decades he made a fortune on inexpensive plate dinners of the best Fulton market beef and poultry. Without a backward glance, he then proceeded to plunge all of his profits into building a hotel. Reporters were flabbergasted by its statistics; twenty-five parlors, a hundred and sixty-five rooms, a lodging capacity of three hundred and a dining capacity of a thousand, with water supplied by a steam engine from a three hundred-foot well. Furthermore, the same engine could be used for hoisting baggage to all floors, grinding knives, and turning enormous spits in the kitchen. The very vastness of the scheme dwarfed all that went before. It was said that the U. S. Hotel killed an ox every day, and could roast seven hundred pounds of meat at a time on its spits to provide no less than twenty-five hundred meals every twenty-four hours.

One final statistic would still give anybody pause: Mrs. Holt had somehow contrived to handstitch all the sheets and pillowcases in the house and to run up a patchwork quilt for each one of those three hundred beds. All of this was made ready for the opening in 1832. But despite Mrs. Holt's economies, and although he charged a dollar and a half a day plus extras for such diversions as promenading on the roof, Holt was unable to balance his budget and was forced to sell out at a hundred and seventy-five thousand dollar loss in 1835.

The only serious competition to Holt's came from the American Hotel which, during the 1830s, drew sharp appraisals from two fastidious Europeans sharing the name of Fanny. The first to arrive was the nineteen-year-old English actress Fanny Kemble, whose devastating dark eyes, ethereal pallor, and sepulchral voice reminded one playgoer of Miss Kemble's aunt, the celebrated tragedienne Mrs. Siddons, as seen through the wrong end of an opera glass. Her father, Charles Kemble, an equally noted actor, had arranged an American tour to recoup the family's financial losses at London's Covent Garden, of which he was the director. It may have been the crassly commercial nature of their visit that put Fanny out of sorts (she had been the toast of England ever since her debut one year before), but out of sorts she chose to be. Anxious to satisfy her slightest whim, the management of the American Hotel had a piano and a tray of lemonade and cake awaiting her arrival in one of its family suites; but she was not long amused, declaring the hotel a cheerless place with bedrooms as shabby as the old Shelbourne in Dublin. She insisted on dining in her own quarters and had this to report in her *Journal:*

Our dinner was a favorable specimen of eating as practised in this new world; everything good, only in too great a profusion, the wine drinkable, the fruit beautiful to look at; in point of flavor it was infinitely inferior to English hot-house fruit. Everything was wrapped in ice, which is a most luxurious necessity in this hot climate; but the things were put on the table in a slovenly, outlandish fashion; fish, soup, and meat at once, and pudding, tarts and cheese at another once; no finger glasses, and a patched table cloth—in short, a want of that style and neatness which is found in every hotel in England. The waiters, too, remind us of the half savage highland lads that used to torture us under that denomination in Glasgow, only that they were wild Irish instead of wild Scotch.

Worse still, there were not nearly enough servants and these had no bedrooms, sleeping anywhere around the hotel they could find. Fanny complained further that the rooms were an unsatisfactory mixture of French finery and Irish dirt and disorder, that the wallpaper was peeling, and that she was ingloriously attended by clouds of mosquitos. There was no bathroom, but she had had to put up with that in Paris and shortly discovered that the Arcade, a bathing establishment behind City Hall, met her exacting standards.

Despite both professional and personal triumphs in New York, Fanny continued to snipe away at the natives in her journals; this proved an unfortunate mistake, and when her observations were published three years later, she incensed every New Yorker who had shown her kindness. These included Philip Hone, the city's urbane mayor. Although the son of a carpenter, Hone was probably the most literate mayor in New York's history and one of its most articulate. He had gone out of his way to introduce the Kembles to society, giving a large dinner party in their honor three days after their arrival. When he later read her offensive report of that particular evening he found he had a few things to enter in his own *Diary,* and while quoting Miss Kemble, offered some parenthetical replies.

"The dinner was plenteous" *(that is the word)* "and tolerably well dressed" *(Peter Van Dyke ought to make her half a bow for that compliment),* "but ill-served; there were not half enough servants to do the work" *(John Stokes is not very ornamental, but tolerably useful and the others are rather smartish, I think, but I have no servants in orange-coloured inexpressibles with tinsel epaulettes; when she comes again I will endeavor to procure a bevy of them from Colonel Berkely, or some other of her distinguished countrymen)* "and we had neither water glasses" *(in this I think she is mistaken, we are never without them),* "nor, oh, horror! that absolute indispensable—finger glasses. Now, though I don't eat with my fingers—except peaches—whereat, I think, the aborigines" *(Oh, for shame, Miss Kemble, to compare Mrs. Davis, General Fleming, and Dominick Lynch to wild savages!)* "who were peeling theirs like so many potatoes, seemed to me, rather amazed. Yet I do hold a finger-glass, at the conclusion of my dinner, a requisite almost to digestion. However, as it happened, I digested without it." *(With all submission I disagree with my fastidious guest. I don't eat with my fingers, and therefore do not require finger glasses. We have them in the house, but do not frequently use them. I think it unseemly to see a company at the dinner-table, particularly the female part, washing their hands, rinsing their mouths, rubbing their gums with the finger, and squirting the polluted water back into the vessel, as was formerly the fashion in this country, a fashion which prevails yet in England in the higher circles.)*

Mayor Hone's knuckle-rapping put Fanny's invective in its proper perspective—the American Hotel could not have been as seedy as she insisted and still have attracted a faithful international clientele, a fact which is attested to by the temperamental Viennese ballerina Fanny Elssler, who, when she arrived in America in 1840, had every claim to being equally spoiled. "I must confess my first dinner astonished me outright. The table was most elegantly garnished with fine linen and beautiful glass—and would you believe it?—I was so positively assured by those who had been here that a napkin was not to be found in the country that I had brought some dozen with me.... There is 'Science' in the kitchens of the hotel, and I begin to think America is not quite so barbarous as fine folks have assured me."

The fine folks of Fanny Elssler's acquaintance were obviously unfamiliar with New York's first authentic Grand Hotel, the Astor House. In 1830 John Jacob Astor, obsessed with building the most luxurious hotel in America, bought up all the property between Vesey and Barclay streets on Broadway at inflated prices. This was to be a true rival to Boston's Tremont, and he hired the same

The Honorable Philip Hone, who refused to dress his servants in "orange-colored inexpressibles with tinsel epaulettes." *Courtesy of the New York Public Library.*

Miss Fanny Kemble, who held a finger-glass, at the conclusion of her dinner, "a requisite almost to digestion." *Courtesy of the New York Public Library.*

architect, Isaiah Rogers, to build it and the same management to run it. The cornerstone was laid on the Fourth of July, 1834, and reposited within it were a silver tablet and a portrait of Lafayette, who had died only a short time before. The plans were magnificent. The structure was to be two hundred and one feet by a hundred and fifty-four feet, virtually twice the size of Holt's Folly, and when completed it was to accommodate six hundred guests in three hundred ninety rooms. Built around a handsome inner court that provided light for every room, the Astor House was extravagantly equipped and ostentatiously furnished. It incorporated all the modern improvements of Holt's as well as those of the Tremont, a one hundred eighty-room palazzo that introduced free soap, the services of a French chef, bellboys, individual room keys, and a peculiar means of guest-to-management buzzer communication called "the announciator." Hot water pumped by steam to the upper floors was a novelty all Astor's own (although bathtubs and water closets remained in the basement).

The Astor House was lit totally by gas lamps, another innovation but not an unmixed blessing. Such large amounts of gas were consumed that the supply might give out in the middle of a ball, and guests unused to the new form of illumination were constantly blowing the flame out, with disastrous results. One visitor noted that "the door of our room is full of holes where locks have been wrenched off in order to let the coroner in."

By opening day, May 31, 1836, Mr. Astor's behemoth was solidly booked. The *New York Mirror* declared it the most commanding edifice ever raised in America with the exception of the Capitol in Washington, and further eulogized: "It quite makes one patriotick and gives us a glimmer of an idea of what might have been the enthusiastick love of country among Roman citizens, who had spent all their lives in the imperial city, while regarding with pride and triumph their baths, acquaducts, their Parthenon [sic] and Coliseum."

As one of the first guests, Davy Crocket was less impressed by the architecture than by the fortune Astor had drained from his fur business to finance the new hotel. Before his departure for the Alamo he snapped, "Lord help the poor b'ars and beavers. They must be used to being skun by now!"

The Astor House managed to do everything for itself except grind corn and fatten beef. On a printing press in the basement, elaborate menus were produced bearing the insignia of the house

—a dignified engraving of itself, a screaming eagle with the national shield, a temple of Liberty, a Phrygian cap on a staff, and a rising sun. It was suggested by one disoriented guest that it might not be amiss to print a map of the hallways as well. Grumblings could be heard about the lack of aesthetic unity in the shop signs on the ground floor, and the façade of blue Quincy granite with its austere pillared entrance on Broadway failed to please everyone. It bore an embarrassing resemblance to the local prison only a few blocks away, and one wag remarked that when he said goodbye to his friends at the door he could not suppress the feeling that he was seeing them for the last time. Matters were not improved by the fact that, like the Boston Tremont, it stood directly opposite a graveyard.

The Astor House was an overnight social success, regardless. Visitors from abroad voiced their approval, particularly the British —its courtyard was compared to an Oxford quad, its granite to London Bridge, and the bustle in its corridors to the Manchester Exchange. Even men of fashion loitered on its steps, and before and after mealtimes a handsomely dressed crowd pushed into the public rooms to see and be seen. Countless celebrities stayed at the Astor House; during its first two decades alone the register bore the arresting names of Thackeray, Jenny Lind, Black Hawk, the Grand Duke Alexis of Russia, and the Prince of Wales (Edward VII), to say nothing of such national luminaries as Clay, Pierce, Van Buren, Buchanan, Taylor, Seward, Choate, Douglas, Jefferson Davis, and Abraham Lincoln.

Guests were summoned to the dining room by what Dickens described in *American Notes* as that "awful gong which shakes the very window frames as it reverberates through the house and horribly disturbs nervous foreigners." Breakfast was served at seven-thirty, dinner (still the most elaborate meal of the day) at three, tea from six to nine, and supper from nine until midnight. The entire table d'hôte was no longer plunked down at once on the communal tables in the "good old English style," but was served from the sideboard according to individual preference from the menu card, in the "good old French style." (Which was, in fact, the relatively new Russian style. Not until 1810 were French tables cleared of all but flowers and sugar fantasies during dinner, and then only at the chic example of the Czar's ambassador to Paris, Prince Alexander Borisovich Kurakine. Rather than spreading out the entrees, roasts, vegetables, and fowl like a gluttonous picnic, the Prince ordered each dish presented individually to his guests, then removed to the pantry. Within two years eating *à la russe* was practiced by all Paris and by 1815 had become a national custom.)

The waiters were put through their paces by the captain with such brisk military precision that a visiting British officer, Colonel Maxwell, declared them as thoroughly drilled as his own men and added that a regiment of such well-trained employees would scare most small European princes to death. The robot-like waiters' drill was initiated, again, by the Tremont, but soon caught on at all the better New York hotels. The typical procedure was as follows:

1. To the fierce beating of a gong, the guests file into the dining room.

2. A platoon of waiters in glistening white gloves, aprons, and jackets form a military escort and smartly march the diners around the table to their places.

3. The grand and patronizing headwaiter, in dress coat and white tie, transfixes his staff with a stare that could blight fruit and portentously raises his official bell.

4. First clang: Waiters retreat in double time to kitchen and reappear bearing large soup tureens with silver covers. They place themselves at intervals behind the guests.

5. Second clang: Tureens are raised on high and slammed down in unison upon the table with such a bang that the chandeliers rattle and the ladies jump.

6. Soup is uncovered and served; at no time does the silver cover leave the waiter's other hand.

7. Third clang: Waiters slam down covers, sweep tureens up over heads like booty, and march back into kitchen.

The ensuing courses were mercifully less dissonant as waiters glided to and fro taking orders and whisking away used plates and cutlery. At the sideboard food was carved carelessly and wastefully; in some instances, a whole fowl or goose would be presented to a diner and he in turn would hack off either side of the breast and send back the mutilated remains untasted.

In season, game was still prolific. One October Astor House menu included black duck, lake duck, meadow hen, short neck snipe, doe witches, cedar birds, grouse, plover, rail birds, mallard duck, robin snipe, surf snipe, and venison.

Americans no longer bolted their food as rapidly as they had, but took a stately thirty minutes to get from soup to dessert. Only a handful of men would linger for another half hour over their port or Madeira. At the Astor House the delights of the Ladies' Parlor and the Smoking Room were criticized by two dismayed English visitors. One, Thomas Grattan, uninitiated into the mysteries of the American rocking chair, was alarmed by the spectacle of what he took to be "two feeble-minded creatures constantly swaying back and forth in rhythmic motion." In his 1848 *A Tour of the United States,* Archibald Prentice was appalled by the gentlemen, many of whom retired after lunch to the lobby "where they sat smoking segars, their feet up on the white marble balustrade and spitting on the marble floor. Four were in the handsome drawing room with a spittoon before them, and as each spat once a minute, on the average, there was a bending toward the spit-box every quarter of a minute. Others had their feet as high as their heads on the arms of chairs, and one who could not find such accommodation had his firmly planted on the wall."

To purchase cigars, a guest at the Astor House had only to step into the tobacconist's on the ground floor. Frequent and unnecessary trips were made in the hopes of becoming more closely acquainted with the shop's special attraction, a ravishing salesgirl named Mary Cecilia Rogers. Mary wisely let it be known that she dwelt under the eagle eye of her old mother at the latter's boardinghouse on Nassau Street. Consequently, her clientele was surprised to read in the papers that Miss Rogers had mysteriously disappeared. A week later, just as the police were about to begin an intensive investigation, she rematerialized, reciting a vague story of having taken the airs for a spell at the home of relatives in the country. There was gossip, all the same, and she was embarrassed into giving up her lucrative post at the hotel. This transpired in February of 1842. Five months later, the press had much more sensational and tragic news to report. Again Mary had disappeared, but this time her battered body had been found floating in the Hudson in the vicinity of Weehawken.

Amazingly, the only valid solution to her murder was suggested by a struggling young Southern journalist, recently arrived in New York. Too poor to frequent the Astor House, his involvement with the incident came no closer than reading the daily reports in the papers. Changing only the names of the principals and places connected with the case, Edgar Allan Poe was able to reconstruct the crime and deduce the identity of Mary's actual murderer in his chilling *Mystery of Marie Roget.*

Revolutionary steps in hotel dining were taken anew by S. B. Monnot, a French chef who opened his New York Hotel at Broadway and Waverly Place in 1844. They caused a furor. The first was the adoption of the "European plan," by which guests paid for food and services separately as desired. The second was Monnot's introduction to a hotel of the "à la carte" menu, whereby a guest would pay for each dish separately, rather than a set price for all he could eat of a vast menu. This system was used by the few existing French restaurants. What set off the greatest controversy, though, was his initiation of "room service." Formerly only celebrities or people of great wealth could command this luxury, but now it was made available to all. The public was shocked. It was considered undemocratic, against the American spirit. Finally, it was suspected to be a bore. Outspoken against the very concept was Nat P. Willis, the New York poet and journalist:

Tell a country lady in these times that when she comes to New York she must eat and pass the evening in a room by herself, and she would rather stay home. The going to the Astor and dining with two hundred well dressed people, and sitting in full dress in a splendid dining room with plenty of company—is the charm of going to the city! The theaters are nothing to that! What good are hotels if people tolerably well dressed and well behaved could not rub elbows and pick their teeth at a public table.

The importance of hotels on the New York scene cannot be overestimated. Citizens were known to pass their entire lives in them. Families tended toward hotel life, and it was said at mid-century that had there been more simple and reasonable establishments no one but the extremely rich would have bothered to operate their own households. Builders could not keep up with the expanding population and private houses were prohibitively expensive. What is more, well-trained servants became increasingly difficult to find in a land of opportunity. Therefore, it is not surprising that fully half the apartments, even in the large commercial hotels, were taken by permanent residents.

A typical guest list at one of the smaller Broadway hotels would include lawyers, editors, opera singers, professional men, brokers and merchants, men of independent incomes, fast rich youths, families from the country, widows looking for husbands, wives who liked society, and old bachelors. Efforts by such hotels to provide a suitably cozy atmosphere for family living were heartlessly ignored by the guests and left to wither by the barroom door or outside the ladies' parlor where, at any hour of the day or night, vivid character assassinations of absent residents held the company spellbound.

Thomas Gunn, a contemporary critic, was dismayed: "The man who sits beside you at dinner is as much a stranger as he who jostles past you in Broadway. He may be either a senator or swindler and you are as little surprised three days hence, to learn that he is a millionaire, as that he's going to be hanged." Conditions for children were also less than perfect:

The boys and girls at our family hotel have an especial table set for them, about an hour earlier than their seniors' dinner, where they indulge in hot, unctious soups, highly spiced French cookery, stale pastry, rewarmed dishes made indigestible with melted butter, cakes, tarts, comfits, pickles and sweetmeats. Their breakfast comprises strong coffee, hot rolls and molasses. As for exercise, their mothers can't be troubled with them while going out shopping, so they're confined to a room devoted to that purpose, or allowed the opportunity of running about the house, being children, or entering saloons, and listening to the oaths and improving conversation of the waiters, or hearing them make love to the chambermaids. If they don't thrive under this treatment, but become excitable, nervous and sick, the doctor is called in to remedy matters.

Boardinghouses took care of the less extravagant, worldly, or purely exhibitionistic souls who found domestic management as distasteful as did the birds of brilliant plumage roosting in the luxury hotels. Walt Whitman was once asked to define the nature of American Yankees. "They are," he replied, "a boarding people."

An extensive study of the various classes of New York boardinghouses was made by the tireless Gunn in his 1857 classic, *The Physiology of New York Boarding-Houses*. Gunn distinguished two major types, the Expensive and the Inexpensive. The first was usually furnished with marble tables, gorgeously bound volumes artistically arranged, thousand-dollar pianofortes, and mirrors capable of abashing a modest man to utter speechlessness; there "he will tarry the advent of stately dames, whose dresses rustle as with conscious opulence. He will precede them—they being scrupulous as to exposure of ankles—up broad staircases to handsome apartments." At the second, he is sure to be met by "Irish girls with unkempt hair and uncleanly physiognomy," and inducted into "sitting-rooms where the Venetian blinds are kept scrupulously closed, for the double purpose of excluding flies and preventing a too close scrutiny of the upholstery."

Gunn deals with a number of subcategories; the Cheap Boardinghouse on a Large Scale is examined at some length. The landlady is a woman of approximately fifty-five summers and married —"the union originated in the male party's running deeply in arrears for board, and honorably compounding the same by matrimony." Her long, whitewashed dining room is furnished with two long tables ringed by fifty or more uncomfortably high stools.

Here we have known the beef of tougher consistency and more veiny construction than desirable, and the potatoes to exhibit as many eyes as Argus; but on the whole, the diet was endurable. Our chief objection applies to the cuisine of such boarding-houses. The meals were universally served up "neither cold nor hot"—a state of which St. John did not approve in Laodicean Christians. The soups, too, might have been improved by a less liberal allowance of grease and unground pepper, of which latter there always remained a deep sediment—as of small shot—in each plate.

The Fashionable Boardinghouse where you Don't Get Enough to Eat is also subject to criticism. Its decorations were of that "peculiar French-New York order which displays more of gilding than of good taste, and more of plate glass than either." One is made to sense the ponderous delicacy of aristocratic exclusiveness, the accommodations are the very model of refinement, and a full-grown man may toss and turn in a bed that is "small and snow white—like a snow drift on a child's grave." No crass bells here. Rather, a servant taps discreetly on your door. But he is not summoning you to much:

At breakfast the ladies are very lively and chatty—so much so, indeed, that a cynic might suspect the existence of a design to keep the boarders' jaws otherwise employed than on breakfast, which is light, tasty, and unsubstantial. There are very small mutton chops, pâtés, nick-nacks, and French bread and coffee—made also à la français. Lunch, served at 1:00, is a paltry affair—pie, delicate shavings of cold meat, and coffee. Dinner at 6:00 consists of five courses commencing with a thin whity-brown soup and concluding with dessert of which water melons form the staple in summer and frosted apples in winter. A cup or two of green and very weak tea, served in the adjourning parlor, after the lapse of a half an hour, concludes the repast.

By sharp contrast, Gunn makes a foray into The Dirty Boarding-house where the food is plentiful, but porky.

Porgies—purchased in their decadence from perambulatory fish-vendors —sometimes varied this anti-Hebraical peculiarity. The coffee tasted like diluted molasses, flavored with roast peas, chicory, Flanders-brick and dirt. The hashes were tallowy. The buckwheat cakes partook equally of the characteristics of flannel and gutta-percha—and sometimes had insects (known as Croton Water bugs) in them. But pork was the universal dish. Everybody was over-porked. Boarders had rashes brought out upon them in consequence, and we remember one consulting a doctor under the impression that he had contracted varioloid.

Then there is The Serious Boardinghouse, displaying this cunning verse prominently on the staircase:

I'm a little Temperance boy,
*   Twelve years old*
And I love Temperance
*   Better than gold!*
Every little boy, like me
The Temperance-pledge should sign,
For God loves little boys
Who don't love wine!!!

This is mercifully counteracted by The Theatrical Boarding-house, where the landlady as well as the cast of residents lies in wait to put the bite on the unwary for just-a-little-something-dear-to-tide-me-over-the-next-few-days. Meals are served at no set time, if at all, but beer can be sent out for at any reasonable or unreasonable hour. On one murky Sunday afternoon, to the author's delight, three gentlewomen of the stage were surprised in the back parlor in the consumption of cigars and hot brandy and water.

A growing horde of single men in New York could neither endure boardinghouses nor afford hotels, but rented furnished rooms and took their meals where and how their funds permitted. Fulfilling their needs were a huge number of eating houses, indispensable as well to the entire downtown business population.

The New York businessman had a positive commercial advantage over his opposite number anywhere in the world: he could hang up his "Out to Lunch" sign, eat heartily for next to nothing, and be back at work within a quarter of an hour. This system was evaluated in 1848 by Geoge G. Foster of the *Herald Tribune.*

"Nowhere else, either in Europe or America, does anything like it exist. It is the culmination, the consummation, the concentration of Americanism; with all its activity, perseverance, energy, and practicality in their highest states of development. In this view the eating-houses of New York rise to the dignity of a national institution."

Foster noted that there were three distinct and separate orders of eating houses, the model of which Linnaeus would classify as Sweeneyorum, Brownivorous, and Delmonican. The first prototype referred to Daniel Sweeney's frenetic short-order house on Ann Street, the lowest on the economic totem pole and a direct descendant of the plate house. Sweeney's was known as a "sixpenny house," serving breakfast, dinner, and tea. The main room was laid out like a church with benches and table replacing the pews, and as no printed bill of fare was ever available, in keeping with the management's low-overhead policy, a barker was stationed at the door keening like a muezzin:

Biledlamancapersors
Rosebeefrosegooserosemuttonantaters—
Biledamancabbage, vegetables—
Walkinsirtakeaseatsir.

Once inside and safely seated, a sightseer could witness "a nice bit of rosegoose . . . flying down the aisle, without its original wings, followed closely in playful sport by a small plate of bile beef and vegetables, until both arrive at their destination; when goose leaps lightly in front of a poet of the Sunday press, who ordered it probably through a commendable preference for a brother of the quill; while the fat and lazy beef dumps itself down with perfect resignation before the 'monstrous jaws' of one of the b'hoys who has just come from a fire in 49th Street. . . ."

Brown's, on the next rung up the ladder, reigned for many years as the heir to the English-style chophouse. There on Water Street, a meal cost the double of one at a genus Sweeneyorum, but the pace was more leisurely, the waiters actually passed within hailing distance every once in a while, and the bill of fare appeared on a printed card with fixed prices, making it impossible to be cheated, as many luckless diners had been at Sweeney's upon ordering an extra side dish of bread, pie, or pickles. Brown's would have resembled an old English-style tavern, but for the fact that two thousand customers charged in and out of it daily.

Candidates for the Delmonican category were discouraged by the prestige of this superstrata's namesake, Delmonico's. The Delmonicos have been credited with introducing haute cuisine to America, and this may be so. They certainly introduced socio-gustatorial snobbery.

New York got its first taste of Delmonican finesse as an indirect result of Lafayette's farewell visit. John Delmonico, founder of the dynasty, had spent the greater part of his adult life, not in the finest kitchens of Europe, but as captain of a three-masted schooner plying its trade between Havana, the West Indies, and New York. Resting at anchor in New York harbor during Lafayette's reception, he was able to observe the preparations for it and came to the conclusion that the city's catering facilities left much to be desired. The following year he gave up the sea for good and bought a wineshop on the Battery. It succeeded well enough to finance a business trip to his native canton of Ticino; there he persuaded his elder brother Peter, a talented confectioner, to return to New York with him. In 1828 the Delmonicos plunged into the world of grande cuisine, announcing in both French and English that the public might henceforth be served with Continental confections at 21-23 William Street. The bill of fare, though bilingual, was familiar to any patron of Corré, Delacroix, or Contoit—chocolate, coffee, bonbons, pastries, liqueurs, pâtés, and the ubiquitous "confectionery of all sorts."

With the addition of hot dishes to the menu two years later, the Delmonicos' reputation spread rapidly. (The employment of a lady cashier, a previously unheard of arrangement, did nothing to hamper business.) The rank and file of the Anglo-Saxon carriage trade were now introduced to the cuisine that the local French had been keeping more or less to themselves; items such as "macaroni and filets" were served to customers at six simple pine tables by one of the brothers, impeccably aproned and crowned with a paper *toque blanche*. Not all New Yorkers were immediately converted; although renowned for his eclectic tastes, Mayor Hone complained in his diary: "Moore, Girard, and I went yesterday to dine at Delmonico's, a French *restaurateur*, in William Street, which I had heard was on the Parisian plan, and very good. We satisfied our curiosity, but not our appetites. . . ." Nevertheless, John sent for his nineteen-year-old nephew Lorenzo to assist him, and shortly thereafter three other nephews, Siro, François, and Constantine, joined them.

A move from the cramped quarters on William Street was hastened by the great fire of 1835. The building was reduced to cinders. When the question of reconstruction arose, the Delmonicos decided to abandon the original site and build on a larger plot at the corner of Beaver and William streets. Secure in their prosperous patronage, the first of the dynasty's grandiose schemes was realized by John with the assistance of Lorenzo, who even at an early age gave indication of his visionary genius as a restaurateur. In the new house, built at an acute angle and standing three stories high, the Delmonicos revealed an unsuspected streak of Italianate romanticism by flanking the entrance with two authentic Pompeian columns, but in their attention to the most tedious of details they remained inflexibly Swiss; Lorenzo and Siro themselves did the shopping every morning at the Washington Market, and in 1840, denouncing the variety and quality of regional farm produce as hopeless, they purchased a tract of land in Brooklyn and personally directed the cultivation of their own.

In the splendidly equipped kitchens, haute cuisine, in its true French definition, was honored in a manner rarely to be found beyond the Pyrenees and the Rhine, and never in America, not even at the Astor House. The Delmonicos, none of whom had been classically trained in their field, must have imported their chefs directly from France.

The concept of the "restaurant" itself originated in France. One Monsieur Boulanger, a Parisian dispenser of soups, advertised his dishes as "restoratives," or *"restaurants."* The word became popular, and a few years later, in 1782, the famous Beauvillier, aware that a traveler had no chance of a decent meal at a Paris inn, organized the first proper "restaurant," with a selection of respectable dishes. After the Revolution, the idea spread quickly, and Baleine opened the trend-setting Rocher de Concale. By the early nineteenth century Paris teemed with restaurants. For haute cuisine there were Very's, Hardi's, and the Quadron Bleu, to say nothing of Tortoni's, the center of romantic intrigue and the traditional site of the duelist's last breakfast, a bracer of pâtés, game, fish, broiled kidneys, and iced champagne. As Arthur Bryant, the British historian, concluded in *The Age of Elegance:* "With his *piquante* sauces and *petit-plats,* his gilded mirrors, bright lights and marble tables—so different from the smokey, wainscotted chop houses of London—the restaurateur was the residuary legatee of the Revolution."[1] It only remained for the Delmonicos to introduce a more or less genuine French "restaurant" on a large scale to New York to complement the American "eating houses."

The menu, under the heading "Carte du Resturant [sic] Francais, des Freres Delmonico," included three hundred forty-six entrees in French and English. Some of the translations seem curious: *mariné* became "pickled," while *meringues à la crème* became "French kisses," *soufflé au riz* a "rice soufflay," *macédoine* a "jumble," and *coquille de faisan* a "scalloped partridge." *Hors d'oeuvres* was translated as "side dishes."

Eleven soups from "consommé" to "diet" *(de santé)* were listed; forty hors d'oeuvres, including such items as lamb's kidneys in champagne sauce, Westphalian ham, mayonnaise of lobster, artichokes à la poivrade, oysters scalloped in their shells, chipolata sausage, and fried ham and eggs; twenty-nine variations on beef (filet with olives, or tomato sauce, mushrooms, Madeira sauce, etc.); twenty-seven veal dishes (breaded with truffles, fricasseed with gravy, "spinage," or sorrel, sealed in oiled paper, and a carbonade with green peas or herbs); fourteen poultry dishes (deviled turkey leg, broiled capon with jelly and rice, compote of pigeon, chicken aspics, and pâté de fois gras—mundanely translated as "Goose liver pie of Strasburg.")

Following the game (quail, pheasant, partridge, woodcock, venison, hare, wild duck, and "Welsh rabbit") came twenty roasts (beef, veal, lamb, chicken, turkey, Guinea fowl, snipe, plover, etc.) and forty-eight fish specialties (turbot pie, salmon with caper sauce, "toasted shad," mackerel with black sauce, eels stewed or *"Tartar fashion,"* fried smelt, turtle steak, pike à la Genoise, "muscles" stewed or aux fines herbes).

"Entremets" included endive with sauce, "jumble" of vegetables, white beans prepared in five different manners, "crusts of mushrooms," truffles with champagne sauce, and an omelet "Celestine."

Pastry, cakes, and desserts embrace "Puddin au riz," Gruyere cheese, cream cake, and sugared raspberries; these were followed by coffee and punches, twenty-four liqueurs, and fifty-eight wines.

The wealthier members of the business world appeared daily at Delmonico's, the French importers among them bringing an air of easy European elegance. Yet the midday crowd was not limited to their ranks alone. In the words of George G. Foster (*New York in Slices, by an Experienced Carver,* W. F. Burgess, 1848), on a typical afternoon there could be encountered an elderly French gentleman dining with his son. . . . At the next table Richard Grant White, the music critic—tall and striking with a ruddy beard. . . . He has evidently just finished breakfast; for there stands the gigantic chocolate cup, deep enough for him, long as he is, to drown himself in it, and before him lie the delicate remains of the *oeufs en miroir* with which he has been trifling. . . . To the right yonder, nestled among piles of preserve-jars, scalloped oranges and *meringue à la creme* are four persons somewhat worthy of notice. . . . Parke Godwin, an Associationist, a Swedenborgian, a Homeopathist, a Hydropathis, a—we know not what. Opposite him, in almost every respect, is a tall, shapely man, daintily dressed . . . and in a style that lets peep out here and there an all-pervading and inextinguishable love for the beautiful—Willis (Nat P.), . . . a sort of human julep, composed of spirits and sugar, with a pungent flavor of the mint which makes the very air spicy about him. . . . At his right hand is Charles Fenno Hoffman, the song-writer of America. The fourth of this group is only an Editor, and not worth wasting time or foolscap upon.

Of the other dawdlers "With the exception of moustaches inconveniently long, which seem to be dragging the soup spoon for dead bodies, they have no salient nor remarkable features."

John Delmonico's own dead body was discovered in November by friends in a thicket near Islip, Long Island, after a day's hunting. He was a novice shot, but that day beginner's luck was with him —up to a certain point. A stag crashed through the trees within close range, John fired, and the quarry fell, followed shortly by the hunter himself who, carried away by the excitement of it, succumbed to an apoplectic fit. His remaining loved ones could not have been accused of excessive crepe hanging; as a matter of fact their impressive resilience to grief stirred some sardonic memories for Mayor Hone. In his *Diary* he mused that

"Business is business" as some man says in some play. The following notice, which was published the day after the funeral of poor Delmonico, is very much in the style of the inscription on a tombstone in Père-Lachaise, which runs somewhat in this form: "Here lies the body of Pierre Quelquechose, who died so and so. This monument is erected to his memory by his widow, who takes this occasion to inform her friends and customers, that the pastry-cook establishment is continued at such a number Rue Saint Honoré, where she will be happy to receive their orders."

This is the counterpart. A card: "The widow, brother, and nephew of Lorenzo, of the late much respected John Delmonico, tender their heartfelt thanks to the friends, benevolent societies, and the Northern Liberty Fire Engine Company, who accompanied his remains to his last home.

The establishment will be reopened to-day, under the same firm of Delmonico Brothers, and no pains of the bereft family will be spared to given general satisfaction."

Lorenzo felt that the Beaver Street location was too far downtown to compete with the fashionable new houses above Bleecker Street and opened yet another restaurant at Broadway and Chambers Street on the site of the notorious Colt Murder.

On Friday, September 17, 1841, Samuel Adams, a printer, appeared at the Broadway Chamber Street office of John C. Colt, an accountant and professor of ornamental penmanship, to collect a sum he felt Colt still owed for the printing of a text on the science of bookkeeping. His manner was undiplomatic, there was a certain amount of grappling around, and Colt, normally a courteous fellow, brained him with a hammer. Colt then nailed Adams into a box and had a cartman haul it off to the docks, where it was transferred to the hold of a boat bound for Brazil. The authorities closed in at the last moment, reclaimed the victim, and sentenced Mr. Colt to the gallows.

Four hours before his execution, however, Colt was granted a sentimental last wish and was married in his cell to a Miss Caroline Henshaw, the ceremony being witnessed by John Howard Payne, author of "Home, Sweet Home." All were amazed at the groom's good cheer; this became more understandable two hours later when an elaborate plot for his escape, involving the substitution of a corpse and the igniting of the jail's cupola, was pulled off without a hitch. Colt's subsequent fate remains a mystery to this day.

The constant uptown movement of society, rather than any ectoplasmic harassments on the part of Mr. Adams, left Lorenzo dissatisfied with the Chambers Street location. In 1861, the Grinell mansion on the northeast corner of Fifth Avenue and Fourteenth Street was put on the market, and he snapped it up. A year was spent on extravagant renovations. When it finally opened, there could be no question that this was the Delmonicos' crowning tribute to gastronomy. It appeared to have been not so much a question of the Delmonicos keeping up with New York as New York keeping up with the Delmonicos. By the middle of the century, the city was running apace, the Croton waterworks were gushing effectively, Central Park was mapped out, the Hippodrome loomed over Madison Square, and suggestions for an elevated railway were being met with howls.

Forty centuries looked down on New York in the form of the truncated Egyptian pyramid that encompassed the Forty-second Street reservoir. Promenades could be taken along its ramparts, and around its base, of a Sunday, crowds jostled one another and nimbly side-stepped a new nuisance, the sandwich-board men, those perambulatory advertisements who, along with painted omnibuses, garish stamped signs, hawkers, and handbills, were giving New York the frantic atmosphere of a fire sale. News reports became increasingly difficult to unearth in the daily muddle of commercial exhortations screaming from every page of the papers. Lydia Maria Child, a Bostonian by birth and New Yorker by adoption, diarist, moralist, and food authority, in her *Letters from New York* points out that "ingenuity exhausts its resources in every variety of advertisement. These articles are in such demand that the writing of them is a profession by itself, sufficiently profitable to induce men to devote their time to it for a living." Testimonial verse was in great demand. Messrs. Pease and Co.'s Hoarhound Drops, for example, were presented to a waiting world not merely as a penny candy of merit, but as a universal panacea.

See where the victim of Consumption sighs—
With hectic cheeks and spirit blazing eyes—
Her frame all wasted by disease and pills
From quacks received, in vain to cure her ills,
Now look again! as buoyant as the breeze,
Behold her bounding under yonder trees!
What miracle is this? What! worn away
Like a lone sunbeam at the close of day
Thus dance along! Yes; Pease has kindly brought
The CANDY here, and thus the magic wrought.

Commerce acquired new prestige with the inauguration by President Franklin Pierce of New York's Crystal Palace at the country's first World's Fair in July of 1853. The palace, a slightly smaller model of the famous exhibition hall in London, stood directly behind the reservoir on what is now Bryant Park. Visitors could admire everything from statues of Parian marble to the recently invented Parisian tin can. Among the native contributions was Cyrus McCormick's reaper.

The previous year a clever entrepreneur named Latting put up a tower on the corner opposite the future Crystal Palace as an observatory and ice-cream parlor. An architectural fore-runner of the Eiffel Tower, it stood approximately twenty-seven stories high, was equipped with a temperamental "vertical railway" or steam elevator and a ground floor café, and did a landslide business with the fair-goers.

To accommodate a raft of visitors from all parts of the world, many of the older hotels had their faces lifted. Two magnificent ones, the Metropolitan and the Saint Nicholas, were built from scratch. The Metropolitan rose up on the site of Niblo's Garden. The theater still operated under Niblo's attentive management, and remained a city landmark; rather than tear it down, the Leland brothers deftly embraced it in the crook of their huge L-shaped hotel which was equipped with a brownstone façade, twelve miles of water and gas pipes, cloth napkins at every meal, but no dinner gong. Two years later the Metropolitan was given a run for its money with the debut of the St. Nicholas Hotel, on Broadway between Broome and Spring streets. The St. Nicholas intended to knock New York's eye out with its worldly decor. The interior was conceived as a procuress's dream and a charwoman's nightmare of beveled mirrors, cut glass, marble, and bronze *objets d'art*. One English visitor refused to leave his shoes outside his bedroom in fear they would be gilded in the night. Another guest noted that every dish in the dining room appeared to have a powerful spirit lamp beneath it and that the gold brocaded curtains had set the management back forty-five dollars a yard. Small wonder that the St. Nicholas was rumored to be the favorite stamping grounds of "Shoddy," that amorphous, glamorous sect of tycoons, social climbers, international adventurers, and professional beauties from which eventually descended lobster palace society, café society, and the jet set.

The final word in stylishness came with the building in the late fifties of the Fifth Avenue Hotel at the juncture of Fifth Avenue and Broadway, regarded as so far uptown it could succeed only as a summer resort. Constructed of white marble with an aura of "restrained elegance," it was considered worth a visit, if only to catch a ride on its passenger-conveying elevator, with a station on each floor. Also on hand were a house detective and his lieutenants. Personal possessions of the rich and the privacy of celebrities had to be protected by all first-class hotels.

Many distinguished Europeans bore home breathtaking tales of the American grand hotels, particularly those in New York; they were cities within cities, completely self-contained, with billiard

The dining room of the Fifth Avenue Hotel, where as late as 1859 "people tolerably well dressed and well behaved could . . . rub elbows and pick their teeth at a public table." *Courtesy of the New York Public Library.*

rooms, hair salons, telegraph, news, and ticket offices, paper and cigar stands, libraries, florists; the dining rooms were open from 6:00 A.M. to 3:00 A.M. and at breakfast, luncheon, dinner, tea, and supper if he pleased, a guest could eat everything on the bill of fare, in large or small sampling portions, and was never expected to tip the waiter. Already, the grand hotels seemed to embody the spirit of American democracy—they were literally "Palaces of the People," where any man, with a few extra dollars to spend, could live like a king, at least for a few days.

The Astor House, renovated for the World's Fair, served over a thousand pounds of beef loin and ribs a week, four hundred pounds of crabs, perhaps eight thousand oysters, and ten thousand eggs. An enormous iron grate in the kitchen could support thirty roasting turkeys at the same time while a five-foot-square gridiron over coals was used for broiling steaks, chops, cutlets, and birds.

Officiating over the steaming cauldrons, pastry tubes, and intricate salmis was Charley Roux, master chef, a product of the Royal Palace kitchen in Paris. And sitting on his right hand was the chief confectioner, furiously concocting "set pieces" of spun sugar, gold leaf, ice cream, candied fruits and flowers. ("Set pieces," or *pièces montées*, reached their zenith in France with the dazzling constructions of Carème, chef to kings and to Talleyrand. Carème devoted every free moment of his apprenticeship to architectural drawing with the intention of reproducing the great edifices of Paris in sugary detail; these creations were sometimes said to surpass their models. His technique was widely copied both in France and abroad.) The success of the Astor's kitchens is born out by a tribute to one of its desserts, unpromisingly called "Soft Vanilla Custard."

It came in a tall glass as slim as a Greek vase. Its color was palest gold, cunningly flecked with infinitesimally small moles of nutmeg, and its top was of whipped cream whiter than ivory. Investigation with the spoon disclosed the fact that down the middle of the custard ran a fine spine of ice, whence radiated tiny spiculae that tickled the palate already swooning with delight. I had six of them for dessert the first time—and stopped then only as one should always arise from the table a little bit hungry.

It is interesting that it was from this time forward that losses sustained by the sale of food were recouped by sales from the hotel's splendidly stocked bar. A first-rate bartender of one of the great hotels of the last century was expected to maintain a working repertory of at least one hundred and fifty cocktails, shrubs, slings, punches, cobblers, juleps, smashes, flips, and nogs. These were consumed at all hours of the day and night; "cocktails," however, were first ordered only as eye-openers.

The supreme creative genius behind a New York bar was Professor Jerry Thomas, and it was he who raised the lowly matutinal cocktail to its glorious estate. His earliest eastern headquarters were at the Metropolitan Hotel, where he presided until 1859. The Professor had a palate of almost neurotic sensitivity, and he concluded that the bitters commonly used in cocktails were solely responsible for the drinks' limited appeal. He perfected a secret formula, bottling it under the trademark "Jerry Thomas's Own Bitters," and history was made overnight. He then published the country's first authoritative manual on the art of imbibing, *The Bon Vivant's Companion, or How to Mix Drinks.* New and updated editions were printed throughout the next twenty-five years, but all of them included the Professor's two earliest inspirations, the Blue Blazer and the Tom and Jerry. It has been claimed that the latter drink is British and named for the heroes of Pierce Egan's 1821 novel, *Life in London, or Days and Nights of Jerry Hawthorne and His Elegant Friend Corinthian Tom;* but no sound evidence has ever been given to support this theory and the recipe remains the Professor's own:

Tom and Jerry

Use punch bowl for the mixture.

5 lbs. sugar	1-1/2 tsps. ground cinnamon
12 eggs	1/2 tsp. ground cloves
1/2 small glass of Jamaica rum	1/2 tsp. ground allspice

Beat the whites of the eggs to a stiff froth, and the yolks until they are as thin as water, then mix together and add the spice and rum, thicken with sugar until the mixture attains the consistency of a light batter.

To deal out Tom and Jerry to customers: Take a small bar glass, and to one tablespoonful of the above mixture, add one wine-glass of brandy, and fill the glass with boiling water; grate a little nutmeg on top.

No order for a Tom and Jerry would be filled until the first snow whitened the window sills. On this the Professor stood adamant, going so far as to smash a bowl of batter at a rival's bar one balmy September evening. The Blue Blazer was as much an exercise in prestidigitation as an antidote to bitter winter weather; heated whiskey and boiling water were combined in equal measure, ignited and poured back and forth like a blazing comet's tail from one silver mug to another until the flames subsided. Here protocol was even more sternly observed; not until the mercury dipped to ten degrees above zero did Professor Thomas deign to perform his incendiary ritual.

With or without the World's Fair, mid-nineteenth-century New York teemed with visitors the year round; but the population was perpetually curious, and anyone of any importance would find it difficult to remain anonymous. Mrs. Trollope, quite probably to her untiring vexation, went unnoticed in America, that is until the publication of *Domestic Manners of the Americans*, at which point her name became anathema. As mentioned earlier, her fancy was caught by the city in 1832, yet there were typical reservations about mint "julaps," chewing tobacco, and queer combinations of food—eggs and oysters, ham and apple sauce, beefsteak and stewed peaches, salt fish and onions. It took almost twenty years for her to relent on the matter of juleps. In her 1849 novel, *The Old World and the New,* a pick-me-up was prescribed for the characters, languishing in the heat of the Cincinnati countryside:

It would, I truly believe, be utterly impossible for the art of man to administer anything so likely to restore them from the overwhelming effects of heat and fatigue, as a large glass filled to the brim with the fragrant leaves of the nerve-restoring mint, as many lumps of the solidly pellucid crystal-looking ice, as it can conveniently contain, a proper proportion of fine white sugar, (not beet root), and then—I would whisper it gently if I knew how—a whole wine-glass full of whiskey poured upon it, to find its insinuating way among the crystal rocks, and the verdant leaves, till by gentle degrees, a beverage is produced, that must create a delicious sensation of coolness, under a tropical sun, and a revival of strength, where strength seemed gone forever.

In *Domestic Manners* she claims the abundance of ice in New York delighted her ("I do not imagine that there is a house in the city without the luxury of a piece . . . to cool the water, and harden the butter"), but informal behavior at the theater and chewing tobacco did not:

I observed in the front row of a dress-box a lady performing the most maternal office possible, several gentlemen without their coats, and a general air of contempt for the decencies of life, certainly more than usually revolting . . . we saw many "yet unrazored lips" polluted with the grim tinge of the hateful tobacco, and heard, without ceasing, the spitting, which of course is its consequence. If their theatres had the orchestras of the Feydeau, and a choir of angels to boot, I could find but little pleasure, so long as they were followed by this running accompaniment of *thorough base.*

The reply came in the *New York Constellation* of July 14, 1832. It started off with the lines, "Mrs. Trollope is commendably bitter / Against the filthy American spitter. . . ." From that day forward, offenders against etiquette in New York's theaters were hooted down with lusty cries of "Trollope! Trollope!" (The chewing and spitting of tobacco was the great American practice, or rather vice, that most upset the sensitivities of European visitors. The country seems to have been a maze of spittoons at ankle level. During the nineteenth century every member of the Senate and House of Representatives had his own "spit box," and there was a giant spittoon at the door of the White House itself.)

A decade after the visit of Mrs. Trollope New York was to have its hand bitten again by a deceptively domesticated literary lion, the great "Boz" himself. Charles Dickens, in *American Notes*, ill-paid his effusive hosts by taking Mr. Tony Weller's advice to Mr. Pickwick to escape to America and "come back and write a book about the 'Merrikins' as'll pay all his expenses and more, if he blows 'em up enough."

Dickens, thirty years old, arrived in the city by way of Boston on February 13, 1842, with his wife, a sackful of imposing invitations, and a cold. There had been no stinting in the preparations for his reception, which provoked a snide editorial in the local French journal, *Le Courier des Etats-Unis*. The *Evening Post* took exception and dismissed the reasons suggested by the French for such national enthusiasm.

The first of these is the instinctive desire which Americans have to refute the accusations of coldness, self-love, money-making, and puritanic strictures brought against them by foreigners, and to do this they resort to unusual displays, just as a man charged with avarice will indulge in some splendid extravagance to retrieve his reputation. The second is that they suppose the author has come to study their character and institutions, and are anxious to conciliate his good opinion, even to the extent of bribing his judgment. Like a young miss who expects a new lover, they take care to make their toilette with pains. And the third is the austerity of our religious tenets and forms, preventing the people from those everyday social amusements to which other nations are accustomed, so that they take these occasions to give vent to their natural love of hilarity and excitement. Six days are employed in the counting-house, and the seventh in the church.

We do not think our French critic has gone to the bottom . . . we have the secrets of the attentions that have been showered on Mr. Dickens. That they may have been carried too far is possible; yet we are disposed to regard them, even in their excess, with favour. We have so long been accustomed to seeing the homage of the multitude paid to men of mere title, or military chieftains, that we have grown tired of it. We are glad to see the mind asserting its supremacy, to find its right generally recognized. We rejoice that a young man without birth, wealth, title, or a sword, whose only claims to distinction are in his intellect and heart, is received with a feeling that was formerly rendered to only conquerors and kings. It is but a fair return.

Ex-Mayor Hone mused that never had a taller compliment been paid such a little man. Hone was a member of the reception committee and instrumental in planning the Boz Ball at the Park Theater on Monday, February 14, as well as the dinner to Dickens at the City Hotel on Thursday of the same week.

Plans for the ball were elaborate, and the decorators ran amok. Lobbies, halls, saloons, boxes, and greenrooms groaned under a burden of gilt, muslin, medallions, festoons, wreaths, silver stars, and tassles. One platform was erected so that the guest of honor and his hosts could be scrutinized at leisure by a milling crowd of three thousand; another was set up for the presentation of tableaux vivants, among them "The middle-aged lady in the double-bedded room," "Mrs. Bardell faints in Mr. Pickwick's arms," "Little Nell leading her grandfather," "The red-nosed man discourseth," and "Washington Irving in England and Charles Dickens in America." Refreshments were catered by Downing, the keeper of an oyster saloon who dared charge twenty-two hundred dollars for his services.

Dickens and wife arrived a few minutes after nine and an aisle was immediately cleared for them. The novelist wore a black suit with a flashy waistcoat, long yellow sideburns with tufts at the jawline, "rings and things and fine array," and a general dandified air that seemed somehow out of joint with his sober reputation. Mrs. Dickens was dressed in a white figured Irish tabinet trimmed with blue flowers, a pearl necklace and earrings, and a wreath wound around her mop of Nell Gwynn ringlets. Hone was not overcome, flatly describing her as a "little, fat, English-looking woman, of an agreeable countenance, and, I should think, 'a nice person.' "

Irving, Cooper, Halleck, and Bryant were in attendance as well as the flushed and beaming General George P. Morris. T. L. Nichols later recalled in *Forty Years in America:* "We tried to dance. Mrs. General Morris honoured the . . . author with her fair hand for a quadrille, but the effort to dance was absurd. I remember being in a set with two young army officers who were afterwards heroes in Mexico, but their prowess could do little toward carrying their partners through the galop in such a rush . . . it was like dancing in a canebrake." Despite the crush, all hands agreed the next day that the ball had been an unqualified success, even Hone, who prefaced his glowing account with a relieved "The agony is over."

What Dickens thought of the food may be gathered from his generalization that Americans ate "piles of indigestible matter." This was probably a little unfair, if we may judge the state of English cooking from a book coyly entitled *What Shall We Have for Dinner,* by the mysterious Lady Maria Clutterbuck, in stark reality Mrs. Charles Dickens. Prominent among her recipes are batter pudding, suet dumplings, and sole in brown gravy, a far cry from the menu served to Dickens on his second visit to New York in 1867. Of course New York fare had improved during the interim, and Delmonico's had taken a firm hold. Still, it must have come as a pleasant surprise.

April 1867
Dinner at Delmonico's in Honor of Charles Dickens

MENU

Huitres sur coquille

Potages

Consommé Sévigné   Crème d'asperges à la Dumas

Hors-d'oeuvre chaud
Timbales à la Dickens

Poissons

Saumon à la Victoria   Bass à l'Italienne
Pomme de terre Nelson

Relèves

Filet de boeuf à la Lucullus   Laitues braisées demi-glace
Agneau farci à la Walter Scott   Tomatoes à la Reine

Entrées

Filet de brants à la Seymour
Petits pois à l'Anglaise
Croustades de ris de veau à la Douglas
Quartier d'artichauts Lyonnaise
Epinards au velouté
Côtelettes de grouses à la Fenimore Cooper

Entrées Froid

Galantine à la Royale   Aspics de foies-gras historiés

Intermède

Sorbet à l'Americaine

Rôtis

Bécassines   Poulet de grains truffés

Entremets Sucrés

Pêches à la Parisienne (chaud)
Macédoine de fruits   Muscovite à l'abricot
Lait d'amandes rubané de chocolat
Charlotte Doria
Viennois glacé à l'orange   Corbeilles de biscuits Chantilly
Gâteau Savarin au Marasquin
Glaces forme fruits Napolitaine
Parfait au Café

Pièces Montées

Temple de la Littérature   Trophée á l'Auteur
Pavillion international   Colonne Triomphale
Les Armes Britanniques   The Stars and Stripes
Le Monument de Washington   La Loi du destin
Fruits   Compote de pêches et de poirs   Petit fours
Fleurs
Dessert

Faced with this kind of menu, modern readers may wonder how
much was actually eaten. The answer is, a good deal. The guest
would be offered everything, he could take as much or as little of

each dish as he liked, and the eating often continued for most of the night. Moreover by 1867 the era had arrived when a man prided himself on his girth.

While the fuss made over Dickens in 1842 showed an awakening cultural interest on the part of New York, the pandemonium on the arrival in 1850 of P. T. Barnum's Swedish star, Jenny Lind, indicated an incipient madness. Barnum had circulated such beatific tales of Jenny that she was met by twenty thousand fervid disciples, some of whom were so transported as to topple off the pier. Her legend alone was quite enough to stir the Victorian heart. Not only was she hallowed for her generosity to the distressed, but she had eschewed for life both the city of Paris, because of its vicious character, and the whole of Grand Opera, because the heroines she might be asked to portray were frequently guilty of lapses in chastity. Hysteria over the thirty-year-old nightingale (blonde, broad-nosed, pious of glance) reached an insensate pitch and a quarter of a dollar was charged merely for the privilege of bidding on tickets for her American debut. The first one went for two hundred twenty-five dollars and the business of scalping was born.

When Miss Lind finally sang at Castle Garden, which had been roofed over to create the largest auditorium in New York, an audience of six thousand breathlessly awaited her opening number—Bellini's "Casta Diva," suitably enough. This was followed by several other selections, including her tortuous Norwegian cow-girl song, "Kom Kiyra," and concluded with an ode to America, composed especially for the occasion and sung, we are told, with reluctance . . .

> . . . *the land of the West,*
> *Whose Banner of Stars o'er the world is unrolled,*
> *Whose empire o'ershadows Atlantic's wide breast;*
> *And opes to the sunset its gateway of gold!*

In 1851, crowds gathered to hail the arrival of the Hungarian patriot Louis Kossuth. The refugee President of a vanquished republic was welcomed with tremendous admiration; thousands cheered him when he addressed the public at Castle Garden; and still more gathered outside the windows of the Astor House during a testimonial dinner in his honor. Sponsored by the New York press, the occasion was marked by noisy rounds of toasts and an unusual collection of *pièces montées*—Grecian Towers at Athens,

Left. Mr. Charles Dickens, who studiously recorded that Americans ate "piles of indigestible matter." *Courtesy of the New York Public Library.*

Right. Lady Maria Clutterbuck, alias Mrs. Charles Dickens, who was able to tempt her husband with "batter pudding, suet dumplings, and sole in brown gravy." *Courtesy of the New York Public Library.*

Mrs. Trollope, who took almost twenty years to relent on the matter of juleps. *Courtesy of the New York Public Library.*

The Swedish Nightingale, Miss Jenny Lind, who restricted her criticism to the city of Paris and the whole of Grand Opera. *Courtesy of the New York Public Library.*

Madame Lola Montez, Countess of Lansfeld, who introduced to the New York housewife recipes "for the purpose of giving elasticity and sprightliness to the animal frame." *Courtesy of the New York Public Library.*

a Hungarian Cottage, and a Turkish Temple with a statue of Abdul Medjid. Unfortunately American aid to the cause of a free and liberal Hungary never went further than spun sugar, and Kossuth returned to Europe empty-handed. At the time of his desperate mission to the States yet another exiled patriot bided his time in New York. Unknown and unfeted, Giuseppe Garibaldi was barely surviving as a candlemaker on Staten Island.

International expositions had stimulated tastes for the beautiful, rare, and costly. At A. T. Stewart's Broadway Emporium, matrons could satisfy their whims for two-thousand-dollar paisley shawls, cashmeres, Brussels carpets, Lyons silks, embroidered linens, and Paris gowns. Flounces of Valenciennes lace, half a yard deep, could be had for six hundred dollars, and a matching collar for two hundred. Some of the extraordinary Oriental brocades were embroidered in gold to half-an-inch thickness. That Stewart's six-story white marble palace staffed by four hundred was perpetually thronged with customers seemed to come as a surprise only to foreigners.

A. T. himself arrived in New York in 1823, fresh out of Trinity College, Dublin. Throughout the years he pursued an undeviating road to riches, staring down any opposition with an icy gray eye. Redheaded, gaunt, parsimonious, gravel-voiced, A. T. was so thoroughly unendearing that when at the end his body was kidnapped from the undertakers, his widow mulled it over for nine days before paying the ransom.

After a hard hour's shopping at Stewart's, most of his crinolined patrons found themselves in need of a little resuscitation at Taylor's and Thompson's gilded ice-cream saloons directly across the street. The forerunners of these two establishments, at which an unescorted lady could order a light lunch without sullying her reputation, were Guerin's, the French confectioner on lower Broadway, where tea sandwiches and sardines were served along with the usual coffee and chocolate, and Palmo's Café des Mille Colonnes at 307 Broadway, where one could play dominoes on marble-topped tables while enjoying ices and chocolate "in accommodations and appointments . . . far in excess of any previous essay in this country." (This last was named after the palatial Parisian "Mille Colonnes," famous for its imposing patronne, who stood over her cash box in blazing red velvet and diamonds, and was whispered to have been a mistress of Napoleon himself.) Unfortunately, both served liquor, and both went the same way. Guerin's cake and pie counter was eventually curtailed in favor of its nemesis, the bar, which, by dispensing both absinthe and maraschino to the Latin element of the city, caused the ladies to flee in alarm. Palmo sold out to a shady operator and the Mille Colonnes shortly became known as Pinteaux's saloon. It came to a final ignominious end in 1848 when it was raided by the police for its titillating "tableaux vivantes," ("The Birth of Venus," "Lady Godiva's Ride," "The Peeping Tom of Coventry").

Taylor and Thompson nobly denied themselves the easy profits realized from the indiscriminate sale of alcoholic beverages, and concentrated on wooing the sort of women who would rather have starved to death than enter a restaurant alone. Creating the proper atmosphere was of major importance; the decor of both establishments gave true meaning to the term "ice-cream palace." Both were huge, with vaulted ceilings, marble tile floors, blooming conservatories, white walls gleaming with gold leaf, gigantic beveled mirrors in gilt frames, and fittings of walnut and thick red plush. Both presented charming displays of fruits, bouquets, and fancy candies on counters of "pure statuary marble," supported by bronze figures of allegorical persuasion. Taylor's held one hundred tables, Thompson's seventy-six, with yet another room upstairs. Both at first served little more than ice cream, pastry, and oysters, but as the feminine carriage trade became more brazen about lunching out, a broadening fare appeared—beefsteak, boiled ham, sandwiches, poached or boiled eggs, boiled chickens, omelettes, coffee, chocolate, toast and butter.

While protesting violently to the contrary, the Victorian woman did not peck at her food. One incurable romantic, and a reporter at that, was moved to eulogize on the sensitive subject of the female digestive tract:

If you take a sly glance at the plates of these delicate customers, or listen a moment to the low-voiced orders entrusted confidentially to the waiters, you will see that ladies have stomachs as well as the rest of creation. Indeed we have sometimes thought that Othello must have been troubled with dyspepsia and envious of the digestion of the gentle Desdemona, when he exclaimed—

—oh curse of marriage
That we can call these delicate creatures ours,
And not their appetites!

Throughout the nineteenth century, a woman had to eat hearty to maintain a rounded silhouette; Englishmen preferred their celebrated "roses" full-blown, pink, and of the cabbage species. They considered New York belles inferior blooms, "so pale and sickly that in Britain they would be regarded as consumptives." Thomas H. James, an unchivalrous visitor of 1845, crassly remarked that the "prominent point of female loveliness which the whole English race so much excell in, is entirely wanting in the American ladies; they are as flat as their own horrid seacoast."

Deficiencies might be remedied in any number of ways. Dentists could fill out unseductive cheeks with "plumpers," hard composite pads inserted on each side of the mouth and costing upward of fifty dollars; bosoms could be artistically stuffed, as could sleeves to flesh out skimpy arms. False calves were deemed appropriate solely for "actresses and women of pleasure." Additional subterfuges might be gleaned from a manual published in New York in 1858, *The Arts of Beauty; or Secrets of a Lady's Toilet, with Hints to Gentlemen on the Art of Fascinating* by Madame Lola Montez, Countess of Landsfeld.

Spunky Lola, undaunted by a series of international scandals that would have unhinged a lesser gentleman's plaything, crashed the city in 1851. She had scarce known an idle moment in her thirty-three years, what with horsewhipping a crowd of indignant Bavarians demanding the abdication of her lover, King Ludwig, subduing both Franz Liszt and India, and fleeing a London bigamy trial. Except for a brief sell-out season at the Broadway Theatre upon first arriving, her balletic talents went unappreciated in New York and she was forced to repair to the lecture platform to eke out a living. Her topics were Beauty, Politics, and the Comic Aspects of Love. These illuminating talks came to the attention of Messrs. Dick and Fitzgerald, who subsequently commissioned Lola's book. In it were revealed the secret recipes of "Europe's greatest beauties," many of which indicated frequent trips to the larder and liquor cabinet as well as the chemist's.

There are some artificial tricks which I have known beautiful ladies to resort to for the purpose of giving elasticity and sprightliness to the animal frame. The ladies of France and Italy, especially those who are professionally, or as amateurs, engaged in exercises which require great activity of the limbs, as dancing, or playing on instruments, sometimes rub themselves on retiring to bed, with the follow preparation:

Fat of stag or deer	8 oz.
Forence oil (or olive oil)	6 oz.
Virgin wax	3 oz.
Musk	1 grain
White brandy	1/2 pint
Rose Water	4 oz.

Put the fat, oil, and wax into a well glazed earthen vessel, and let them simmer over a slow fire until they are assimilated; then pour in the other ingredients and let the whole gradually cool, when it will be fit for use.

Lola also asserted that pilfering the fruit bowl could lead to a sparkling glance. "The Spanish ladies have a custom of squeezing orange juice into their eyes to make them brilliant. The operation is a little painful for a moment, but there is no doubt that it does cleanse the eye, and impart to it, temporarily, a remarkable brightness." She did draw the line however at another Spanish practice whereby women slept with their arms chained to pulleys on the bedposts, to awake with hands of unearthly whiteness and immobility.

The Arts of Beauty was Lola's swan song, and she died just as a homegrown femme fatale, Adah Isaacs Menken, first shook polite New York to its foundations. Adah was the first girl to bob her hair, the first to smoke in public, and uncontestably the first to charge up a theater runway strapped naked to the back of a horse. For good measure, she wrote passionate verse. If she paused to catch her breath, it was at Charley Pfaff's dingy cellar restaurant on Broadway and Bleecker, where a rebellious crew of artists and intellectuals, calling themselves "The Bohemians," gathered at any hour of the day or night to rail at each other and refuel themselves with bean soup, beefsteak, and pfannekuchen. Pfaff owed much of his popularity to Henry Clapp and his staff of the *Saturday Press*, a short-lived satiric journal which announced its collapse with the notice: "This paper is obliged to discontinue publication for lack of funds; by a curious coincidence, the very reason for which it was started."

Like John Jacob Astor before him, Charley Pfaff was a native of Baden. His first New York restaurant was opened in 1855 right in the heart of the theater district; almost immediately the "Bohemians" took over, bringing with them a mixed bag of literati and a coterie of hangers-on. Their routine intellectual hassles were refereed by a rumpled, tieless, emotional Walt Whitman.

On June 13, 1861, the most crucial day in her life, "The Menken" appeared at Pfaff's and took her last dinner in obscurity. It was the opening night of *Mazeppa*. Adah had little reputation among theater audiences and all depended upon its reception. At twenty-five Adah had known dizzying ups and downs, as everything from a prostitute in Havana to the wife of a highly respected Cincinnati businessman, and an hour before curtain time, she wolfed down a platter of clams on the half shell, a tureen of chicken soup, a trencherman's sirloin steak, and a deep-dish pie of mixed fruit. Watching her in silent admiration were her companions, Walt Whitman and the poet-journalist Fitz-James O'Brien. Edwin Booth and Ada Clare, the "Queen of Bohemia," passed by to give their best wishes, and Charley Pfaff trotted out of the kitchen with a thick turkey sandwich to tide her over during intermission.

Adah, whispy black bangs curtaining china-blue bedroom eyes, her sensuously athletic figure costumed to advantage in the roles of both heroine and hero, was a sensation, and provoked literal hysteria by slinking menacingly around the stage in boots, pants, and a man's shirt with most of the buttons undone. The climactic gallop up the runway, in which she portrayed the hero chained to a wild horse, came as the *coup de grâce*.

A hastily arranged celebration supper was whipped up by Pfaff for Adah and her escorts, Whitman, O'Brien, James Murdoch, her director, and the editor Newell, in fact all of Pfaff's menagerie. Adah, who was hellbent on becoming a lady of letters, prattled on about a new novel by Mrs. Henry Wood called *East Lynne*. She ate three helpings of New Orleans shrimp and oyster gumbo, immediately renamed in her honor, and topped it off with a slab of roast beef. Even priggish old Horace Greeley congratulated the new celebrity and Charles Edmond Burke, a poet and novelist, drank a toast from her slipper.

Or at least attempted to do so. Adah, who had recently endured the most anxious poverty, wailed that it was her one and only pair and tried to grab it from him. There was a scuffle which grew more obstreperous as the combatants dove beneath the table and continued to thrash around amidst hoots and cheers of encouragement. Heaving and panting, choking with laughter, they were finally separated, only to have Adah haul Burke home as her trophy for the evening. It was a tribute to Pfaff that Greeley would remain two minutes in such impure surroundings.

By 1860, the New York theater had suffered through a long, convulsive upheaval. Gone were the patrician days of the Park Theatre when a Cooper, Kemble, Kean, Tree, or Wallack would regularly grace the boards. It was gutted by fire in the eighteen forties and its shell was left to rot. By 1858 it presented such a sorry spectacle that George G. Foster lamented, "We never pass the old Park walls, standing grimly defended by troupes of paupers, peddlers and thieving beggars, and letting the sunshine or the moonlight through its windows on wide crannies, like a dismantled fort, without a sigh and a keen memory of the past."

Other less genteel theaters sprang up to take its place. The most flamboyant of these was the Broadway, which tended to book outlandish vehicles on the order of *Mazeppa*, leaving the old guard to pine for earlier days. An occasional breath of fresh air would be provided by an actor of Edwin Forrest's caliber; it was at the Broadway that he played a record-breaking engagement of sixty-nine Shakespearean performances in the roles of Macbeth, Hamlet, Lear, Othello, and Richard III—this hot on the heels of his embroilment in the Astor Place riots and at the very moment of a picaresque divorce suit brought against him by his wife.

The Broadway was considered socially acceptable and would have been regarded as downright smart had it not been for the number of "abandoned women" patrolling the vestibules. As it was, respectability was maintained only at the Academy of Music, where opera was rapidly catching on, and the glamor of possessing one's own box had become the ultimate symbol of social rank.

The Bowery Theatre was unquestionably the liveliest. The audience here was composed of "butcher boys . . . the mechanic with his boisterous family, the b'hoy in red flannel shirt sleeves, the chop woman, the sewing girl, the straw braider, the type rubber, the paper box and flower maker, with a liberal sprinkling of undercrust blacklegs and fancy men—and their thirst for drama was unquenchable."

The opening of James W. Wallack's spectacular house on Broadway and Thirteenth Street gave the theater a much-needed boost. Overnight it became as fashionable as the opera, and its stars, Lester Wallack, Mrs. John Hoey, and Mary Grannon (most theaters functioned essentially as stock companies), were the idols of New York. So was its renowned guest artist, Charlotte Cushman, when she returned to entrance New York audiences with her repertoire. In *Henry VIII* she appeared alternately as Queen

Katherine and Cardinal Wolsey, accoutred for the latter role in a bushy beard.

But Shakespeare went into a deep decline as reality and the present took over in the form of Dion Boucicault's topical dramas and the drawing-room comedies of T. W. Robertson. The New York stage even went so far as to produce Sardou. Moreover, the public's greedy taste for the sensational met its apogee in 1866 with the performance of *The Black Crook* at Niblo's Garden. Although it had come to depend on rope-dancing and pantomimes to stay in business, Niblo's retained a tenuous respectability by allowing no unescorted ladies into the theater. Now it came back into its own with an extravaganza combining dancing, singing, acting, and bare legs (halfway up the thigh) that ran nonstop for five hundred performances. Everyone was horrified and no one missed it.

Mark Twain reported: "The scenery and legs are everything. Girls—nothing but a wilderness of girls—stacked up, pile on pile, away aloft to the dome of the theatre . . . dressed with a meagerness that would make a parasol blush." The story line was as fuzzy as a Chinese opera's, but some idea of it was divined by that master of plotting, Charles Dickens, who witnessed it during his trip to New York in 1867.

The Black Crook . . . is the most preposterous peg to hang ballets on that was ever seen. The people who act in it have not the slightest idea of what it is about and never had; but, after taxing my intellectual powers to the utmost, I fancy that I have discovered Black Crook to be a malignant hunchback leagued with the Powers of Darkness to separate two lovers; and that the Powers of Lightness coming (in no skirts whatever) to the rescue, he is defeated.

The theatrical haunts of the sixties resisted change and clung closely to the English chophouse tradition; contrary to the usual New York practice, they all placed emphasis upon the kitchen rather than the bar. All were operated by men of warmth and charm, and their interiors were similar in that each was jammed to the bursting point with fascinating clutter.

For elegance and originality, the De Soto won hands down. Located at 71 Bleecker Street, three doors off Broadway, the De Soto was named after a packet-steamer belonging to the New Orleans and Havana Lines for which the English owner, William G. Garrard, had previously acted as caterer. The main dining room was long and narrow, fitted out much like the first-class saloon of an ocean-going steamer; a magnificent chandelier with simulated candles fed by gas swung from the ceiling and a padded velvet band ran around the black walnut wainscoting as a headrest for those who could not light up a corona without tilting back their chairs. The bar, though small and trim, was laden with bric-a-brac, and from it flowed some of the best liquor in the city. Ale, on draught or by the bottle, was ordered as an accompaniment for a late supper of broiled kidneys or Welsh rarebit.

Rowdier devotees of the drama caroused in oyster cellars or, as they were later known, after gilt and plush poisoning had set in, "oyster saloons." The trademark of all oyster cellars, no matter how fancy or fetid, was a "balloon" of solid red or red-and-white striped muslin stretched over a wire frame and illuminated within by a candle. Some of them burned until dawn. Cellars were always just that, and entered a few steps down through swinging double doors. Lower Manhattan was catacombed with them.

Oyster cellars served a shifting cast of characters depending on the hour. Breakfast was dished out to flights of businessmen and tradesmen on the wing, and lunch could be downed even more speedily than at the genus Sweeneyorum. By evening, after the honest burghers had gone home to supper (of oyster stew, most probably, since that menu appeared nightly on many New York tables), the cellars would begin to ring with shouts for ale and brandy punch, the coy squeals and giggles of hard-eyed shopgirls, the insinuating sniggers of dandies, the clump and bang of firemen's boots. Blood ran high. A police officer recalled the night in 1848 when, as the climax of a tense political chat in an oyster cellar, Tom Hyer the pugilist thrashed Yankee Tim Sullivan within an inch of his life. This indicated a certain lack of foresight on the part of Hyer, since Sullivan's gang was one of the city's most lethal, and the officer, as in all good Victorian potboilers, arrived just in the nick of time to save Hyer, optimistically loading his gun, from being slaughtered.

During the thirties and forties the greatest concentration of oyster cellars could be found on Canal Street, which gave its name to the "Canal Street plan"—in essence, all the oysters one could reasonably eat for six cents, the "reasonably" being a matter of interpretation to the management; a bad oyster would miraculously turn up to curb a glutton's appetite. Some oysters grew so large that they had to be cut up into three or four pieces (Thackeray claimed that eating his first American oyster was like swallowing a baby). Served raw, they were accompanied by a simple array of condiments: salt, pepper, oil, mustard, lemon juice, and vinegar. For fifteen cents, a customer expected a modest bowl of stew to be teeming with at least three dozen oysters, along with a generous slab of bread and butter, salad, and a relish or two.

Not all cellars encouraged derring-do, nor did they necessarily limit their menus to raw, fried, or stewed oysters. Downing's on Broad Street, for example, was the very model of comfort and prosperity with its mirrored arcades, damask curtains, fine carpet, and chandelier. Downing, a Negro, drew a regular clientele from the nearby banks and customs house augmented by a distinguished group of businessmen and politicians. (It will be recalled he catered the Boz Ball.) His was the only oyster house to attract the aristocracy as well as ladies in the company of their husbands or chaperones, and the menu listed unusually elaborate dishes—scalloped oysters, oyster pie, fish with oyster sauce, and an unusual specialty, poached turkey stuffed with oysters.

By 1850, over six million dollars worth of oysters were sold in New York each year; this still included the "R"-less months. Local beds supplied a huge number, but cellars for the connoisseur, such as Dorlan's and Libby's near Fulton Market, served a choice variety of eastern seaboard specimens: Cape Cods, Chesapeakes, Blue Points, Lynn Havens, Mattitucks, Peconics, as well as Little Neck and Cherrystone clams. By 1880, however, most of these local beds had been exhausted, and the simple inexpensive oyster cellar fell into the realm of nostalgic recollection.

But even at their rock-bottom cheapest, oyster cellars were beyond the means of the ruling class of street arabs, the newsboys. For practically nothing they could sustain themselves on practically nothing at an all-night coffee shop. The Delmonico's of these flyblown snack bars was Butter-cake Dick's on Spruce Street, down the block from the *Herald* offices. Here an army of sharp-faced adolescents gathered every midnight, hoarse from news-hawking, to consume a buttercake, "a peculiar sort of biscuit with a lump of butter in its belly," and a cup of coffee for three cents. Their demigods, ex-newsboys who had somehow scrounged and saved enough to become newsdealers, occasionally graced the premises, commandeering the cherrywood tables, puffing on cheap cigars, and further turning the air blue with oaths and tales of lust and riches. They in turn idolized the Olympian b'hoys, those red-flanneled, jack-booted, steaming supermen who galvanized the company every time they staggered in from a late fire, reeking like fresh smoked hams.

More than mere firemen, the b'hoys comprised an elite, and rivalry between the fiercely independent volunteer companies amounted to gang warfare. Once an alarm had been sounded by the fire watch in the cupola of City Hall (a specific number of bells indicated the ward), all hell broke loose. As the laurels logically went to the first company on the scene, New Yorkers learned to nimbly flee oncoming fire wagons, or to just lie low until the rumbling monsters had safely passed. More often than not, if two companies arrived at a blaze simultaneously, they would fight one another instead of the fire.

The b'hoys were for the most part Irish. By 1860 48 percent of the city's population were foreign born of which 27,000 were English, 120,000 German, and 204,000 Irish—virtually half of Ireland's population had emigrated during the famine years of the 1840s, the majority to America. Many arrived after a horrendous voyage ill and unfit, and until 1855, when Castle Garden was finally converted into an immigration depot, there was no federal or municipal processing of immigrants beyond the sketchy reports of individual ship captains. Slums mushroomed in various quarters of the city, especially around the wharves and along the fast-decaying Bowery, and a shantytown of impoverished German and Irish squatters disfigured the banks of the Hudson from Fortieth to Eightieth streets. Those who arrived with sufficient funds to see them on their way were often cheated by extortionate boarding-house proprietors or swindled outright by agents selling fake train tickets.

As a result of the unwieldy influx, jobs reached a premium; the newcomers were so hotly resented that a Native Party was founded to assure the Yankees their privileges. Gangs, such as the "Dead Rabbits" and the "Bowery Boys," proliferated and fanned the flames of Irish-versus-native-born hostilities. A Bowery Boy could be distinguished at a glance; in sartorial exaggeration he fell somewhere between the Parisian "Incroyable" of the 1790s and London's "Teddy Boy" of the 1950s; the overall effect was burlesque-house Prince Albert—cinched frock coat, vibrant plaid bell-bottoms, a Micawber top hat, and an organ-grinder's neckerchief. He was decidedly less comic when engaged in active warfare, swinging heavy clubs and heaving pavement stones.

New Yorkers resented the immigrant's political power even more than the unemployment crisis he created. Suffrage had been extended to the newly arrived in 1826. Their vote was often bought and misused, and citywide corruption spread like dry rot. Tammany Hall ceased to function as a private political club and developed into a well-oiled machine. The wolves were gobbling up the sheep at such a clip that in 1855, when Fernando Wood (of Tammany's rival, Mozart Hall) was elected mayor, the cool and upright Philip Hone was forced to remark, "instead of occupying the mayor's seat, he ought to be on the rolls of the State Prison."

"Vote early and vote often" was the ward heeler's friendly advice as he doled out rewards to the civic-minded. And two minutes out of the polling booth, his hard cash could be converted into hard liquor at a corner "grocery" or placed on a dog fight or cock fight. These usually took place behind closed doors in the back rooms of a grog shop or gin mill. In appearance, grog shops were depressingly the same—whitewashed windows to frustrate prying eyes, dirty sand or sawdust on the floors, crude plank bars set on horses and littered with shot glasses and bowls of shriveled lemons. The "whiskey" cost the management twenty-five cents a gallon. Compounded of anything from adulterated alcohol to oak juice, it was frequently sold as "cognac," or if the distiller had neglected to add coloring, "gin."

The corner liquor groceries differed only slightly. During the day, alcohol and basic foodstuffs such as ham, molasses, flour, and butter were sold to the neighborhood poor. The quality of the merchandise was abysmal, but as the locals were craftily permitted to buy in minuscule quantities, outrageous profits could be exacted by the owners. At night, with the shutters closed, the scene changed; greasy decks of cards appeared on the counter and a scruffy collection of cabmen, groomsmen, b'hoys, pickpockets, and other nocturnals would drop in for a hand or two of poker and a belt of redeye. On a good night the tone of the company might be lifted by the presence of hearty girls in gorgeous attire whom contemporary moralists described as "those wretched female outcasts whose breath is pollution and touch leprosy."

Police protection was an easy matter, and many liquor grocery operators reaped fortunes. One in particular bought a box at the opera, moved his wife and daughter into a mansion, and became a familiar figure on Wall Street. The moralists squawked about the fate of his victims: "As to his customers—from whose madness he amassed a fortune—they have not fared so well. Some are in the Hospital, some in the Almshouse; and others in the Penitentiary or State Prison—while a few of the more fortunate among them are in their graves."

For more serious entertainment, "pretty waiter girls" were popping up at concert saloons all over the downtown area, raising their hemlines, lowering their bodices, and shaking the bells on their red kid boots. For the weary businessman they offered sweet companionship and a welcome respite from workaday care; for the thirsty traveler they provided succor by the glass; for the lonely and bleak of heart, an occasional dance or whatever else they cared to work out between themselves after hours.

Needless to say, the concert saloon was yet another institution to drive Victorian reformers frantic. Its arrangements were simple: a crowded room with tables at which drinks were served, something by way of a band, and an area cleared for dancing. The rules of the houses were brief and to the point: drunkenness, unseemly behavior, and discourtesy to women were forbidden; "all must call for refreshment as soon as they arrive; the call must be repeated after each dance; and if a man does not dance he must leave." A habitué's sole requirements were a stout set of lungs, resilient metatarsals, and a bulging wallet.

The Louvre reigned as the undisputed queen of New York's concert saloons. With its thirty "pretty waiter girls," columned billiard room, and drinking hall, its jewel-colored frescoes, gilded archways, and marble fountain stocked with gold and silver fish, it looked like Thompson's and Taylor's jaded, painted, cigar-puffing sister, hung with diamonds and blowing smoke in the very eyes of the Fifth Avenue hotels.

In *Night Side of New York*, a compilation by members of the New York press, a typical siren of the Louvre is described in the act of bamboozling a tableful of English playboys:

She is clad in a dark green dress—sitting very elegantly on her well-defined bust and falling in graceful folds, its sombre material sets off well the dazzling whiteness of her round throat and snowy hands; her chin is dimpled, her lips very red, and wear a luscious and habitual pout, her nose is thin and straight, her large eyes dark hazel, with a long shadow of black lashes; her hair is raven, very heavy and glossy; a golden band binds her forehead, and down her back fall clusters of pendent curls and as she tosses a well-shaped hand, very conscious of her winning charms, there is a queenly grace that in hearts less steeled than the major portion of the youths about town would work fearful havoc. The portly old figure at the bar, and his busy staff, look very much pleased as they cast a frequent glance at the table, for this Hebe is a great coiner to them of a golden vintage.

"Kant yer drink some more of the wine, my dear?" says the first cockney to her, as he peers with bloodshot eyes at her yet untasted glass. . . .

Hebe looks at the bottle in the bucket (it's nearly empty), then at the cormorants at the bar who express an unmistakable "go ahead." Hebe says she likes fresh wine; she's afraid that in the bottle's flat.

"Gwacious! What a chawming creature," says another son of Britannia, and the soft youth calls for a fresh bottle of "Madame Cliquot."

At the other end of the courtesan's pole were the "dolly-mops," or sailors' tarts, who slank and jigged about the dance halls on the wharfs to the accompaniment of a fiddle and a tambourine. Another slumming correspondent reported:

They are all dressed in a style of tawdry finery that is repulsive. Immense waterfalls, made apparently of horse's tails, sit like poultices on the napes of their necks. These appendages are profusely decorated with faded ribbons and sown all over with spangles. Short skirts, frequently of bright tartan patterns, appear to be quite the thing here, showing a good deal of soiled petticoat and cotton stocking. Paint is laid on an inch thick, but it cannot conceal the expression of bestial depravity that characterizes these wretched creatures in every case.

As far as drinking with these maids was concerned, "it is absolutely necessary to patronize the bar. Not exactly necessary to drink, indeed; for having had one taste of the liquor poured out for you, nothing is less likely than that you will be in a hurry to drain the maddening cup. It is a wonder how people can drink that stuff three times and live."

Between the *grandes cocottes* of the Louvre and the maritime dolly-mops fell the girls at Harry Hill's on Houston Street. Harry did not supply a resident stock of charmers, but "ladies" were admitted free of charge, and as long as they wore fashionable street dress and drank reasonably well, Harry welcomed them with open arms. However he insisted on keeping a "respectable house" and banished any doxy found soliciting under his roof or passing out under his table. Whether they knew it or not, the girls were tolerated only so long as they pleased Harry's well-heeled hard-drinking patrons; despite the nightly presence of Five Points thugs, policemen out of uniform, soldiers, sailors, and firemen, a free-spending crowd of judges, lawyers, officers of the army and navy, merchants, bankers, and editors made up the backbone of Harry's favored clientele. Profits ran as high as fifty thousand dollars a year.

Unquestionably the lowest concert saloon—not the cheapest and not the most derelict, but the *lowest*—was Billy McGlory's Armory Hall on Hester Street. McGlory made no claim to protecting his clients from bloodshed, robbery, or murder, but this was possibly unnecessary as most of his customers were hardened criminals to begin with. To those who were not, knockout drops were administered with regularity; a victim might wake up in an alley naked and shivering, robbed of his small-clothes as well as his money, or he might not wake up at all. Following one of McGlory's girls home was tantamount to suicide.

Just for fun, McGlory supplemented his gang of waiter girls with a bevy of lithe-limbed waiter boys, likewise outfitted in short skirts and high red boots. Mingling with the crowd at McGlory's might easily prove disastrous to the innocent uptowner in search of new thrills; so as not to discourage such lucrative trade, private boxes were made available above the colorful dance floor at a considerable fee. Following the Civil War, it was McGlory who, aided and abetted by such volunteer hostesses as the Poll Parrot and Big-Mouth June, was to stage one of the most hair-raising episodes in the history of New York catering.

The only place as morally ambiguous as the Armory Hall was John Allen's concert saloon on Water Street, by the river, although this was so for different reasons. Allen began his illustrious career as a divinity student, and to his second calling brought much of the first—a customer, upon retiring to an upper chamber with one of the pretty waiter girls, could rely on his sanctimonious host to have a Bible piously displayed beside the bed.

Meanwhile, in Kit Burns' nearby Sportsmen's Hall, rough wooden benches and a pit provided the setting for fights in which a starved gray waterfront rat was set against a terrier, or, on certain occasions, another starved rat. One of the attractions of the Sportsmen's was Kit Burns' own son-in-law, known simply as Jack the Rat. Jack's parlor trick was to bite the head off a mouse, and he was happy to perform for a ten cent wager. For a quarter he would try his talents on a rat.

Generally speaking, Water Street was a thoroughfare of vice and iniquity to challenge the imagination of the most graphic

Victorian preacher. Many of its tenements had a dance hall, house of prostitution, or saloon-type dive on every floor. These were interspersed with rooming houses where sailors were regularly robbed, murdered, or, if they were lucky, just plain shanghaied.

Water Street's most popular night spot was definitely the Hole-in-the-Wall, at the corner of Dover Street, run by One-Armed Charley Monell with the assistance of two ladies, Gallus Mag and Kate Flannery. Gallus Mag was a six-foot-tall English woman who kept her skirt up with "galluses," or men's suspenders. For use in her role as bouncer she kept a pistol in her waistband and a bludgeon strapped to her wrist. She knew how to use both. Like so many thugs of the period, she had her own entertaining "turn." When she had clobbered an undesirable customer, she would drag him to the door by his ear, which she clutched between her teeth. If he gave any trouble, she summarily bit it off, and placed it in a jar of alcohol which she kept behind the bar for that very purpose. She was generally regarded by the police as the most savage living female.

There is a touching tale about how Gallus Mag finally made peace with her rival, Sadie the Goat. It was Sadie's technique to butt a likely victim in the stomach, throwing him off his guard, so that her accomplice could slug and rob him. Sadie enjoyed considerable success in the neighborhood until she became involved in a fight with Gallus Mag and her ear ended up in the bottle. At that point, fleeing the district, Sadie took up piracy. During the mid nineteenth century it was occasionally the habit of New York waterfront thugs to row up the East or Hudson rivers and terrorize the countryside, robbing farm houses and mansions alike, and taking men, women, and children for ransom. Sailing in a sloop under the Jolly Roger, Sadie the Goat and her Charlton Street Gang were particularly successful, and it is even said that Sadie obliged uncooperative hoodlums to walk the plank. But when the farmers took to meeting her forays with gunfire, Sadie and her gang retired, and she was forced to return to her waterfront quarters and arrive at some kind of peace with her former rival. When Sadie acknowledged Gallus Mag as supreme on the waterfront, Mag was so touched she returned to Sadie her ear, which she in turn, as a gesture of sentiment, preserved in a locket hung around her neck.

It should not be thought that despite their soubriquets and penchant for eccentricity, there was anything particularly quaint or charming about the muggers of New York's ninteenth-century waterfront. They were criminals as vicious and deliberate as any in the history of the city, and there was an atmosphere of braggadocio that led to murder for the sake of murder. It was at the Hole-in-the-Wall that Slobbery Jim murdered Patsy the Barber over the division of the spoils of a robbery. They had waylaid a German immigrant walking along the sea wall at the Battery. After beating him unconscious with a club they robbed him, and threw him into the river to drown. It was a more or less typical crime. The take was twelve cents, all the immigrant had in the world, and quite sufficient to set the two muggers to murdering each other.

The Five Points, named after a brewery-cum-tenement at the crossing of Baxter, Worth, and Cross streets, on the site of what had once been the Collect Pond, was the most fearful slum in New York, possibly in the world. It was said that a murder was committed there every night for fifteen years, but the inhabitants were not so much criminal as desperately poor. Charles Dickens, sociologically fascinated by the seamier side of Victorian civilization, compiled a report on the scourges of Manhattan. (Geographically he was less acute: "one day, during my stay in New York, I paid a visit to the different public institutions on Long Island, or Rhode Island. I forget which.") Of the Five Points he demanded,

What place is this, to which the squalid street conducts us? A kind of square of leprous houses, some of which are attainable only by crazy wooden stairs without. What lies beyond this tottering flight of steps? . . . Open the door of one of these cramped hutches of sleeping negroes, Pah! They have a charcoal fire within; there is a smell of singeing clothes, or flesh, so close they gather round the brazier; and vapours issue forth that blind and suffocate. From every corner, as you glance about you in these dark retreats, some figure crawls half-awakened, as if the judgment-hour were near at hand, and every obscene grave were giving up its dead. Where dogs would howl to lie, women, and men, and boys slink off to sleep, forcing the dislodged rats to move away in quest to better lodgings.

Dickens wished to visit a Five Points dance hall; his escort steered him to a basement door where a scruffy shingle proclaimed one word: "Almack's" (the name, ironically, of the most exclusive club of Regency London). Proceeding down the steps and through the door, they were warmly greeted by the owners:

The Five Points, with its hub of Corner Whiskey "groceries." *Courtesy of the New York Public Library.*

Heyday! the landlady of Almack's thrives! A buxom fat mulatto woman, with sparkling eyes, whose head is daintily ornamented with a handkerchief of many colors. Nor is the landlord much behind her in his finery. How glad he is to see us! What will we please to call for? A dance? It shall be done directly, sir. "A regular breakdown" . . . the dance commences. Every gentleman sets as long as he likes to the opposite lady, and the opposite lady to him, and all are so long about it that the sport begins to languish, when suddenly the lively hero dashes in to the rescue. Instantly the fiddler grins, and goes at it tooth and nail; there is new energy in the tambourine; new laughter in the dancers. Single shuffle, double shuffle, cut and cross-cut; snapping his fingers, rolling his eyes, turning in his knees, presenting the backs of his legs in front, spinning about on his toes and heels like nothing but the man's fingers on the tambourine; dancing with two left legs, two right legs, two wooden legs, two wire legs, two spring legs—all sorts of legs and no legs—what is this to him? And in what walk of life, or dance of life, does man ever get such stimulating applause as thunders about him, when, having danced his partner off her feet, and himself too, he finishes by leaping gloriously on the bar-counter, and calling for something to drink, with the chuckle of a million of counterfeit Jim Crows, in one inimitable sound.

These impressions were gathered by Dickens on his first extended trip in 1842 and included in *American Notes.* The reaction to this journal was less vitriolic than to Mrs. Trollope's tirade of the previous decade; Dickens was respected by Americans as an avid social reformer and courageous critic of Britain's brutal economic and social systems. Nevertheless they had not expected to have their hospitality repaid with the same sort of subjective accusations that they applauded at a comfortable distance. Native reformers were quick to take up the banner; Harriet Beecher Stowe wrote her shocking best-seller *Uncle Tom's Cabin,* and charitable organizations protecting children, immigrants, and runaway slaves burgeoned in New York.

Of all the reformers, none were more shrill or dire in their prophecies than the temperance crusaders; whether it be a casual Tom and Jerry in the barroom of the Metropolitan Hotel, a ladylike finger of anisette at Taylor's, a convivial mug of ale at Pfaff's, or a shot of nameless alcohol between polkas at Harry Hill's, all led the same way, and the soul of the nation quailed at the thunderous voice of doom.

In these decanters lie coiled up the venomous serpents that have stung the life of so many hundred thousand noble hearts, corrupted such myriads of pure souls to the capacity and similitude of demons—beggared so many wives and daughters—made desolate so many homes and hearths —and spread desolation and despair through all ranks of society. Taste, drink, riot, grow mad, and die! 'Tis only sixpence a glass! Cheap enough — yet oh, rash tippler, only to be purchased at last with the life of thy immortal soul. In those dainty decanters, smiling amber-hued and rosy— flashing the spiced air with purple beams and glowing with golden and fiery essence—lie the spells that once conjured to the lip, will lead thee to ruin, madness, blear-eyed idiocy, and untimely death.

Dare and die!

Chapter IV

1860-1914

Little Birds are dining   *Hid, I say, by waiters*
Warily and well   *Gorgeous in their gaiters—*
Hid in mossy cell: *I've a Tale to tell.*

"Pies kill us!" . . . "On to Glory!" . . . "Blood!" . . . "Zwie Lager!" . . . "Lincoln and Seward Assassinated!" . . . "Grant for President!" . . . "Brodie Leaps from Brooklyn Bridge!" . . . "Mrs. Astor requests the pleasure . . . !" "Don't come back till you get the recipe!" . . . "Paint my face and I'm off to Rector's and to hell with the rest!" . . . "Will you listen to that elevated!" . . . "There's death in the pot!" . . . "God, Nell, ain't it grand!" . . . "What this town needs is a little hatchetation!" From the Civil War until the turn of the century New York behaved as if everything it did or said or even repeated should be followed by an exclamation point. Feelings had been decidedly mixed about the prospect of war. New Yorkers vacillated between apathetic procrastination, abolitionist rallies, and pro-Southern commercial panic—large debts owed by the South to New York would be canceled by the event of war. There was even illogical talk of the city's secession from the state of New York in order to maintain neutrality. But war jitters were briefly dispelled by the arrival in 1860 of two extraordinary social side-shows from abroad—the first from Japan, the second from England.

On June 16, six thousand soldiers and hordes of gaping New Yorkers lined the Battery and welcome route for the first mission from Japan to a Western power, only seven years after Commodore Perry, a New Yorker and father-in-law of August Belmont, had steamed into Uraga Bay and bestowed free sewing machines on the Imperial Court, thereby cementing trade relations once and for all. The delegation, led by First Ambassador Simmi Boojsen No-Kami, was composed of seventy-seven kimonoed Japanese ranging in rank from prince to secretary to an alert little boy who was to capture public affection and the nickname "Tommy." They were met by an imposing roster of American dignitaries, as well as a posse of hospitable aldermen all wearing, for some unexplained reason, yellow kid gloves.

What impressed New Yorkers most about the Japanese visitors was the ball given in their honor at the Metropolitan Hotel. As the astronomic price of thirty dollars had been extorted for each of the three thousand tickets of admission, homage payers to Nippon were determined to get their money's worth; ten thousand bottles of champagne were consumed and by dawn the ballroom floor was literally awash. The Japanese may have thought this simply a local custom; in any case, what impressed them most was the city's broad boulevards and miraculous gas illumination which, as one of the entourage recorded, "is so wonderful and such a surprise to us that I cannot describe it."

"Dear Bertie," alias Albert Edward, Prince of Wales and the green apple of Victoria's eye, sailed into New York harbor on October 11, 1860, and thereby precipitated a social crisis ending in unparalleled exultation and despair. Slim, pop-eyed, and nineteen years old, the prince was remarked upon, by New Yorkers in the know, as being handsome and unusually frisky. (It was later claimed by residents of the Fifth Avenue Hotel, where His Highness resided, that Bertie was in the habit of playing leapfrog with his retinue before bedtime.) Apparently these insights had not been conveyed to a committee of city elders led by the aged and august Peter Cooper, and their proposed testimonial banquet was rejected by Bertie's Minister, the Duke of Newcastle, as not quite the royal idea of an amusing evening; a ball, on the other hand, might be just the thing. This suggestion was met with dour disapproval by the old guard, but with squeals of delight from their wives, daughters, and nieces. The question of who should be invited to the ball and who should be singled out to dance with the Prince was treated with Solomon-like detachment by the committee despite the machinations and intrigues of the socially ambitious.

The ball was finally held at the Academy of Music, and referred

"The glory of America—how it is crowded." The Japanese Mission's impression of Fifth Avenue, with women in paisley shawls and six-story skyscrapers. *Courtesy of the New York Public Library.*

"Dear Bertie"—Albert Edward, Prince of Wales; the social pandemonium precipitated by a ball in his honor caused the dance floor to collapse. *Courtesy of the New York Public Library.*

Opposite. Never before had New York Society schemed so perfidiously. *Harper's Weekly,* October 13, 1860. *Courtesy of the New York Public Library.*

THE PRINCE AND THE LADIES.

EMILY reads: "The number of British Princes who have married commoners is very large indeed, and any virtuous and amiable girl may become the wife of a member of the Royal Family. The Duke of Clarence married Mrs. Jordan, George the Fourth married Mrs. Fitzherbert, James married Ann Hyde—" Now Mary, I ask you, why should not—?

CLARA has heard that the Prince is rather violent in his style of dancing, and throws his legs about à l'anglaise. She practices accordingly with her brother Tom, who learned the "pas," he says, at Paris.

SOPHY is not going to be taken by surprise by any future occurrences, and rehearses carefully before a glass—"Yes, dearest Albert Edward, if I only believed you were sincere—"

Our sweet friend ARABELLA thinks she's captivating, and that the Prince will be sure to notice her; but Mary feels confident that the $100 she has just laid out in that new bonnet has dashed Arabella's hopes forever. If the Prince is not struck by that bonnet, the sooner he goes home the better.

As the Prince passes, brother George will cry "Fire!" and I'll faint.

If it's graceful dignity he seeks, here must he bend his knee—eh, Chloe? Yes, 'um—yes, 'um; any where 'bout heyah.

to thereafter by awed guests with hushed respect. A mob attacked the buffet catered by Delmonico's with such ferocity that John Jacob Astor III and several others were deputized to guard the supper-room door. The crowd was so dense that a few moments before the opening quadrille, the floor languidly collapsed. The Prince stood placidly by with his hands folded until it was repaired. Dauntlessly, he then glided out on it first with Mrs. Governor Morgan, then with the Misses Fish, Mason, Butler, and other post-debutantes, including one sober-eyed Massachusetts heiress with camellias in her chestnut hair who had been introduced to him by a contemporary wit as the "Princess of Whales." Bertie never did catch her true name which was Hetty Robinson (later Green). The evening gave rise to a gaggle of spurious romantic legends: the Prince had squeezed his hands into a pair of kid gloves sent by a "lady present in the house" (assuredly not Hetty), the Prince had caroused till dawn in the city's most gorgeous bordello, etc., etc., etc. . . .

Six months later, less frivolous events took the limelight. With the attack on Fort Sumter, New Yorkers were shaken to positive action and immediately committed themselves to the Union cause.

On April 19, 1861, barely six days after the fall of Sumter, the Seventh Regiment answered Lincoln's call to arms, and after gathering in front of the Astor House, marched down Broadway to the cheers of an awakened public. They were followed shortly by Colonel Elmer Ellsworth's brigade of theatrically uniformed Zouaves and the Fighting Irishmen's Sixty-ninth Regiment, hastily recruited from the Bowery. New York further supported the Union by privately advancing the government $210,000,000 to defray military expenses and by the founding of the Committee for Union Defense, the Women's Central Association of Relief, and the tireless U.S. Sanitary Commission, progenitor of the Red Cross. While the city never suffered attack, it was constantly reminded of the crucial situation by innumerable regiments marching through its streets on their way from the North to the front lines.

Yet in retrospect the attitude of the citizenry seems disturbingly paradoxical, as, time and again, New Yorkers permitted their attention to wander from national tragedy to passing follies. The year 1863, for example, was made memorable for the citizenry by the arson, looting, lynching, and atrocious anti-Negro attacks associated with the Draft Riots, caused by the slum toughs' determination to avoid a draft call for three hundred thousand men. It was also the year of the widely publicized wedding at Grace Church of General Tom Thumb to the equally minuscule Miss Mercy Lavinia Bumpus. Breaching an ice-cream fortress overnight became the fashion at all the best dinner parties, and, at the end of the day, dedicated volunteers of the U.S. Sanitary Commission would drop their bandage rolling and rush off to Taylor's to pick one up before closing time. The news that a Southern plot to simultaneously burn down all the central hotels had been foiled was received with relief, for it meant that the Astor, missing a charred room or two, would still be able to hold its traditional New Year's Eve Ball. Worse still, selfless efforts to support a war of moral principles were too often canceled out by cynical profiteering on all levels. Speculators made ruthless killings in the market, while some quite innocent investors such as Lorenzo Delmonico, who had bought unreliable stocks just prior to the war, were saved by the inflationary upsurge in business.

Peace came at last with the tidings of Appomattox on April 9, 1865, a glorious peace with hugging and kissing in the streets and miles of brilliant bunting and tearful prayers of thanksgiving. Three weeks later New York sank to its knees under layers of black crepe weeds; a coffin passing on its way from Washington to Springfield was detained for a day so that mourners could pay their stunned respects to the man they had only fitfully supported. And when the boys came marching home during the succeeding months, the gaps in their ranks were appalling.

The Civil War marked the end of the Age of Innocence. The philosophical tenets of Jefferson and Paine had somehow got lost in the shuffle. In New York money became king. Mary Eliza Tucker, a sheltered poetess from the South, clarified the situation for future social historians in her epic, *Lewis Bridge, a Broadway Idyl.*

> *Each eager face in passing seems to say—*
> *"Chasing a dollar, comrades, clear the way!*
> *I am ambitious and I fain would win,*
> *Would gain the dollar, even if I sin!"*
> *And oft, alas, in raging lust for gold,*
> *Life's cup is broken, and a soul is sold!*

A freshly minted class emerged, bright as a silver two-bit piece, and no aristocratic epithet—"parvenu," "shoddy"—could slap it down. The great robber barons rode roughshod through Wall Street and drawing rooms alike. Suddenly, thanks to a union with the ancient Dutch Schermerhorn family, the name of Astor took on the old-guard patina of a Van Rensselaer; but even in exclusive aggregate, the city's princely ranks found it difficult to stave off the bumptious assaults of a Jay Gould or a Commodore Vanderbilt, whose attitude was summed up in his famous remark "What do I care about the law? Haint I got the power?" When criticized for breaking the railroad connection between the banks of the Hudson, so that passengers were obliged to cross on the ice, Vanderbilt was outraged—"Can't I do what I want with my own?" These were the days of totally unbridled enterprise, and even in New York it was an era of barefaced frontier "might makes right," a philosophy best summed up by Frederick T. Martin in *The Passing of the Idle Rich:*

. . . we are the rich; we own America; we got it, God knows how; but we intend to keep it if we can by throwing all the tremendous weight of our support, our influence, our money, our public-speaking demogogues into the scale against any legislation, any political platform, any Presidential campaign, that threatens the integrity of our estate.

Under the epicine maneuvering of pasty Ward McAllister, a native of Georgia, the Patriarchs and the Four Hundred arose to bar the gates against intruders; for years McAllister reigned as their general factotum, their pampered acolyte, their finger in the dike, living proof that, in the words of William Dean Howells, "Inequality is as dear to the American heart as Liberty itself." McAllister picked their guest lists as he picked their carpets—with an eye to elegance and authenticity—and all listened gratefully to his shrill admonition that truffles should never appear more than once at dinner.

A short quotation from McAllister's manual, *Society as I Have Found It,* reveals that his fine-honed snobbery was tempered all the same with a healthy respect for money: "If you want to be fashionable, be always in the company of fashionable people. As an old beau suggested to me, if you see a fossil of a man, shabbily dressed, relying solely on his pedigree, dating back to time immemorial, who has the aspirations of a duke and the fortunes of a footman, do not cut him; it is better to cross the street and avoid meeting him." McAllister does not mince words: "It is well to be in with the nobs (aristocrats) who are born to their position, but the

Ward McAllister, acolyte to the Mystic Rose, receives appropriate adulation in "Puck." *Courtesy of the New York Public Library.*

support of the swells (nouveau riche) is more advantageous, for society is sustained and carried on by the swells, the nobs looking quietly on and accepting the position, feeling they are there by divine right; but they do not make fashionable society, or carry it on. . . ." He advised Mrs. Astor to invite "professional men, doctors, lawyers, editors, artists and the like," *only* on New Year's Eve.

The trend to exclusivity in this era resulted in the almost endless multiplication of men's clubs. The ancient and revered Union Club (founded in 1836), literally pupped. The peculiar multiple birth is perhaps best described by Cleveland Amory in *Who Killed Society:*

The Century . . . which dates from 1847, was formed in the belief that the Union was slighting intellectual eminence. "There's a club down on 43rd Street," said one Union Clubber, "that chooses its members mentally. Now, isn't that a hell of a way to run a club?" The Union League, a Republican club dating from 1863, was formed in answer to the fact that a Confederate Secretary of State was allowed to resign from the Union Club, when, according to Union leaders, he should have been expelled; the Manhattan, originally a Democratic Club, was formed a year later in answer to the answer. The Knickerbocker (1871) was formed because its members felt the Union was taking too many out-of-towners and wanted a club limited to men of Knickerbocker ancestry; the Metropolitan (1891) was formed because the elder J. P. Morgan could not get a friend of his into the Union, and thereupon, in the Morgan manner, built his own club; the Brook (1903) was formed because two young Union Clubbers had been expelled for having attempted, unsuccessfully, upon the bald head of the Union's most revered patriarch, to poach an egg.[2]

In the precincts of these institutions, the club member found, in the words of Mr. Amory, his four freedoms: "Freedom of speech against democracy, freedom of worship of aristocracy, freedom from want from tipping, and above all, freedom from fear of women."

The new Grand Hotels buried themselves hermetically under an avalanche of consoles, tapestries, ormolu, gilded wood, tesselated marble floors, etched mirrors, and enough brocade to maintain the entire Orient in purdah. The Hoffman House prided itself on a pair of "2,000 year old" blackamoors supporting baskets of metallic grapes on their heads, and in the dining room of another hotel the walls revealed allegorical representations of the Spirits of Music, Art, Coffee, Tea, Tobacco, and Bouillon. The overall scene was tempered somewhat in the eighties by the sobering introduction of a sort of Victorian-Queen-Anne-Revivalist style, and minstrel galleries began to usurp areas formerly enhanced by gilt mouldings and floral frescoes.

For all the concentration on beauty within, the city's growing pains without were agonizing both financially and visually; rather than appearing, as it had to Mrs. Trollope, as an American Venice rising from the sea, New York with its countless excavations and demolitions resembled nothing less than a delinquent biblical metropolis enduring divine vengeance. Spirits withered within a cage of electric poles and wires. The elevated trains provided speed and comparative comfort, but no beauty. They were necessary, however. Rush Hour had become an institution and, as one visitor put it, "Neither the boulevards, the Strand, nor the Corso of Rome in Carnival time can give an idea of this tumultuous movement."

Advertisements were plastered everywhere. The success of suspense campaigns had been discovered, and for six months the public was kept in the dark as to the meaning of "Sozodont," which seemingly glared from every available surface. (It was eventually unmasked as a dentifrice.) The following year, an eerie graffito gave pedestrians a nasty turn. For months curbstones all over the city were imprinted with the red letters "B-L-O-O-D." And, by the end of the century, it is said the entire city as seen from Brooklyn Heights seemed to cry out "Castoria."

Growth was tremendous. By 1870, streets numbering up to One Hundred and Fiftieth had been laid out, although undeveloped areas still existed south of Central Park. The population then stood at nearly a million. By the end of the century, building enveloped the park, and outlying areas such as Harlem were devoured by the encroaching city. In 1898 the "counties" were finally consolidated into the "boroughs" of Greater New York, bringing the census up to three and a half million.

With the refinement of the elevator and the steel frame building, New York's distinctive tendency to spread upward rather than outward was given free rein. The five stories of the Equitable Building in 1870 were soon outreached by the Western Union Building and the Tribune Building, which rose to two hundred eighty-five feet. In 1888 the Tower Building climbed to thirteen floors, and by the close of the century, the Park Row Business Building towered twenty-nine breathtaking stories over the city. The most outstanding attempts at beautification had a fantastical, seven-wonders-of-the-world quality: the Brooklyn Bridge and the female colossus in the harbor, the Statue of Liberty. Rudyard Kipling, however, huffily dismissed the whole of Manhattan as a "long narrow pig trough" and took himself off to live in Vermont.

Fifth had become the avenue of the rich and a parade of splendid residences—Moorish chateaux, Venetian manor houses, Tudor palazzi, and Georgian monasteries—inched steadily up from Thirty-fourth Street to "Millionaire's Row," which sprawled majestically above Fifty-ninth Street facing the Park (the inhabitants of this Eldorado were familiarly known as "Avenoodles").

Central Park itself looked ravishing. An open competition for park designs was won in 1858 by Frederick Law Olmsted and Calvert Vaux, who incorporated many of the most appealing aspects of European public gardens into their overall plan: drives and rambles, rustic shelters and Oriental bandstands, a grand mall and fountains, lakes and reservoirs, grottoes and commemorative statuary, and a half-million plants and trees. After Central Park's basic completion in 1876 the drive from Fifty-ninth Street and Fifth Avenue to the mall, between the afternoon hours of four and five, became nationally famous for its extraordinary display of equipages, fashion, and horseflesh. Everyone of note—the old guard with their conservative broughams and loathing of ostentation, Mrs. August Belmont in her stylishly eccentric demi-d'Aumont, beparasoled opera stars in their gracious victorias, bright young things racing their smart-as-paint phaetons or lolling in the barouches deemed chic by Empress Eugenie—indeed everybody who was anybody joined in the daily excitement of a vehicular point-to-point. All this did not prevent alarmists in the press from voicing dire warnings, however: "Perils of the Park—a Ruffian's Refuge where Ladies, Children and the Unprotected Generally are at the Mercy of Villains."

When the final bills were totted up, the park was said to have cost the city seven million dollars, the lion's share having instinctively found its way into Boss William Marcy Tweed's bulging pockets. This was nothing new; by the time the public caught on and the law caught up, Tweed and his friends had managed to siphon off an estimated thirty million dollars of municipal funds. A violent newspaper campaign led by George Jones of the *Times*, Horace Greeley, and cartoonist Thomas Nast brought the Boss plummeting earthward in 1871. The descent was painful: temporary imprisonment, trials and retrials, and an unlikely escape to

Spain, where he was identified by means of a Nast cartoon, which was shown to the authorities. In it, Tweed was depicted beating a child representing Justice. The Spanish police swiftly located "The Boss" and charged him with kidnapping. His final incarceration in the Ludlow Street jail proved fatal.

Tweed and the Tammany politicians employed the freewheeling, every-man-for-himself ethics of the new millionaires, which they frequently became. George Washington Plunkitt, for forty years a Tammany "sachem," presented his *apologia* in the form of a lecture on the distinction between "honest" and "dishonest" graft:

I've made a big fortune out of the game, and I'm getting richer every day, but I've not gone in for dishonest graft—blackmailin' gamblers, saloon keepers, disorderly people, etc.—and neither has any of the men who have made big fortunes in politics.

There's an honest graft, and I'm an example of how it works. I might sum up the whole thing by sayin': "I seen my opportunities and I took 'em."

Just let me explain by examples. My party's in power in the city and goin' to undertake a lot of public improvements. Well, I'm tipped off, say, that they're going to lay out a new park at a certain place.

I see my opportunity and I take it. I go to that place and buy up all the land I can in the neighborhood. When the board of this or that makes the plan public, and there's a rush to get my land, which nobody cared particularly for before.

Ain't it perfectly honest to charge a good price and make a profit on my investment and forsight? Of course it is. Well, that's honest graft.

Tweed and his friends took unconcealed advantage not only of City Hall but of thousands of immigrants whose hasty, illegally arranged citizenships assured him instant votes. Neither the Irish nor the Germans were immune to this sort of political exploitation. They were exploited economically as well, particularly the Germans, who were willing to suffer deplorable abuses from American landlords in order to remain in their beloved *Kleindeutschland,* or "Little Germany."

"Little Germany" represented the first self-contained foreign-language quarter to develop in New York. By the 1840s it had swallowed up a sizable chunk of the native residential area between the Bowery and Houston Street, forcing the previous tenants to flee—to live among immigrants meant total loss of face. The German newcomers in their "long skirted, dark blue woolen coats, high necked and brass buttoned waistcoats, and flat military caps or gray beavers, their wives in many colored scarves and glaring jackets" arrived in such droves that by the time of the Civil War one hundred twenty thousand resided in New York.

Kleindeutschland, by both preference and necessity, became a city within a city, founding its own clubs, libraries, schools, and theaters. It eventually took possession of the Bowery Theatre and Dr. Faustus consorted with the devil where once Macbeth held congress with witches. Of greater cultural significance was the rising tide of lager that, having burst the walls of Bowery grog shops and gin mills, launched a flotilla of Teutonic taverns or "beer halls"—one for every two hundred inhabitants.

Equally populous were small restaurants or eating houses that fed the many Germans who lacked kitchen facilities. A typical meal, costing about fifty cents, might consist of roast veal, fried potatoes, and a pancake. Native Americans were critical, however, and associated small German restaurants with grease and cracked crockery. While the coffee and bread were agreed to be palatable, the butter was accused of being a neat mixture of butyric acid and lard.

The "Continental Sundays" of the quarter were a delight for those stultified by the puritanical Sabbaths of Anglo-Saxon New York. After the most cursory of religious observances, the Germans' day of rest was devoted to picnics, tavern feasts, or a visit to the "Volkstheater," where turns by singers and comedians on a small stage accompanied the universal consumption of cheese and beer (hard liquor was unknown and wine too expensive). Respectable Manhattanites shuddered to hear that the German fraus and frauleins, smiling and in their Sunday best, joined in these unseemly pursuits.

Unquestionably the true spirit of Little Germany was represented by the Atlantic Garden, the beer hall to end all beer halls. Lager beer was the German immigrants' gift to America, and its mass production irrevocably changed the nation's drinking habits. Lager was light, delicious, and, happily, it could be consumed in quantity. The brewing and aging techniques employed by the Germans were more refined and time-consuming than the old-fashioned English methods brought to the New World by the Pilgrims (a one-room brewhouse was constructed during the first winter at Plymouth); the principal difference lay in the "bottom fermentation" process characteristic of lager as opposed to the

Lager and "tangled hair." *Courtesy of the New York Public Library.*

"top fermentation" process for stouts and ales. After the addition of yeast as a fermentive agent to a brew, the Germans would maintain it at a low temperature for a period of a week or ten days; when the yeast gradually settled to the bottom of the vat, the clear amber "bier" would then be siphoned off into other kegs, where for at least a month or two it would undergo a secondary phase of after-fermentation. With "top fermentation" no pains were taken to regulate the temperature; consequently only three to five days were necessary for the completion of the fermentive process; the yeast rose in an ungainly mass to the top of the barrel and was then skimmed off, the remaining liquor being bottled or kegged for immediate consumption, if so desired.

Many German brewers in New York opened beer halls as well as breweries and often owned magnificent teams of dray horses. Thatched hooves pounding on the Belgian paving-stones, these beautifully groomed animals were kept clopping back and forth between the supply and the demand, while behind them might be seen thundering wagons swaying and creaking beneath pyramids of copper-girded barrels, surmounted in turn by blue-smocked drivers puffing on white clay pipes.

The Atlantic Garden, located on the Bowery near the theater, owned four such conveyances. This cavernous hall, seating more than a thousand, was a "garden" only in the sense that a rural mural had been painted on the walls. Upon closer inspection, however, this scene proved to be a graveyard represented in the minutest detail. A pianist, violinist, and a drummer performed in a box suspended from the ceiling; interminable games of cards, dominoes, and dice progressed below, and at a rifle range customers could fire popguns at the flaming bodies of ferocious Turks and Zouaves. Teenagers wearing the traditional New York "pretty waiter girl" uniform of short skirts and red-topped boots, but speaking German and of unquestioned virtue, served beer at five cents a stein. Beer sales were so profitable that the Atlantic Garden competed with the various other neighborhood beer halls for the patronage of organizations holding all-day "picnics," and paid up to five hundred dollars for the privilege of catering to an unquenchable party.

In *The Night Side of New York* one member of the press recounts gingerly setting foot in the barroom adjoining the main hall. He was thoroughly impressed by what he saw: "a large low ceiling, square room with a bar on either hand, groaning with the weight of dripping kegs, piles of crystal glasses of all dimensions and variety; loaded with sausages, those famous snake-looking, American 'Bolognas,' with cakes and bread, brown and shiny, contorted and twisted, patted and moulded as only the German brain could devise; green, fresh-looking salads dripping with oil, dimpled with red beets and scarlet turnips; salads of fish, salads of meat, salads of herring. . . ."

The Germans were known for their slow, lethargic manner, and the silence in which they drank. Rows in beer halls were infrequent, although temperance fanatics inevitably managed to find fault. In the words of *Harpers New Monthly Magazine* in 1866: "Lager bier, though said not to intoxicate, has a decidedly exhilarating effect as its first stage; after that comes a stupor, which is expressed by the term 'tangled hair.' " Only on occasion was the subdued atmosphere, the almost apathetic quiet of the German community, shattered. One New Yorker, W. O. Stoddard, in his *The Bowery, Saturday Night,* recalled his bewilderment on coming into the Atlantic Garden one evening in 1870:

Ordinarily the crowd was so quiet over its beer and wine to seem almost stupid. But you should have been here on Saturday night, September 10 . . . when the details of the great French overthrow at Sedan were being brought in here by telegraph office. They sang the German's Father-land and Luther's Hymn . . . until they were hoarse, and hardly could get breath to drink the health of "unser Fritz" and Von Moltke and everybody else. Nobody was drunk and one could not help feeling that a part of the "Fatherland" was here for they had brought it with them in their hearts.

In another part of town that night, the news from Sedan was greeted with stunned silence. By 1870 the French quarter, if smaller than the German, was as firmly entrenched. French immigrants, largely refugees from severe and repetitious political upheavals in Paris, inhabited the area west of Broadway, south of Washington Square, and north of Grand Street. A frugal lot, they earned little and spent less; but, following their native tradition, they insisted on eating well, and their quarter was peppered with tiny and inexpensive restaurants. One of these, on Bleecker Street, bore the legendary name Restaurant de Grand Vatel. (Vatel was a chef at Versailles who may or may not have run on his sword after an order of fish failed to arrive in time for a banquet.) Its menu was inviting and reasonable in the extreme: café supérieur, three cents; café au lait, five cents; soup plus a plate of meat and bread, ten cents; beef and vegetables, ten cents; veau à la marengo, ten cents; mouton à la ravigotte, ten cents; ragout de mouton aux pommes, eight cents; macaroni au gratin, six cents; celeri salade, six cents; compote de pommes, four cents; fromage Neufchatel, three cents; Limbourg, four cents; Gruyere, three cents. A "gloria" (black coffee with cognac) usually ended the meal.

At the Grand Vatel, sand covered the floors and a pewter cruet centered each small oilcloth-covered table, and it has been reported that the restaurant's proprietess, wearing many petticoats and white wool stockings, would occasionally pause in her knitting to screech "Tranquille!" at two parrots which had fallen into the habit of jabbering uncouth Revolutionary slogans. On the wall flapped a placard announcing a grand festival, banquet, ball, and artistic tombola to celebrate the Anniversary of the Bloody Revolution of March 18, 1871, under the auspices of the "Société des Refugiés de la Commune."

In nearby cafés, out of work and bombs, nihilists and thwarted regicides discussed politics and dispiritedly harangued each other over absinthe and beer; "absinthe houses" such as the Taverne Alsacienne were rumored to be dangerous and outsiders were *persona non grata*. Their regular patrons seemed consumed only by grim desperation and billiards.

Happily, there existed many less depressing resorts of casual entertainment in the quarter, and of all these the most tantalizingly unfamiliar to Americans were the *cafés chantants* on a Sunday night; the curious had only to cross the zealously guarded brownstone threshold to find themselves in Paris. A seat at a long table ringed with babbling French families and cluttered with wineglasses, overcoats, and babies could be taken for a mere ten cents. No English was spoken, but dancing continued beneath the plaster mouldings and clinking chandelier until dawn. Aging singers trilled away atop a midget stage in the corner; after each took a final bow, she would bounce back into the crowd either to shift a drowsing child before it fell in a heap off its chair, pour herself another glass of wine, or ceremoniously applaud a sister soubrette's recital.

Other national tastes could be gratified in scattered areas about the city. As early as 1853 a Spanish hotel on Fulton Street became well known, while on William Street an Italian restaurant featured Bolognese mortadella, the mother of American "baloney."

From 1870 on, hundreds and eventually thousands of Chinese came to live in the vicinity of Mott, Doyer, and Pell streets. Although they were generally considered respectable and law-abiding (except for a tendency to take Irish wives—several at a time), there hung over the Chinese colony an uncertain aura of opium and white slavery. Their unusual cooking habits were also a matter of conjecture, though few New Yorkers could claim to have sampled their eating houses; only the most curious journalists or courageous Bohemians gave them a try.

Damned with champagne tastes and small-beer pocketbooks, the Bohemians began to congregate in aromatic French "holes-in-the-wall" and early Italian trattoria-like establishments around Washington Square, making them the first native-born Americans to discover the cheap, picturesque charms of the "little foreign restaurant." And as the more prosperous of the German population moved to Union Square, the Bohemians took up residence in the Fourteenth Street beer gardens as well.

Union Square presented a scene of popular press romanticism when Bohemians such as the critic James Huneker, Vance Thompson, Winslow Homer, and William Dean Howells, leader of the literary realistic movement, strolled beneath its elms and maples or paused by its splashing central fountain on their way to one of the cozy *weinstuben* that vied with Tiffany's and Brentano's book store for popular attention.

From the terrace of Brubacher's Wine Garden, on the east side of the Square at Fourteenth Street, convivial parties of Rhine-wine tipplers could lay odds on the next victim of Dead Man's Curve, one of the most bloodcurdling traffic hazards in the country. It was said that an early streetcar conductor had arrived at the conclusion, no one seemed to know why, that the curve from Broadway onto the Square should always be taken at full speed, else the cable would be lost. This deadly theory was adopted by the rest of the city's carmen. A ghoulish new spectator sport arose as money was placed on a daring pedestrian's chances, and Brubacher's became known as the "Monument House." "Papa" Brubacher attracted the best German society with his formidable cellar, and the king of the connoisseurs, Baron August Hartmann, was a frequent patron.

There were many *weinstuben* in the Union Square vicinity, and all stocked excellent vintages; the most dependable were Weber's,

The Hoffman House Bar, 1898. *Courtesy of the Museum of the City of New York.*

New York's free lunch, as seen by the Paris press. *Courtesy of the Picture Decorator, Inc.*

Chinatown restaurant, 1896. *Courtesy of the Picture Decorator, Inc.*

Messerschmidt's, Wiehl and Weidmann's, and Muschenheim's Arena. The last was considered too cosmopolitan for some tastes although its list of three hundred thirty-three varieties of vintage wines and liquors never failed to impress true connoisseurs.

Hermann Weber, wine expert and a lieutenant of the Independent Marksmen, not only offered one Rhine wine at twice the price of two quarts of champagne, but also genuine Russian caviar and imported German Handkäse in his gothic-windowed, curio-strewn *weinstube*. Count Alfred von Wiehl, owner of Wiehl and Weidmann's, and a painter and violinist himself, attracted international artists and the most important wine dealers of the day.

In 1879 August Guido Luchow, a Hanoverian of twenty-three, arrived in New York. Three short years later, with the financial assistance of William Steinway of the piano and music empire, he was able to buy out the beer halls of his employer, Baron von Mehlback. The hall was then only one-eighth the size it would shortly become. Mr. Steinway, surrounded by his executives and visiting artists, assured Luchow's reputation by appearing every day for the forty-five-cent lunch. Through the next decades, the exquisite banquets arranged for him on the second floor became the talk of social and artistic circles, and Luchow found himself mentioned in the same breath with the Academy of Music, Union Square, Steinway Hall, and Tony Pastor's (where Nellie Leonard, a scintillating blonde from Clinton, Iowa, made her debut as an English ballad singer, rechristened by Pastor for the occasion as Lillian Russell).

Luchow became the exclusive American agent for Wurzburger beer in 1885, and later for Pilsener; at the turn of the century Harry Von Tilzer composed a song in honor of the lager and the man responsible for popularizing it. The famous lyrics by Vincent Bryan were sung wherever beer lovers assembled:

> *Rhine wine it is fine,*
> *But a big stein for mine,*
> *Down where the Wurzburger flows!*

The champion Wurzburger toper was Baron Ferdinand Sinzig, of the House of Steinway, who achieved a high-beer mark that has never been reached since: thirty-six seidelsfull at one sitting. It is said that he did not rise once during the operation.

Luchow was one of his own best customers, eating and drinking along with the most assiduous diners. After washing down a series of dishes with a lagoon of beer (the Wurzburger townspeople had gratefully presented him with a decorative six-quart stein), he would more often than not require the aid of four busboys to maneuver him upstairs to the quarters he shared with his sister. On one occasion, however, he managed to nip to as adroitly as the rest of his customers who, amidst screams and shrieks and upended hoopskirts, were suddenly seen to leap about over tables and chairs and a welter of hand-carved knick-knacks in a mad scramble to flee the premises. It seems that a lion, fattened on scraps from the restaurant's kitchen, had escaped from his cage next door at the Hubert Museum: barely ambulatory, toothless, he had waddled into Luchow's dining room in search of leftover dumplings and sausage ends.

The position of Union Square as the heart of the music world, with Steinway Hall and the Academy of Music hard by, was presently threatened by lions of a different order. The eighteen private boxes at the Academy were hardly sufficient to accommodate the growing numbers of new millionaires, nor was the old guard in a hurry to relinquish them. The Rockefellers, the Morgans, the Vanderbilts, and the Goulds, meanwhile, resented having to sit in the orchestra beneath the hovering seraphim of New York's ancient aristocracy. A proposal to create twenty-six new boxes failed to meet their rancorous demands, and a group led by the Vanderbilts undertook the construction of the new Metropolitan Opera House at Thirty-ninth Street and Broadway. Its one hundred twenty-two boxes in three tiers looked to one unimpressed critic like "cages in a menagerie of monopolists."

The "Met" opened in 1883, and within two years the Academy of Music closed its doors. The Academy's adherents were forced to migrate to "the new yellow brewery," and within a few years Broadway from Madison Square to Forty-second Street emerged as the new theatrical vortex of New York.

New York was apparently in need of a new theatrical vortex. America's favorite composer was the Alsatian Jacques Offenbach, although his rollicking operettas, featuring "lewd" dances like the can-can, were denounced from the pulpit. Offenbach himself had recently toured the States and conducted his romps, like "Orphée aux Enfers," for clamorous audiences. He felt the country was like a giant, physically perfect, but without a soul—art. America needed a conservatory of music, theaters, museums, art schools, and "when this people, which has given such admirable proofs of will power, activity, and perseverance, decides to win a place among the artistic nations, it will not be long in realizing this new dream." The building of the new Met could only be regarded as a step in the right direction, one of the many made in New York toward the end of the century. (These included the founding, in 1870, of the Metropolitan Museum of Art, and the construction of Carnegie Hall in 1891, which made Fifty-seventh Street New York's perennial musical thoroughfare).

"Society, saith the text, is the happiness of life." With this Shakespearean quotation commences a volume entitled *The Season, an Annual Record of Society,* first published the very year of the opening of the new opera house, 1883. Society was undergoing a sudden expansion—quantity was catching up with quality. There was an excess of spending, an excess of food, an excess of snobbery, and an excess of etiquette. A new phenomenon appeared like a Mercury with ankles trimmed in magpie feathers: the popular press. By the 1870s there were two dozen newspapers in New York alone, and a wide audience panted for tales of bloodshed and sexual scandal. Flitting around the rich, and lapping up every detail of their lives, the "new journalists" then diffused the tidings over the land; by the turn of the century there was not a housewife in Toledo unable to enumerate the stars in Mrs. Astor's tiara, while in Atlanta, hostesses reproduced deftly tossed "Waldorf salads." The outward mannerisms of "polite society" were carefully diagrammed for the willing and worthy in do-it-yourself tomes of inexpressible archness. Over two hundred such books were published between 1870 and 1917. The authors of an 1886 bestseller, *American Etiquette and Rules of Politeness,* four of whom were asserted to be college professors, warned the budding generation that "every young lady and gentleman should cultivate a love for society—not as an end, but as a means . . . to regard it as a means to an end—makes it a constant source of interest and profit." Books of etiquette also covered the rules for refined dining. *Hills Manual of Social and Business Forms* (1879) dictated "Never allow butter, soup or other food to remain in your whiskers."

The excessive sense of etiquette led inevitably to a concept of Royalty and Court in New York society, and when Mrs. William Backhouse Astor, nee Schermerhorn, determined to seize for herself the position of Queen, she was able, with the Machiavellian advice of Ward McAllister, later Lord Chamberlain, to achieve the closest thing to a Throne in the history of the United States: at the end of her Fifth Avenue ballroom (which could contain, as McAllister pointed out, exactly "Four Hundred"), where the Empire State Building now stands, a discreet dias was erected for the accommodation of Mrs. Astor and that year's Privy Council. There, on an outsized, overstuffed couch, the "Mystic Rose" would thoughtfully preside, and, like a marble dial in the center of a formal maze, indicate the zenith of the sun. "Patents of nobility" were conferred, of course, by acceptance at court. In her rise to Absolute Power, she found it unfortunately necessary to edge out her sister-in-law, Mrs. John Jacob Astor III, who had committed the social blunder of exercising her intellect and permitting authors and painters to track up her carpets.

Blatantly conspicuous spending got under way at the very close of the Civil War with the Peto dinner at Delmonico's. Sir Samuel Morton Peto, a visiting Englishman, made his splash by entertaining one hundred American coffee and tea merchants at a twenty thousand dollar dinner amid flowering trees and restrained music. Printed in gold-leaf on satin, the menu ran from oysters to the "ruins of Paestum," and reflected both the finesse achieved by Delmonico's and the limits to which one could go for a two hundred dollar-a-plate dinner in the age of the five dollar-a-week wage.

MENU

Barsac.   Huitres.

                             POTAGES
Xeres F.S., 1815.   Consommé Britannia.
                             Purée à la Derby.

                             HORS D'OEUVRE
          Cassolettes de foie-gras.   Timbales àl'écarlate.

                             POISSONS
Steinberger Cabinet.   Saumon à la Rothschild.
                             Grenadins de bass, New York.

                             RELEVES
Champagne Napoléon.   Chapons truffés.
                             Filet de boeuf à la Durham.

ENTREES

Chateau Latour.

Faisans à la Londonderry.

Cotelettes d'agneau Primatice.

Cromesquis de volaille à la purée
de marrons.

Aiguillettes de canards à la bigarade.

Rissolettes à la Pompadour.

ENTREES FROIDES

Côtes Rôties.

Volière de gibier.

Ballotines d'anguilles en Bellevue.

Chaudfroid de rouges-gorges
à la Bohemienne.

Buisson de ris d'agneau Pascaline.

Sorbet à la Sir Morton Peto.

ROTIS

Clos-Vougeot.

Selle de chevreuil, sauce au vin de Porto
groseilles.

Becasses bardées.

ENTREMETS

Choux de Bruxelles.   Haricots Verts.

Artichauts farcis.   Petits pois.

SUCRES

Tokai Imperial.

Pouding de poires à la Madison.

Louisiannais à l'ananas.

Gelée aux fruits.   Pain d'abricot à la vanille.

Moscovite fouettée.   Gelée Indienne.

Vacherin au marasquin.   Cougloff aux amandes.

Mazarin aux pêches.   Mousse à l'orange.

Caisses jardinière.   Glaces assorties.

Fruits et Desserts.

PIECES MONTEES

Madere Faquart.

Cascade Pyramidale.

Corbeille arabesque.   Ruines de Paestum.

Le Palmier.   Trophée militaire.

Corne d'abondance.   Nougat à la Parisienne.

Needless to say, New Yorkers felt somewhat abashed upon learning that Sir Samuel, possibly the most open-handed visitor to these shores in history (his charitable contributions were many), had been indicted for fraud shortly after returning to England.

The prize for gastronomic aesthetics of the period probably belongs to the "Swan Banquet" of 1873, when Edward Luckemeyer, an importer, handed over a ten thousand dollar tax rebate to the imaginative Lorenzo Delmonico with the request that a memorable dinner be devised. The results were not easily forgotten. A huge oval table filled the entire ballroom; in the center of it glistened a thirty-foot lake surrounded by a landscape of flowers. A society reporter noted that "There were hills and dales; the modest little violets carpeting the valleys, and other bolder sorts climbing up and covering the tops of those miniature mountains." Moreover, floating on the lake in supernal serenity were four swans from Prospect Park, their flight being inhibited by a table-to-ceiling cage covering the entire indoor sea and executed in gold by Tiffany. Above the guests' heads, songbirds trilled in tiny gold cages. Only the very edge of the table was left free for plates, and the guests ate in some comfort, despite the raucous mating of two of the swans and the difficulty of cross-table conversation.

Even still, the Luckemeyer banquet could not claim the distinction of being the most expensive, per plate, of the period. That honor goes to a yachting enthusiast who arranged for a replica of his yacht, complete with satin sails and golden oars, to be moored in a cut-glass basin in front of every plate. This entertainment cost him four hundred dollars per guest.

The eighties and nineties represent one of the most foolish periods in American social history. Mrs. Stuyvesant Fish, who had made insulting repartee fashionable, gave a dinner party in honor of a monkey, and set a style. Thanks to that glamorous event, people were hard put to come up with party ideas more pacesetting than those of their friends. As a great deal of dinner conversation was spent on horseflesh and dogs, what better than to honor them appropriately? One hostess persuaded a number of canine-loving friends to dress as maids and serve their pets off trays at a banquet table. A more squeamish acquaintance restricted her banquet to inanimate guests of honor and requested that all her friends bring their favorite dolls to lunch beside them.

Two equine dinners received ample coverage by the press. The first was the more conservative of the two in that the guests were

Observing the social spectrum, 1882. *Courtesy of the Picture Decorator, Inc.*

seated in chairs throughout the evening. The table undulated in abstract curves and was solidly banked with a repellent combination of flowers and broad ribbons; seen from a balcony, the misshapen outlines of a horse's head, neck, and bridle could be detected dimly over the diners' shoulders. Artificial grass and trees, a painted cyclorama, and a champagne cart decked in flowers, drawn by a pony and driven by an eight-year-old child, almost completed the picture. The maître d' was on horseback.

All the guests dined on horseback at C. K. G. Billingsworth's white-tie banquet for the New York Riding Club at Sherry's. Trays were secured to pummels and champagne had to be sipped through lengthy straws from saddle bags. It was the first dinner party in American history requiring stableboys with dustpans to augment the regular staff.

But the exquisite heights of discomfort were reached at an Easter dinner given by a hostess with a taste for Parisian peephole candy eggs. Her urge to dine within one persisted to the point that artisans were instructed to build a canvas egg large enough to accommodate sixty-odd people. At that time, ventilation as a science was not advanced. Halfway through the entree the company began dropping in its tracks and the evening was cut short as oxygen-starved victims staggered through the orifice of the egg and out to the street, into their carriages, and home to their beds.

Aside from thematic absurdity, formal dinners dragged on interminably, until Mrs. Fish finally put her foot down and kept her guests at table for a smart fifty minutes. An absurd tonnage of food was served while an even more absurd amount of snobbish dogma decreed its selection, preparation, and ingestion. One family emigrated to Europe in disgrace because their friends' chefs surpassed their own: another committed the unforgivable gaffe of serving a rum rather than maraschino or bitter-almond-flavored sorbet, the refreshing water ice produced between meat courses to clear passage for further onslaughts, much like the clear broths presented intermittently by the Chinese throughout a thirty-three-course meal.

The sanest gourmet of the century, whose masterful menus and flawless taste saved a tattered professional reputation, was Sam Ward. He first married Emily Astor, sister of William Backhouse and John Jacob III. Emily died, Ward's banking business fell apart, his second marriage was judged a fiasco, and the Astors froze him out. Through years of wildcat speculation, however, Ward finally managed to disentangle the financial cat's cradle of his life and emerged triumphant as King of the Lobby. He was a kind and intelligent man (he proposed Oscar Wilde for a guest membership at his club despite the sniggers the visitor provoked) and his expert advice on food and wines was sought even by his detractors; in fact, Ward was the only layman allowed into the kitchen at Delmonico's to prepare his own sauces. His superb sense of balance is crystallized in this, one of the innumerable menus of his devising:

Little Neck Clams

Montrachet
Potage tortue vert à l'anglaise
Potage crème d'artichauts

Amontillado
Whitebait, Filets de bass, sauce crevettes

Rauenthaler
Timbales à la milanaise
Filet de boeuf au madère

Pommery sec
Selle d'agneau de Central Park, sauce menthe

Moet et Chandon Grand Cremant Imperials
Magnums
Petits pois, Tomates farcies, Pomme croquettes
Cotelettes de ris de veau à la parisienne
Crêpes à la bordelaise
Asperge froide en mayonnaise
Sorbet au Marasquin
Pluvier rôti au cresson

Chateau Marqaux
Salade de laitue
Fromages variés

Old Madeira Charleston and Savannah
Bombe de glace Fraises Peches Gateaux
Raisins de serre
Café

Cognac et liqueurs

Another noted amateur chef, Skipper Ben Wenberg, sailed the waters of the Atlantic and Caribbean as director of his own line to South America. The side benefits of his voyages would be reconstructed at Delmonico's upon his return to New York. One afternoon he burst in and called for a "blazer," or chafing dish, two large lobsters, a half pound of sweet butter, six eggs, a glass of Jamaica rum, and a pint of heavy cream; he had brought some special hot chili peppers with him. Charles Delmonico delightedly stood by as the Skipper whipped up his discovery; other late lunchers stopped in mid-meal to sniff the extraordinary aroma. Within days everyone was ordering "Lobster à la Wenberg." But some time later Wenberg broke a stringent house law by engaging in loud and abusive political argument and was peremptorily blacklisted by the management. The Delmonicos had a way of letting a patron know he was no longer welcome; he would be seated suavely, but never, never served. The lobster dish continued to gain in popularity, however, and so, rather than banishing it along with its namesake, a simple operation of anagrammatical surgery was performed on the name by Lorenzo, who converted it to "Lobster à la *New*burg."

The city's appreciation of such lofty exercises in finesse as the invention of Lobster à la Newburg failed to register with the recording angel, for Baedecker's Guide to New York stated that "New York is like a lady in ball costume with diamonds in her ears and her toes out at her boots." The Great Robber Barons did nothing to dispell this impression, the murder of Jim Fiske being a case in point.

Fiske, a Wall Street desperado, was Jay Gould's partner in his murkier dealings. "Colonel Jim" (he had gleefully accepted the titular command of the Ninth Regiment) was fat and foppish, a ladies' man, and a physical coward of note. Any woman who attracted his attention could expect her cup to run over. A predatory divorcee from Boston, Mrs. Helen Josephine Mansfield Salor, set her sights on Fiske and managed to secure not only the key to his heart but to his bank account. Following a year or so of amorous dalliance in a baronial love nest on Twenty-third Street, Josie became more and more mysteriously withdrawn and peaked; this was not surprising for, had the truth been known, she was engaged in a little additional dallying on the side with one of Fiske's collection of festering arch-enemies, Edward S. Stokes. Together they trumped up a scheme to blackmail Fiske with letters he had indiscreetly written to his inamorata. Fiske countered by threatening Stokes with ruin, accusing him of blackmail and fraud. It is difficult to distinguish total ruin from the position in which Stokes, feckless playboy son of an eminent family, found himself after Fiske had taken his first vindictive steps. In a blind rage, Josie's "fancy man" set off for the Grand Central Hotel to blast Fiske off the face of the earth. They met on the stairs as Jim was coming in from the street. Stokes drew a pistol, slowly and deliberately aimed, and shot Fiske through the torso: Fiske tried to rise, but was struck a second time and rolled to the bottom of the steps. He died the next morning in one of the Grand Central's bed chambers and Stokes was immediately arrested and arraigned for the first of his three trials. Ironically, Jim Fiske's favorite philosophical bromide, when accused of vicious business transactions, had been "Let each man carry out his own dead."

Fiske could not have been shot down in a more suitable place. The Grand Central Hotel at Broadway and Third was as gaudy and outsized as the victim of the dastardly deed. Opened first in 1854 by La Farge, a representative of Louis Philippe, it was rebuilt after a fire in 1856, and renamed The Southern Hotel, strangely enough, ten years later. Incorporated within its ample confines was the largest theater in New York, first called the Burton, then the Metropolitan, and later the Wintergarden, and it was here that Edwin Booth gave his first performance following Lincoln's assassination. The hotel was bought by E. S. Higgins in 1869, enlarged, and reopened as the Grand Central; its six hundred fifty rooms could cope with fifteen hundred guests padding across seven acres of carpet, and it was even able to live down the Fiske scandal.

Thriving in the same era were the Windsor, where Andrew Carnegie lived; the conservative Brevoort, on Fifth Avenue and Eighth Street, exhibiting huge portraits of Queen Victoria and the Prince Consort; and the Clarendon, featuring female waiters in the dining room. William Ferguson, a visiting Englishman, was altogether impressed by them, noting their smart pink and blue uniforms with white aprons and the "suitable attitudes" in which they stood. Moreover, he noticed that many were quite good-looking. Unfortunately the illusion was shattered when he peered into the kitchen and caught one of these paragons heaving a corn cake at another.

Although the Astor House sank slowly into a decline, many of the older hotels continued to hold their own. Throughout the Civil

War the Fifth Avenue Hotel had harbored the "Gold Exchange" in which Gould was so perniciously active; there too, the Republican party made its New York headquarters. In fact, it was at the Fifth Avenue Hotel that a Republican made the mistake of calling the Democrats the party of "Rum, Romanism, and Rebellion." One area of the lobby, where Boss Platt gave his yes-men their unsealed orders, was known as "Amen Corner." The Metropolitan Hotel retained its standing as the stronghold of the Democrats; in the early 1870s Boss Tweed leased it for his son, who spent half a million dollars on redecoration and all of his animal cunning in discovering new surfaces on which to brand the Tweed monogram.

Inevitably, the machine age took its toll on hotel business. The advent of elevated trains made it possible for New Yorkers to get home quickly, and the hotels of the business section, as well as restaurants, found it increasingly difficult to fill their dining rooms in the evening. But a still more decisive threat to the entire hotel world lay in store.

In 1874, permanently boarding families or bachelors constituted 50 percent of all hotel residents despite persistently nagging complaints of "too much show, too little comfort, too much hurry and excitement." Only the wealthy could afford private houses; the situation seemed hopeless. Then came the solution. Under Parisian influence, young Rutherford Stuyvesant built a "French Flat" on Eighteenth Street, complete with a concierge office. The "flats," more decorously called "apartments," were an immediate success with fashionable young couples and apartment houses went up all over New York, especially in the outskirt areas above Forty-second Street, on Madison, Lexington, and Fifth Avenues in the east, and on Seventh Avenue and Broadway in the west. The new apartments were often roomy and comfortable, and sometimes actually cavernous, with fine woodwork, fireplaces, steam elevators, and maids' rooms.

By 1881 the hotels had lost most of their permanent residents. General bankruptcy would have ensued had not expanding business brought more transients to New York, and had not the vogue for entertaining at home been superseded by the ostentatiously extravagant practice of giving private dinners and parties in grand hotels or "Delmonican" banquet halls.

Hotels still tried to retain something of a family atmosphere, and put themselves out, particularly during the holiday season, to center the activity of the city within their halls. Christmas dinners were sumptuous, and featured menu cards ingenious and marvelous to behold. It was said of the Hotel Vendome's 1888 menu that "the outside was like a sighing swain's Christmas card to his inamorata, and the inside was like the epitaph of one who had died of overeating."

The height of seasonal festivities was traditionally reached on New Year's Day, when hotels vied with one another in producing elaborate free lunches for "unfortunate and fashionably attired men about town, who are comparative strangers to the comforts of domestic life." In 1893 a team of reporters visited most of the free lunches in the city, and their subsequent tales have a "Nutcracker Suite" ring to them. Starting with the simpler downtown establishments around Union Square they found substantial menus: game, salads, pickled oysters, salmon, sandwiches, and pastry. But these hotels, catering to actors, artists, and such, did not find it necessary to decorate their wares in order to attract a New Year's Day crowd. Uptown the situation was quite different. At one grand hotel the reporters discovered a large sugar lion, "copied after one of Gerome's pictures," glaring fiercely at Virginia hams "in the shape of mandolins," and at the Murray Hill Hotel they were abashed to find a "barbecued pig's head, which by the chef's skill had been made to wear a sentimental expression," and which "appeared to be gazing with loving eyes on a tired-looking salmon." The caterers at the Vendome, again striving for distinction, displayed the total diorama of the landing of Columbus, along with "some very natural roses in sugar, some pheasants, a boar's head and a pig with spectacles." It was a good year for Columbus, as he was also represented in the restaurant of the New York Athletic Club, where he overshadowed "Corbett, who was looking quite sweet, and was performing his pleasing act of knocking out Sullivan, who looked sticky and melted." All agreed, however, that the lunch was unsurpassed at the Hoffman House where a replica of Columbus Arch, a sugar fort, and life-sized statues of the Graces hovered over the buffet.

The Hoffman House originally opened in 1864; it took another decade for it to become one of the most celebrated hotels in the city. Chiefly responsible for its metamorphosis was Ned Stokes who, upon his release from Sing-Sing four years after shooting Jim Fiske, bought part interest in the hotel with money advanced him by his family.

Ned's flamboyant and voluptuous tastes soon set the general tone of the Hoffman House. The gentlemen's bar, home of the Manhattan cocktail, was its focal point. An enormous painting, "Nymphs and Satyr," by the French artist Bougereau, much in demand for his overblown nudes, dominated the room. Ned had picked it up for ten thousand and ten dollars at a private auction of the Wolfe estate; a few years later he turned down an offer of three times the original expenditure; the painting had become a national sensation and no one came to New York without stopping in for a prolonged aesthetic evaluation. A reproduction of it appeared on boxes of Hoffman House perfectos, ten cents straight, in almost every cigar store in the country, and a best-selling comic lithograph of the period pictured a hobo mesmerized by the nymphs over the bar with the caption, "I've been looking all over the world for that creek, but darned if I can find it."

Once a week, "Ladies' Day" made it possible for feminine art lovers to broaden their cultural horizons; the management provided a catalogue of museum-like chasteness in which the virtues of the house's collection were extolled in erudite prose. Two great favorites were the "Narcissus" and "The Vision of Faust," a jumbo-sized painting of the German romantic school in which the doctor's dilemma seemed centered on a roseate nudist colony in the act of levitation.

Running fifty by seventy feet and paneled in mahogany, the barroom itself became the archtype for the era. The heavily carved bar was equipped with the customary brass-plated footrest; the floor, flecked with Turkish rugs, was tesselated in shades of cream, putty, brown, and cerulean blue.

Wine sprees were common at the Hoffman House bar in those beautiful, swaggering days when "standing drinks for the house" or "treating" was as much a part of the national character as showboats, red velvet corsets, and the incorrigible "Bet-a-Million" Gates, the Western barbed-wire king who would have taken odds on a sandcrab pushing a freight train. Not everyone could afford to play the game; three or four hundred dollars could be casually spent by a fledgling playboy during those tedious afternoon hours when life held nothing until the time arrived to change for dinner.

However, it was not the smart hotel bar, but the plainspoken neighborhood saloon that would be transformed into an historic institution by future generations of nostalgic Americans. On the surface, saloon life radiated a bleary, beery charm immortalized later in song: "Just mention 'saloon' and my cares fade away." A neighborhood saloon became the hub of a workingman's universe and he visited it daily, sometimes as frequently as morning, noon, and night. A saloon gave a man a chance to air his political opinions; it gave him masculine privacy, a feeling of belonging, easy credit (despite the curt notices on the wall to the contrary), escape from nagging household problems, and what's more, a free lunch. "Swell" bars like the Hoffman House spread out a free banquet seven days a week, the only hitch being that it was necessary to tip the starchy waiter a quarter and refer to the lunch as a "buffet." In solid saloon circles, the free lunch table served a felicitous double purpose—it kept a man on his feet and provoked violent thirst all at the same time. Salt was the common flavoring agent of its principal dishes; spiced hams, sauerkraut, olives, summer sausages, crackers, potato chips, dill pickles, potato salad, spring onions, pungent cheese, cervelat, baked beans, herring, and pretzels—any or all of these were consumed at better establishments between rounds of foaming steins or straight shots of bourbon and rye. To take whiskey any way but neat was considered effeminate, although chasers were allowed: water—branch or otherwise—clam juice, ginger ale, or milk. Scotch meant anything from Scotland, and made no inroads with drinkers until golf was introduced to the United States in 1887 by a Scottish resident of Yonkers, who returned from a trip home with some clubs and balls. For some unknown reason the popularity of the two imports grew in equal ratio.

The prize-winning dehydrator on the free lunch table, sardellen, a crude and caustic cousin of fancy oil-packed sardines, was immediately identifiable by its arresting tang; quite a lot of beer had to be gulped down to rinse the last, lingering traces of it out of the mouth. "Black-Eyed Susans" and "Blind Robins" (dried herring) followed closely as runners-up but they somehow never equaled the sardellen.

Free lunches in the simpler saloons and no-star hotels were drastically abbreviated; bean soup or cheese and crackers might constitute the entire menu. At the Criterion, a theatrical hotel at Fourteenth Street and Union Square, a tippling journalist noted that at 11:00 A.M. precisely "the free lunch, consisting of a wide segment of cheese and a bowl of hard crackers, was placed upon the counters across the room from the bar. The comedians present made funny falls toward the repast while the tragedians advanced with stately tread, and, in an incredibly short space of time, nothing remained of the banquet but the rind of cheese which by some daily miracle still maintained an upright position, and a handful of impalpably fine cracker dust at the bottom of the bowl."

At proper saloons, such gusto went unappreciated when customers poured their own straight drinks. After a man called for whiskey, a black bottle and a two-ounce shot glass were slammed down in front of him and he was expected to pour a "gentleman's drink," in other words, not in excess of an ounce and a half. If etiquette were breached, the bartender invariably demanded, "Will you need a towel?" In barrelhouses a nickel slug meant a brimming shot glass; whether or not the management had thinned its liquor with prune juice and water and then revitalized it with a generous dash of cayenne seemed not to matter; nor did a false-bottomed measure as long as the glass was *full* and you could feel your money's worth burning all the way down.

The "bum's rush" was initiated not as a means of expelling drunks but of repelling daring daylight raids by fleet-footed indigents on the free lunch counters. It was the Golden Age of the bouncer's art. Typical scene: burly, biceps-gartered bouncer leans laconically against stamped tin and imitation paneled wall, his lovelock plastered firmly to a Neanderthal forehead, beneath which flicker small, unseeing eyes. He remains as motionless as the bass-fiddle-shaped nude above the brass-ringed bar; if he is conscious of the constant drone of compatible elbow-benders or the tintinnabulation of brightly polished glasses, he could certainly fool the scruffy vagabond poised by the doorway shuffling his feet in the sawdust, his eyes darting triangularly from lunch to bouncer to door to bouncer to lunch. A moment's hesitation and the intruder steps purposefully toward the bar, hand upraised as if to catch the bartender's attention. Suddenly he veers, as if about to return a tricky backhand tennis shot, and pounces on the lunch counter. Clouds of sawdust. Swinging doors bang open as trajectile object arcs gracefully onto the street; bouncer strolls back to battle station, eyes unseeing, biceps, lovelock, dignity in place. Drone of compatible elbow-benders. Peace.

The bouncers most assiduously to be avoided policed the countless saloons proliferating in the Bowery and the Tenderloin. The Tenderloin, the brash new area of vice, possessed neither history nor scruples. It ran along Sixth Avenue from Fourteenth Street north to Madison Square, and by the nineties had spread as far as

Forty-eighth Street, from Fifth Avenue clear over to Ninth, rubbing shoulders with the spreading slum of Hell's Kitchen, where gangs such as the Hudson River Dusters and the Tenth Avenue Gang made the old Bowery b'hoys look like so many dancing-class students. The Tenderloin got its name from the remark of one Police Inspector Williams, who, when transferred to the district, gloated that he had long been restricted to chuck steak, and told Charles Delmonico that he was now getting a "piece of the tenderloin."

The dives of the Tenderloin could never have survived without the vigorous cooperation of Tammany Hall and the police, who were munificently repaid by the proprietors. Despite its dreadful reputation, the prosperous establishments of the quarter were by no means all cheap brothels, dance halls, and saloons. The French Madame's, for example, was one of the most reputable "houses" in the city and had no truck with the police or Tammany Hall, claiming protection from higher authorities—that is, until the day the French Madame herself, Mrs. Matilda Hermann, was hauled into court and, to her undying shame, accused of common prostitution, an act which she personally had never even contemplated. The Haymarket, a dance hall, developed such flamboyant standards that Diamond Jim Brady often frequented it; and the quasi-Oriental White Elephant seethed with an exotic elegance, with its stained-glass windows, rumbling bowling alley, pachydermic frescoes, and sirocco of musky perfume.

The Bowery, as the century came to a close, fell on hard times. In his study *The Bowery*, Edward Ringwood Hewitt reported its decay from the tough glamor of the days of when it was the favorite home of the b'hoys (in its precincts Billy the Kid was born in 1859) to its final bleak estate: "Still the gardens and saloons and occasional pistol shot at night; still the cheap amusements; more and more frequent the three balls, and the shops offering the cheapest clothing in the world; and the 25¢ and 10¢ lodging houses, with all they mean of homeless men ever increasing. . . . But to most, the Bowery is a half-humorous, half-dreaded portion of the city, strange and foreign." There was a saloon in every doorway, many catering specifically to beggars and thieves. These "distilleries" or "morgues" sold a flask of whiskey for five cents, and some kept a "velvet room," where for a nickel drunks could get a bowl of alcohol to finish them off for the night. Other saloons offered for that sum a barrel with a hose from which a customer was allowed to drink all he could hold without catching his breath. Among these were Harry Cooper's, McKeon's, and Old Man Flood.

The transition from Bowery b'hoy to Bowery bum has been attributed to Chuck Conners, "Mayor of Chinatown." But Conners was actually more symptomatic than casual. He was born George Washington Conners of a respectable Irish family, precisely the type of family that soon left the area. Chuck, however, was heroically unregenerate. Having worked briefly as a locomotive fireman on the elevated until the early death of his wife, he settled down to the less mettlesome career of Bowery philosopher and character. A busily lethargic person, Chuck's laconic sayings were faithfully and constantly reported in the papers, where he was represented as a lovably irresponsible genre figure. He may possibly claim the invention of the "deez, dems and doze" dialect, and his most celebrated advice was "Work? Ah-h fergit it!"

Chuck portrayed himself in variety shows on the stage, rising to the proud heights of Oscar Hammerstein's Victoria Theatre on Broadway. He even managed to get himself shanghaied, and wound up for two weeks in Whitechapel, where he felt quite at home. Upon his return, as self-decreed bellwether of Bowery fashion, he affected the broad sailor pants and pea jacket of the English costermonger; but the style never caught on, and he resumed his initial trademark, an iridescent outfit of black suit and waistcoat set off by rows of pearl buttons and a black bowler with a button on each side.

It was as a political force, however, that Conners was best known. He eventually took over as official ward heeler for the Bowery and Chinatown and formed the Chuck Conners Club, whose "honorable Members" included Al Smith, John L. Sullivan, Mickey Finn, Richard Mansfield, and Oscar Hammerstein.

The Bowery, unlike the Tenderloin, never courted organized vice, nor did it completely lose its sense of neighborhood or charm; at P. J. Kennedy's spotlessly clean eating house, for example, neatly dressed old men and women forgathered nightly for a good ten-cent dinner and the latest local gossip. Irish revolutionaries still met in the back rooms of many Bowery saloons, and another famous "native son," Steve Brodie, promoted his noisily chronicled career under the very nose of the "Chuck Conners Club."

Brodie's first bid for immortality was a well-organized plunge from the three-year-old Brooklyn Bridge. The year before a man had been killed in a similar attempt. Steve connived to arouse publicity by trumping up a bet that he would back down, at the same time keeping the date of his projected leap a closely guarded secret. When the day arbitrarily dawned, "something" was seen to fall from the bridge by Brodie's close friends, and a conveniently idling barge did haul a soggy Steve out of the water. Despite conflicting evidence and hearsay, the young hero was soon appearing in museums and was presently set up in the saloon business by a brewery.

Steve Brodie's in the Bowery attracted the sporting element of the nineties, together with those uptowners who simply wanted to gawk at roughshod celebrities, gamblers, and toughs. In the barroom, among souvenirs and mottoes, hung a painting of Steve in that never-never moment between bridge and water, and alongside it the framed testimony of the barge captain who had rescued him. The pinnacle of his glamorous history was reached when a producer sought Steve to star in a three-act play about himself. He had outdone Chuck Conners once and for all.

Tersely entitled *On the Bowery,* the production arrived at the Bowery Theatre via Philadelphia and Brooklyn on October 22, 1894. It was a great day for the precinct. Above the street, "Welcome to Steve Brodie" banners flapped in the wind, and the many floral tributes commissioned for the occasion included a replica of the Brooklyn Bridge. A painstaking facsimile of Steve Brodie's own saloon, down to the "painted women drinking real beer," provided the setting for the second act, and needless to say, the house broke up when Brodie finally burst into his theme song, "My Poil Is a Bowery Goil." The finale was climaxed by the jump itself, artistic and historic license having been taken by the playwright so that Steve could save the heroine.

The antithesis of Brodie's overpublicized saloon, securely entrenched on Seventh Street between Second and Third avenues, was, and still is, McSorley's Ale House, where German lager is never served and whiskey rarely. John McSorley first opened his house in 1854 and ruled over it as benevolent monarch until his death in 1910 at the age of eighty-seven. He discouraged flashy customers who, although spending freely, disturbed the conservative workingman's atmosphere which McSorley so tenderly nurtured. Activities were limited to ale drinking, clay-pipe smoking, and congenial, *seated* talk. The pipes and tobacco were provided by the house; everyone who trudged through the luxurious turf of fresh sawdust to take a chair at one of the sturdy old tables was expected to fill one on his way and, once seated, strike up a conversation with the neighbor nearest him. As disorder was foreign to McSorley's, a bouncer would have been useless window-trimming.

McSorley looked after himself as carefully as he did his customers, rising every morning at five and marching briskly down to the Battery baths. Drinking steadily and soberly for years, McSorley abruptly stopped on his fifty-fifth birthday; not only did he eschew alcohol for the rest of his natural life, but tobacco as well. Such commendable moderation was unwelcome in most saloons, and consequently, even the most innocuous taprooms found themselves under attack, at first from the speaker's platform, then from the pulpit, and finally "in the field," where they had to be defended from female armies by brute strength alone.

The temperance movement got under way in the 1840s with the formation of the Washington Total Abstinence Society. The Washingtonians were all male and their appeal was to Pure Reason. Speakers for the Society enlivened their lectures with imitations of God's own children under the influence of alcohol, and they enjoyed tremendous popularity. These "drys" claimed shortly to have cut the consumption of hard liquor in the United States by one half and to have sworn one hundred and fifty thousand drinkers to the "pledge." Other temperance societies and many abolitionists soon joined them and by the mid 1850s nine states including New York had actually passed temperance legislation, little of it enforced.

The post-Civil War period saw two important advances in the temperance movement; it was taken up by women and by the church. During the war, women had served as nurses, rolled bandages, sewn uniforms and generally taken an active part in the campaign. There were now many in the nation who discovered they enjoyed campaigning and refused to retire to their hearths. If abolition could succeed, so could other moral crusades against evils standing between Victorian life and utopia. The saloon figured prominently among them. Moreover, nineteenth-century women undoubtedly felt that by attacking the saloon they were, however indirectly, attacking the male establishment that had oppressed and paralyzed them for over a hundred years. The Middle West was suddenly besieged by "The Praying Women" kneeling patiently on the floor of saloons with their eyes to heaven as beer passed over the bar, possibly history's first "sit in" demonstrators. By the 1880s the Women's Christian Temperance Union had sixty thousand members, and the clergy was busily borrowing their thunder to hurtle from the pulpits. In New York, temperance speakers baldly denounced all foreigners as alcoholic reprehensibles.

Not until the nineties did a brimstone-and-fire-breathing exterminating angel appear to goad the W.C.T.U. and sister splinter groups into evangelic action—Carry Nation, the Kansas Smasher herself, looking like a souvenir postcard pugdog in a poke bonnet. Bearing the scars of a disastrous first marriage to a drunk and the grim memory of her psychotic mother who, thinking she was Queen Victoria, had habitually driven about in an open carriage wearing a purple cloak and a glass crown, Carry was a weird little figure behind whom misguided idealists, incipient vandals, and generally repressed Victorians could orgiastically rally in the names of Motherhood and Morality, Piety and Prohibition.

With her symbolic, but useful, hatchet in hand, Carry set out to physically destroy every saloon in the country. The devil's utensils being largely made of glass, this goal lay gratifyingly within the realm of possibility.

Unfortunately for Carry, the law was never really on her side, as may be gathered from her interview with Police Commissioner Murphy when she first arrived in New York. The following accounts appeared in the *New York World* on August 29 and September 2, 1901:

With a two foot hatchet strapped to the girdle under her linen jacket, her beaded black poke bonnet pushed down firmly on her head, her broad jaw set at its most pugnacious angle, the Smasher strode into Colonel Murphy's room at Headquarters at eleven a.m., plumped into a chair close to him, and in ringing tones demanded: "Don't you think New York is an awful bad place."

—"You don't know what you're talking about," said the Colonel, in a rage, "Go back to Kansas. If you want to do something, why don't you do it for your husband?"

—"I have no husband now," said the Smasher (recently divorced from her second husband). . . .

—"Oh, yes," said the Commissioner, with a grin. "All I have to say is that I congratulate Mr. Nation."

After more such badinage, the Commissioner gave his ultimatum: "If you violate the law I'll have you locked up." But Carry had the last word. As she departed, she passed a man smoking in the corridor. "You horrid, nasty man!" she screeched, "Don't you know that your fate will be an eternal smoking?"

Mrs. Nation left that very evening to address a meeting in Ohio, much to the relief of the proprietor of the Hoffman House, who, having been tipped off that she was about to attack his bar, had enlisted all available forces to help protect his priceless "Nymphs and Satyr." He might well have worried, considering Carry's reaction to a nude figure of Diana in the lobby of the Victoria Hotel. "Look at that image," she cried. "She ain't got a thing on; where's the hotel proprietor?" When a clerk was summoned, she ranted on: "There's a woman over there without any apparel on her. It ain't respectable. Would you like your wife to see that? Now I'd like you to put a little something on her right away or there may be a little hatchetation around here."

Carry returned some days later to precipitate the promised riots, but she was stopped in her tracks as she proceeded through the Tenderloin at the head of an army of thousands. Three saloons and two "concert halls" had felt the purgative blow of her hatchet by the time the police finally halted her progress at Twenty-eighth Street and Eighth Avenue. "I am not disturbing the peace," she asserted. "You are raising a crowd and creating a riot, and I arrest you," came the reply, curtailing forever her activities in New York.

Temperance sired the soda fountain, and the ladies of the movement selflessly mothered it in the fond hope that it would someday vanquish the bar. The ice cream soda, composed of carbonated water (introduced in the 1830s), fruit syrups, and ice cream, had evolved by the mid 1870s. The fountains, usually constructed in various shades of colored marble suggestive of their cool, particolored contents, were awesome works of art in themselves, replete with mirrors, filigrees, columns, and entablatures and might cost upwards of forty thousand dollars. One is recorded that could supply the white-coated operator with three hundred different combinations of flavors. The "fountain" itself often stood behind a low counter at which customers were served, and it was generally allied to a drug or candy store or a restaurant. John Matthews, the inventor of soda water, died in 1870 and was laid to rest beneath a monument which was made to resemble a marble soda fountain. By 1891 there were actually more soda fountains in New York than bars, so the temperance spirit would appear to have

Soda-water fountain. *Courtesy of the New York Public Library.*

been victorious. Yet there were those who regarded even the ice cream soda as sinful.

Total Abstinence Cookery, a prophetic tome published in 1841, gave fair warning to lovers of "tipsy cakes" and fried pork chops that heavenly storm clouds were gathering over the kitchen. Temperance was required in food as well as drink. The Reverend Sylvester Graham, immortalized in the cracker bearing his name, remained firmly convinced that the miller's bolting (sifting) cloth was fast becoming the shroud of the American people. He raged not only against white bread but seafood, salt, pork, and butter. Even milk and honey were found suspect and ungodly. Will K. and Dr. John Harvey Kellogg, in tandem with C. W. Post, took up the cry and turned Battle Creek, Michigan, into the cradle of the cornflake.

The dietetic bandwagon militantly rolled on, and through the ensuing years attracted such disparate types as Thomas Edison, Horace Greeley, and Harriet Beecher Stowe. Mrs. Stowe upheld the theorem "the frying pan has awful sins," while Mrs. Horace Mann, mincing no words in her *Christianity in the Kitchen,* bleated "There's death in the pot."

The popular press began firing a few salvos of its own. In 1866 a correspondent for *Harper's New Monthly Magazine,* alarmed by the culinary menace stalking the land, concluded that pies jeopardized the national future.

We cry for pie when we are infants. Pie in countless varieties waits upon us through life. Pie kills us finally. We have apple-pie, peach-pie, rhubarb-pie, cherry-pie, pumpkin-pie, plum-pie, custard-pie, oyster-pie, lemon-pie and hosts of other pies. Potatoes are diverted from their proper place as boiled or baked, and made into a nice heavy crust to these pies, rendering them as incapable of being acted upon by the gastric juice as if they were sulphate of baryta, a chemical which boiling vitriol will hardly dissolve. Life is short, and we have no time to waste in eating. Thus our tables become railway-station counters, and we devour our food as if the conductor were outside ready to cry "All aboard!" We enjoy less than any other people. We have no time for even our pleasures. Pie is at the bottom of all this nervous unrest.

Decrying meat as "cow's corpse" or "butchered hog," vegetarians agitated their chlorophyl-stained fingers at a protein-packed public. The second annual dinner of the New York Vegetarian Society was held in 1894 at the St. Denis Hotel. One hundred and fifty members convened to exchange physiological pointers and banquet on cream of celery soup and haricot beans and bread with curry sauce.

Along with inveighing against fats and meats, diet faddists railed against overeating in general. Here they had a point. The first meal to be successfully reduced was breakfast, and by the 1870s toast, eggs, oranges, and grapefruit were seen routing steaks and chops from the early morning table. This did not please the English journalist George Augustus Sala, who had dearly loved his breakfast at American hotels, where he could order from a menu "a foot-long hominy, iced milk, whitefish, eggs, a cutlet, a prairie hen, corn bread and coffee." National specialties such as raw eggs beaten into a glass of milk and pancakes also appealed greatly to him. As to the last, he described, in a letter, his competition for pancakes at one hotel table with the ladies of the house: "lean and sallow creatures—that masculine yet bony authoress from New England has actually built a monument to Mrs. Beecher Stowe of flapjacks. Now she butters each layer, then pours libations of molasses on the whole, and lo! In less than ten minutes the monument is no more, and the strongminded woman herself has stalked downstairs and gone shopping in defiance of all dietetic laws, human and divine."

But while the forces of righteousness and abstinence were militantly closing ranks, their enemies did not sit idly by. The fate of the Hotel Brunswick, for example, rings out in the annals of the crusade against vice.

Under the management of Messrs. Mitchell and Kinzler, the Hotel Brunswick on Fifth Avenue and Twenty-sixth Street, thrived for over a decade as a resort for gourmets, coaching clubs, horse fanciers, high society, and well-set-up undergraduates from Harvard and Yale. Traveling aristocracy alighted in migratory flights, and Kinzler became noted for the ingratiating way in which he brightened the journeys of Lord Durham, the Earls Russell and Cairns, the Duc de Morny, the Comte de Tocqueville, and King Kalakana of the Sandwich Islands.

The Brunswick's name remained unblemished until one wintry day in 1882 that started out as auspiciously as any other; in the café, the barmen abstractedly polished their crystal ware, fresh flowers were arranged on the reception desk, impeccably dressed guests paced to and fro in the lobby after breakfast, and in the kitchen the chef supervised the paring of truffles into decorative shapes.

Shortly before noon, a magnificently tailored figure, black and white as a dageurrotype, with sandy hair and manicured hands, an educated accent, and divinity-school manners, irritably left Delmonico's, crossed the street to the Brunswick, and introduced himself at the desk as "Mr. Thompson." He had been planning a supper for fifty friends that very night following a sleighing party, and had simply taken it for granted that Delmonico's would have a private dining room free. No such luck. As the cuisine at the Brunswick was considered by many to be of equal merit, would they possibly be able to accommodate him and his party? He would happily pay in advance. The desk clerk hastened to assure Mr. Thompson of the corporate delight occasioned by his patronage. Under no circumstances could an advance payment be accepted from such a distinguished gentleman. Arrangements were quickly and efficiently made for a one A.M. supper and Mr. Thompson quietly withdrew, his departure followed by many admiring glances.

That night the maître d' and his staff hovered over the elaborate table setting in the private dining room, straightening an oyster fork here, filling a salt cellar there. One o'clock struck, the candles were lighted, but no guests appeared. An hour passed. Waiters shifted from foot to foot and stifled cavernous yawns. Suddenly they began to look at each other obliquely, eyes narrowed and ears cocked. What sounded suspiciously like a brass band in full complement seemed to be hideously blaring forth beneath the windows. They converged in the lobby and cautiously peered outdoors. The spectacle that greeted them made their blood run cold and bank around the corners.

Out in the snow, a caravan straight out of a Goya nightmare had come to an ominous standstill behind a twenty-two-piece marching band. Twenty-five carriages of every description began to belch forth their clamorous cargo; from the lead coach emerged Mr. Thompson, who, it turned out, was none other than Billy McGlory, the scourge of Manhattan and proprietor of Armory Hall, that delectable distillation of a hundred Road to Ruins. Still operating sardonically as a "concert saloon," Armory Hall had become nationally known as a sort of underworld Valhalla.

Lurching and cackling, The Poll Parrot and Big-Mouth June bolted their carriage and treated the company to a high-kicking sidewalk reprise of the can-can and cootchy-cootch which had so enthralled the guests at the presupper party at Armory Hall.

Nigsy, Princess Kate, Little Swipes, and other "bright and shining lights of East Side female depravity" raced each other into the hotel. It became immediately clear to the management that many more than fifty revelers were staggering through their doors. It also became immediately clear that something drastic would have to be done to placate the regular tenants who, at that very moment, were clustering furiously at their windows, too stunned to speak a word, too intrigued to avert their eyes.

Into the dining room McGlory shepherded his fifty female guests (forty-seven to be legally accurate: three were in reality men) and gave orders that they were to be served all the champagne they could drink. He and the masculine contingent commandeered the café; one courageous waiter protested the abuses being heaped upon him by the ladies and was promptly pitched through a door and into a coma by the gentlemen.

Back in the dining room the hors d'oeuvres were flying—radishes, celery, pâtés, and olives. The Misses Anderson and Howard had hit upon this method of gaining the waiter's attention and the rest of the crew followed suit. As they had all climbed into fancy-dress costumes for the momentous occasion, the scene was one of riotous color; blue and silver page-boys passed out in the laps of Spanish dancing girls, ladies of Marie Antoinette's court were being ill under the table, and two "pédestriennes" in jockey caps and short skirts, unable to recall all the words to a dirty song, beat out time on their glasses.

The merrymakers were not escorted home by their Spanish grandees and Pierrots until after four in the morning, and then only with difficulty. As the *Herald* reported, "some had to be dragged from the scene of gastronomic carnage and thrown bodily into the carriages—legs of every size and shape bristled from the carriage windows and the drivers reeled so violently they barely kept their seats on the boxes."

The *Herald* was not by any means the only paper to give the story full play, and try though Kinsler might, nothing, not even complete redecoration, could restore the Brunswick's reputation. The decline was a drawn-out one as the hotel limped along until 1896 when it finally, ruefully, gave up the ghost.

Some years later, the Low Life and the High Life met yet again, on the occasion of The Awful Seeley Dinner. This debacle, conversely, was instigated by the High Life, which may explain why society was so much more thoroughly shocked by the festivities.

Herbert Barnum Seeley, a nephew of P. T. Barnum, may not have epitomized the soul of New York aristocracy, but he did have pretensions. Following his brother's engagement to the well-connected Miss Tuttle, he determined to give a bachelor dinner at which no expense would be spared. The scene of the bash was to be Delmonico's, but, possibly because of a house regulation prohibiting service behind closed doors, the location was transferred to Delmonico's up-and-coming archrival, Louis Sherry's. Seeley then signed a one thousand dollar contract with a Broadway agent, it was later claimed, to provide suitable entertainment.

Herbert Barnum Seeley proved to be a showman worthy of his uncle, albeit by default. On the night of Saturday, December 19, 1896, Mr. Seeley's guests assembled at Sherry's, twenty in all, among them representatives of New York's most gelid elite. Around mid-evening Captain Chapman of the Thirtieth Street Station was apprised by an undisclosed informant that something of an obscene nature was transpiring at Sherry's. Captain Chapman, noted for his aggressive whiskers and dauntless courage, held within his jurisdiction the juiciest slab of the Tenderloin. It was not often that he was called to work at Sherry's, which happened to fall within the geographical boundaries of his precinct although it lay well outside its spiritual realm.

Inured as they were to raiding, the captain and his men barged into Sherry's with a wild-eyed fervor more appropriate for invading the French Madame's. They stormed upstairs and through a door, the first they saw, which happened to be the entrance to the dressing room of the girls, who were to perform. Maidenly squeals and piteous protestations followed. The guests crowded in to find out what was amiss, and in the ensuing scuffle Chapman narrowly escaped defenestration. Order was restored and the police, who had found nothing incriminating, withdrew. The bachelors, some of whom were married men, among them the father of the host, readdressed their dinner.

All would have remained quiet had not the Seeleys and Sherry's brought charges against Captain Chapman. At the resultant trial before a police board, matters were looked into which did not bear looking into. The story hit the headlines and remained there. Tales abounded of girls approached by agents with the immodest suggestion that, for a tantalizing sum of money, they dance in the all-in-all. Definitely established was the fact that a certain dancer of vile repute was not only on the premises at the time but had actually performed. This was Little Egypt herself, who had introduced the belly dance to Western man at the Chicago World's Fair three years before. Ultimately Seeley and his agent were indicted for "conspiring to induce the woman known as Little Egypt to commit the crime of indecent exposure." Had their conspiracy succeeded? According to Little Egypt's own testimony, she had danced twice, clad for both divertissements in a Zouave jacket and a pair of lace pantaloons. She had not frolicked in the buff; although she had hoped to.

Fortunately for Louis Sherry, his restaurant's reputation did not suffer the same public condemnation as was heaped upon the Hotel Brunswick, where he had been employed early in his career. In fact society took no punitive measures whatsoever against the caterer of The Awful Seeley Dinner; by the turn of the century Sherry's was superseding Delmonico's in fashionable popularity.

Louis Sherry was French by descent only, having been born and raised in St. Albans, Vermont. After working in his teens in the Brunswick dining room, he moved on to a better job as maître d' at an exclusive hotel in Long Branch. There he developed his social skills and contacts, and became the trusted friend of Henry Clews, F. R. Coudert, and William Dinsmore. In 1881 they backed him in his first confectionery and restaurant venture in an unassuming building at 662 Sixth Avenue. Once installed, Sherry went after society with his biggest guns.

His first windfall came through the Metropolitan Opera. Having catered a "kirmess" brilliantly at the new opera house, Sherry was rewarded by extensive newspaper coverage. Soon after, in 1885, he captured a catering plum when chosen by Paran Stevens to arrange a gala dinner for the Badminton Assembly at his town house. Herman Oelrichs commissioned a "different" kind of party aboard his new tug, the "Rob 2"; Sherry was also tapped to arrange a luncheon in honor of Bishop Henry Codman Potter, "a man of the world, a lover of rich food and rich houses" (according to Harry Lehr who should certainly have been able to spot one). As the craze for clergy at dinner parties ran unabated, Bishop Potter became a prize catch for any hostess. "Stately and magnificent," he invariably incanted grace before the meal; at the table of a Mrs. Fish he would portentously clear his throat, assume a kind of ghost-of-Hamlet's-father tone of voice, and launch into a "Bountiful Lord, we thank Thee for all Thy blessings" form of incantation.

It was maliciously noted, however, that in less elaborate households where the food might have been boiled for hours by an Irish woman of all work, the Bishop would start off more hesitantly, his expression one of abject self-effacement, with "Dear Lord, we give thanks for even the least of these Thy mercies. . . ."

In 1890 the Goelet mansion at Fifth Avenue and Thirty-seventh Street fortuitously appeared on the market and Sherry wasted no time in snapping it up. Its confines provided sufficient space for large and small ballrooms, a small restaurant, and a confectionery shop. But patrons began to rely on Sherry for all manner of pampering and he soon realized that the new quarters was still inadequate in size. He commissioned Stanford White to design a splendid twelve-story building for the corner of Fifth Avenue and Forty-fourth Street. It was finally opened on October 10, 1898. Society was duly impressed; so was Charles Delmonico, who had moved into his own new quarters the previous year—on the opposite corner.

Delmonico might well have been incensed. Mrs. Astor canonized the new Sherry's by giving a ball on its premises, and the luminous likes of Richard Harding Davis, Ethel Barrymore, and J. Pierpont Morgan, known informally to the staff as "Uncle John," established immediate squatter's rights. There was plenty of space for all of them. The main dining room, instantly approved by the women, occupied the Fifth Avenue frontage; the superb men's café was entered from Forty-fourth Street. Upstairs lay the ballroom and less palatial rooms for private parties, and above them, residential apartments; one bachelor kept a suite of twenty-seven rooms.

Louis Sherry's most inflexible rule was "Never disappoint a customer!" Catering commissions were no longer limited to the environs of New York. When J. P. Morgan heard the call and made a California pilgrimage to attend the Triennial Convocation of the Protestant Episcopal Church, he turned to Sherry rather than his pastor for guidance. Sherry sent agents to San Francisco to scout around for a fit mansion in which Morgan might entertain on his brief and pious mission. This achieved, they set about redecorating it from garret to wine cellar, installing antiques, electricity, paintings, and other unmonastic niceties. Everything was running as smoothly as a Wall Street messenger by the time Uncle John arrived. Not until three days after his departure, when the house was completely dismantled, could Sherry's agents breathe freely once again.

Delmonico's had maintained its pre-eminence well into the 1880s under the tireless supervision of Lorenzo—Lorenzo who faithfully rose at dawn each morning to visit the Washington and Fulton markets, arriving at his restaurant by 8 A.M. There Rimmer, the manager of the café, brought him a cup of strong coffee and a Figaro cigar, after which Lorenzo padded home to East Fifteenth Street. The same steadfast soul reappeared at dinnertime and remained until midnight, chatting with friends, and overseeing generally while nibbling at *aspic de canvassback*. Following the uptown movement of wealth, in 1876 Lorenzo had herded his entire staff from Fourteenth Street to new premises on the corner of Fifth Avenue and Twenty-sixth Street, where it formed a glittering triangle with the Brunswick and the Fifth Avenue Hotel. This was the biggest and most elaborate Delmonico's yet to be seen, with café, main dining room, and private dining rooms each attracting their own devotees.

On Twenty-sixth Street, during the late seventies and early eighties, Delmonico's enjoyed its greatest popularity. It was here that Ned Stokes, who couldn't seem to learn his lesson, broke the nose of James Gordon Bennett (not at the table, but fair and square in the hall, as Lorenzo suggested), and it was here that George Augustus Sala tasted his first baked Alaska and Oscar Wilde, in town for the opening of *Patience, or Bunthorne's Bride,* in which he himself was satirized, complained about his bill.

But in the early 1880s tragedy befell the Delmonico family. Lorenzo died in September of 1881, leaving the bulk of his estate to his widow, and the business to his nephew Charles. Siro was left no share of the restaurant and died shortly thereafter—of a broken heart, it was said. Charles, affable and charming, ran the business with success for two years, but made the mistake Lorenzo had made some time before: he gambled on the stock market. He did not invest in Brooklyn oil wells, as Lorenzo had done, but succeeded in losing his shirt nonetheless and suffered a severe nervous breakdown as a consequence. Despite the surveillance of round-the-clock nurses, one bitter day in January he mysteriously disappeared. The police and detectives were called in and large rewards were offered to anyone able to supply information as to his whereabouts. No one, not even the household servants, could recall seeing or hearing anything unusual. As the hours crawled

Lady's luncheon at Delmonico's, 1902. *Courtesy of the Museum of the City of New York.*

Stag dinner at Sherry's, 1906. *Courtesy of the Museum of the City of New York.*

Above. The kitchen at Delmonico's. *Courtesy of the Museum of the City of New York*.

Right. Easter sugar fantasy under confection at Delmonico's. *Courtesy of the Museum of the City of New York*.

by, the possibility of kidnapping or murder grew in the minds of Charles' frantic family. Finally a commuter stepped forward to say that he had noticed a well-dressed gentleman, bareheaded and obviously distracted, pacing the deck of a New Jersey-bound ferry a few days before. The physical description fit, search parties were alerted, and soon the riddle was solved; Charles, in his pathetic confusion, had set off to visit a retired friend in the suburbs and eventually had found his way by train to Orange. Leaving the station on foot, he strode purposefully forth into a blizzard and tumbled into a deep ravine. There he was eventually discovered by the authorities, a snow-whitened bundle on the frozen ground, a Delmonico dead of starvation and cold.

After Charles' unlikely death the succession of the house of Delmonico passed to his late sister's son, young Charles Crist, who wisely appended Delmonico to his name. But able though he was, the empire he inherited was soon to be challenged by the forces of Louis Sherry. Battle was not actually joined until Delmonico's moved to its final site on Forty-fourth Street and Fifth Avenue, with Sherry's just opposite. It was the largest Delmonico's of them all, with a main dining room of yellow and green Louis XVI euphony. Nevertheless, Sherry's, of Sanford White design, was thought to be more distinguished architecturally. Delmonico's still appealed to the conservative, but Sherry's had definitely bagged the fast set. It was at Sherry's that Billingsworth's uncomfortable horseback banquet took place, and of course, The Awful Seeley Dinner. But business fluctuated. One year Delmonico's was in favor, the next year Sherry's—the social locomotive of New York was on a seemingly perpetual shuttle between the two. Still, in the matter of cuisine, Delmonico's sovereignty went undisputed until after the death of Charles Crist Delmonico in 1901.

The publication in 1903 of *The Epicurean*, by Charles Ranhofer, chef at Delmonico's since 1861, caused a furor. This comprehensive and unique cookbook, compiled with the blessings of his employers, dealt a mortal blow to the mysteries of the gastronomic art. When told of its contents, the fraternity of chefs expressed both rage and dismay. Ranhofer's defection now made it possible for absolutely anyone to prepare petits-fours, roast suckling pig, half-glaze, salmis, *rognons au madère, veau à la portugaise,* or any other dish in the professional repertoire. The recipe for veal in port is one of the briefest in the huge collection:

Chop one shallot very finely and put it into a saucepan with half a pint of port wine and a finely shredded orange peel, a pinch of cayenne pepper, lemon juice and half a pint of espagnole sauce, boil and reduce to the consistency of sauce. Heat some cooked veal fillets cut up into slices, dress them in the center of a circle of poached eggs laid on oval-shaped croutons, and place on top of each egg a piece of tongue cut the shape of a crouton; pour sauce over all.

That the average housewife would find such recipes too complicated for everyday use did not interest the professionals in the least; what principally alarmed them was the insurrections the book was sure to cause in their own kitchens. Until *The Epicurean's* appearance, a full-fledged chef might labor twenty years practicing, learning, tasting, listening, and experimenting before reaching his exalted position. Now any upstart could apprentice himself to an acknowledged master for a year or two, pick up the basic tricks of the trade, buy the book, and open a restaurant of his own.

Social upheaval in the restaurant world could by no means be attributed solely to the publication of Ranhofer's *Epicurean;* toward the turn of the century, the scepter of fashion was stolen from the *haut monde* denizens of Delmonico's by the "lobster palace" society. The men of that enclave were largely self-made, and the women as well. Quite literally shining as a star among them was James Buchanan Brady, known as Diamond Jim, originally a railroad-equipment salesman, and the set included such disparate types as Oscar Hammerstein, "Bet-a-Million" Gates, Gentleman Jim Corbett, the prizefighter, and Stanford White, the brilliant architect with a fatally roving eye. The story goes that White threw a birthday party for Diamond Jim at which the guest of honor was presented with the end of a red ribbon that led innocently to a Jack Horner pie. When the ribbon was pulled, a souvenir chorine burst from the confection covered by "nothing but the ceiling" and was wound in by Jim, who would probably have preferred something more genuinely edible. The other guests, voicing their dissatisfaction with this inequitable gesture, were quickly appeased when, at a clap of the host's hand, a corps of identical mementoes bounded in and onto their laps. It is more likely that the activities of these lads rather than those of the unfortunate Seeleys gave rise to the myth of the high-life dinner attended by men in white tie (a sartorial denomination equalizing "Bet-a-Million" Gates and his deadly enemy and social superior,

J. P. Morgan, in the eyes of impressionable newspaper readers) and girls in Jack Horner pies.

Lucius Beebe regarded "lobster palace society" as the forerunner of "café society"; it might more reasonably be considered the last blooming of the Romantic period, an American extension of the demimonde of Dumas *fils* with its cast of actresses, courtesans, and assorted admirers. But the New York lobster palace set was altogether more robust than the consumptive heroines and heartless libertines of the Second Empire. The old guard, rather than looking askance, seemed intrigued by them, enough, in any case, to dine at the same lobster palaces in hope of catching a glimpse of another, presumably lower world. It was with many thanks to the tastes of these attractive untouchables that the scheme of New York dining and entertainment went through a series of revolutions, the sum total of which became known to later generations as "The Gay Nineties."

Lobster palace society indulged itself in a healthy exhibitionism which led to its most characteristic ceremony, the "entrance." The scene is Rector's, or Shanley's, or Martin's. The time is somewhere between eleven thirty and midnight. The orchestra is playing, when it is suddenly called to a halt. The leader has caught sight of a star just about to enter (if she is not a star recognizable on sight, he has probably been tipped off in advance as to her identity). There is a pause of silence during which all conversation ceases. Then the orchestra strikes up the song currently associated with the star who, blushing faintly, glides swanlike to her table, skin dazzling, diamonds winking, profile at the proper tilt. Her escort, probably hidden behind a blanket of violets, her evening's tribute, knows his name will go down in history.

The first chorus girls to cut a wide swath in New York were the members of Lydia Thompson's company of "British Blondes." They were bigger, blonder, pinker-and-whiter than anything before seen outside the British Isles. They kept in shape on midnight suppers of anchovies on toast, sirloin steak and potatoes, tripe and onions, and wedges of Stilton. The more petite natives of the American stage preferred what became traditionally known as a "bird and bottle." The Britishers were replaced in quick succession by Lily Langtry, Anna Held, Lillian Russell, and ultimately Eva Tanguay, Billie Burke, the Dolly Sisters, and Annette Kellerman, the mermaid who dove from the Hippodrome roof.

The lobster palaces themselves gained their name from what

was regarded as their most characteristic dish, the expensive and modish lobster. The menu for the Café Martin (price fixed at $1.25, $1.50 on Saturdays, Sundays, and holidays) conveys the general concept of those dishes associated with luxury, although this early-twentieth-century menu is definitely less elaborate than the menus of a quarter of a century earlier, and the diner is distinctly limited to a *choice* from each course. Lobster represents but one of seven courses to which the diner was entitled:

CAFÉ MARTIN

Dinner April 15, 1904

All	*Hors d'oeuvres*— Norwegian Anchovies, Radishes, Celery–Root Parisienne	
Choice	*Potages*— Cream Parmentier, Consommé Printanier	
Choice	*Poissons*— Cutlet of Lobster Victoria, Broiled Kingfish	
Choice	*Entrées*— Sirloin with stuffed tomatoes, potatoes Colinet, Timbal of Blackcock Grand Veneur, purée of Chestnuts	
Choice	*Légumes*— New asparagus, sauce Hollandaise Spaghetti à l'Italienne	
Choice	*Rôtis*— Spring Lamb, mint sauce Roast capon Truffled Paté of imported Hare à la gelée	
Choice	*Entremets*— Baba au Rhum Genoises Glacees Coupe aux Cerises Amandines Souvenir	
Cheese	Dessert: Assorted fruits	Demi-tasse

In 1899, Louis Martin, who had successfully operated a small hotel on Ninth Street popular with French visitors, leased the site of the recently abandoned Twenty-sixth Street Delmonico's. He made an exploratory trip through Europe to study new trends and returned with a grabbag of fresh ideas. Most sensational was his Art Nouveau decor. He introduced the intimate "banquette" to more expensive restaurants, and his murals of ravishing females afloat in space by William de Leftwich Dodge were admired by "art lovers, food lovers, and plain lovers." Another beguiling feature was his dining terrace, which was placed just above the street and covered with a brightly striped awning. Seated behind shrubs, flowering plants, and palms, guests could admire the splendid view of Madison Square without being seen. Finally, Martin hired an orchestra for his café and allowed women to dine there if escorted, and even to be served drinks ("cafés" normally operating as masculine preserves). It was an ideal rendezvous for the Stanford White clique, that lived in the area around Madison Square. White himself kept an apartment atop Madison Square Garden. Near at hand were not only the mansion Jim Fiske had given Josie Mansfield, but the splendid residence that Freddy Gebhardt, a local playboy, had bestowed upon Lily Langtry, the first "society actress." Other neighborhood hangouts were Mouquin's, very like a middle-class restaurant in France with its excellent onion soup and *vin ordinaire,* and Cavanaugh's on West Twenty-third, perhaps more stuffy and Anglo-Saxon, but a favorite with the "fast set" just the same.

For decor, however, neither Cavanaugh's, Mouquin's, nor even the incredible Martin's could hold a candle to Murray's Roman Gardens, a pleasure temple of questionable tone that appeared on Forty-second Street. Murray's daringly essayed a Pompeian room (it was assumed at the time that that unhappy city had met the same fate as Sodom and Gomorrah and for the same reasons) as well as an Egyptian room. A painting of Goujon's "Fontaine des Innocents" stood between the doors of the entrance; just inside it there hung a huge mosaic global chandelier of shattered colored glass. An ambiance of D. W. Griffith abandon was achieved by mirrors suspended at every angle, lights rising through pink-clothed tables, fountains, temples, Roman galleys, caryatids, views of the Bay of Naples, Libyan tigers, Egyptian peacocks, and Greek Phrynes and Aphrodites. Martin somehow managed to get back in the running when he opened his Café de l'Opéra, which achieved a Nebuchadnezzar's palace effect in a controlled color scheme of blue and gold set off by black marble. A heathen atmosphere was hopefully established by a huge bas-relief of a winged lion with a human head at the base of the main staircase which swept up to a landing where hung a vast painting of Babylon falling. As Julian Street, author of *Welcome to Our City,* saw it: "To place such a picture in a New York Lobster Palace would be daring if Lobster

Palace Society had brains or used them. But why have brains? Aren't sweetbreads just as good? So the elite of Lobsterdom have sweetbreads, and, eating, fail to see the writing on the wall."

Forty-second Street became the hub of the new theaters and the great "entrance" restaurants. Principal among these were Shanley's, on that part of Times Square then called Longacre Square, run by the Shanley brothers; Churchill's, known for its impeccable food and popular owner, "Cap" Churchill, an ex-policeman; and Bustanoby's Café des Beaux Arts, on Fortieth Street and Sixth. Jacques Bustanoby introduced the first and possibly last women's bar, at which only women could order, and for which his license was almost revoked. The lawyer who defended him on this occasion was none other than the young James J. Walker. Dancing between courses proved a more felicitous innovation and, to cut the ice, the host himself executed exhibition numbers with musical comedy stars. Bustanoby also featured *"soirées artistiques,"* at which Lillian Russell, Anna Held, Maxine Elliot, Douglas Fairbanks, David Warfield, Vernon and Irene Castle, and Isadora Duncan all performed. Reggie Vanderbilt staged his horse-show parties at the Beaux Arts, clearly the right place for cutting up, and on one historic evening he led in a hansom cab horse to be guest of honor. But the star of Times Square, Rector's, was to other lobster palaces what Lillian Russell was to the Floradora Sextet— a shade more lush, more vibrant, more exciting. Rector's personified an era even dearer to American hearts than the Roaring Twenties—the Nineties and the turn of the century, a period which more than being simply "Gay" or "Naughty" was in turn obstreperous and sentimental, gorgeously vulgar and vulnerably innocent.

Charles Rector, son of the owner of several frontier taverns in upper New York state, and his young wife, into whose family home on Tenth Street in Washington, D. C., the assassinated Lincoln was carried to breathe his last, moved to New York after the Civil War. Rector took a job as conductor on a Second Avenue horse-car; this did not stimulate him so he moved to Chicago to work for George Pullman. An inherited talent for organizing kitchens became apparent as Charles busily dreamed up innovations for Pullman dining cars; the first transcontinental journey was duly put in his charge. The time spent with Pullman proved invaluable, for excellent contacts could be made every trip. By 1884, Rector was ready to open his own café, complete with the sign of the Griffon, in downtown Chicago. Two of its features put him instantly on the map: one was the long, dramatic staircase descending into the restaurant that women of fashion with their Turandot-length trains found indispensable; the second was the importation of oysters, live lobsters, and three-hundred-pound green turtles from the West Indies. In no time at all Rector's garnered a national reputation as the first restaurant of the second city.

To express its gratitude in no uncertain terms, Chicago granted Rector the only restaurant operator's license for the 1893 fair. At the resultant Café Marine, first-comers to the fair were treated to the gargantuan *prix fixe* dinner that was to become one of the exhibition's greatest drawing cards. Princess Eulalie of Spain, upon hearing that Rector served whiskey in porcelain teacups to ladies, ordered a soup bowl full. This performance, coupled with a lot of stern cigarette-smoking, eclipsed the restaurant's fourteen-course special; Her Highness, exciting the goggle-eyed indignation generally reserved for Little Egypt, really gave the tourists something to tell the folks back home as she made her unsteady exit, parasol menacingly atwirl, her hat cocked over one eye.

Toward the last years of the century, Rector heard of an ideal location in New York for a lobster palace, and left on the first train. Situated on Broadway between Forty-third and Forty-fourth streets, the building stood only two stories high; nevertheless, with the electrification of its Greco-Roman façade, it assumed a theatrically imposing air. Rector spent two hundred thousand dollars on transforming the interiors into a mirrored paradise of green and gold. Linen especially woven in Dublin and hand-stenciled silver covered a hundred tables on the ground floor and seventy-five on the second. Four private dining rooms completed the architect's plans for reconstruction.

Just before the grand opening on September 23, 1899, a last-minute raid was made on Delmonico's staff: the *saucier*, Charles Parrandin, the maître d'hôtel, Paul Perret, and the business manager, Andrew Mehler, were all wooed away. Rector believed in battalions of help (one hundred sixty-five in all) and preferred his waiters to have graduated from professional schools in Switzerland. By today's union standards, their hours seem incredibly brutal: 10 A.M. until 3 A.M., with only three hours off in the afternoon. Yet a job at Rector's was much sought after, and there were several ways in which to augment the twenty-five dollars weekly salary; the house gave a percentage on "extras" that a waiter "pushed," such as relish trays, and wine agents paid fifty cents corkage for every bottle a customer might be wheedled into buying. A waiter could also look forward to phenomenal tips, particularly on New Year's Eve or Election Night. Frank Gould once left a hundred-dollar bill for a bromo, and the maître d' never realized less than twenty-five thousand dollars in Yuletide gratuities.

Opening-night festivities at Rector's were delayed two hours by a glorious new toy at the entrance, the first revolving door in New York. The guests refused to get out of it. Like a spinning Fabergé menagerie, it twirled and sped, sending off sparks and blurs of jewels, egrets, sable, otter, birds of paradise, and flashing teeth. Finally, Gentleman Jim Corbett and his wife were catapulted into the entrance hall followed by other reeling celebrities.

Preparations for the first dinner were made under the beady eye of the head chef, Emil Hederer, a temperamental artist whose terrapin à la Maryland had inspired Queen Victoria to send him a snapshot of herself with royal admiration scribbled all over it. (He subsequently never moved without this accolade and carried it in his hat.) The menu was cosmopolitan indeed; it offered Egyptian quail, English pheasant, African peaches, and Southern European strawberries.

Rector, like Louis Sherry, would go to any extreme to please a favored patron, even to the point of taking his own son out of college and packing him off to Paris as a kitchen apprentice for two years. Diamond Jim Brady became obsessed by filet of sole Marguery while in France, and could not rest until the dish could be duplicated for him in New York. George Rector, the son of Charles, was then studying law in a rather desultory manner at Cornell. He had spent many summer months in the paternal sculleries and hankered to cook despite it; his father's decision to send him, like Parsifal, on Brady's sacred mission could not have delighted him more. George eventually talked his way into the Parisian cathedral of haute cuisine, the Café Marguery, and by dint of diligent sipping, sniffing, and snooping, solved the riddle. Chef Marguery, rather than boxing his plagiarist's ears, loaned him out to the Palais des Champs-Elysées as chef for a state dinner to King Oscar of Sweden. For this triumph, George was decorated by the French government with the Cordon Bleu. Returning to New York, he was met at the dock by Brady, who, chubby little hands cupped to quivering jowls, could be heard bellowing above the crowd, "Did you get the recipe?"

Diamond Jim's legendary eating bouts were scarcely exaggerated. He ate prodigiously from the moment he rose until early the next morning when, wheezing and groaning, he would collapse like a beached whale on his long-suffering bed. Jim, an inveterate first-nighter, dined at Rector's both before and after the theater. The first session would usually begin with two or three dozen outsized Lynnhaven oysters, personally reserved for Brady by Maryland seafood wholesalers aware of his Pantagruelian appetites. These would be followed by six or seven boiled crabs and a double portion of turtle soup. Six or seven mammoth lobsters would then be dispatched in time to get through two orders of terrapin, two canvasback ducks, and an enormous steak accompanied by mounds of vegetables. All of this, plus a tempting array of desserts and an entire tray of French pastries, would be washed down with a gallon of fresh orange juice. Jim enjoyed company almost as much as he did eating, and when the last crumbs of fruit tart had been brushed from his chins and stomachs, he would call for a two-pound box of candy to be passed around the table. The smile never deserted his face if one of his guests actually accepted a piece, but if anyone did, a second box would shortly appear, untouched by greedy human hands, for Brady's delectation at the theater. Shaw, he explained, was more tolerable with bonbons, while Ibsen sat better with glacé fruit.

Lillian Russell was Jim's great friend, his natural-born soul and stomach mate. A pin-up for thirty-five years, Lillian, or "Nell" as she was known to Brady, let her weight fluctuate as irrationally as her love-life. As time passed, the control of both began to bore her. An evening with Jim, on the other hand, suited her right down to the ground. After a queen-size dinner, a good show, and multiple bottles and birds, they would stand arm in arm and take in the lights on Broadway, Jim peering at her affectionately with big brown eyes and croaking, "God, Nell, ain't it grand?"

Together they were one of the unnatural phenomena of the era —Lillian with her fabled figure and jewels, Brady with his fabled figure and jewels—as they sat down to a marathon banquet. Once in Saratoga she vanquished him at his own game by slipping into the ladies' room and out of her corsets, wrapping them in a tablecloth, and handing them to a waiter for safekeeping with the admonition "Don't look!" Needless to say, the day came when Nell's fans would have no more of it and public criticism of her avoirdupois forced her into temporary retirement. Jim tenderly

Above. The Mystic Rose, Mrs. Astor. *Courtesy of the New-York Historical Society.*

Below. Lorenzo Delmonico. *Courtesy of the New York Public Library.*

Above. After the heroic plunge—Steve Brodie. *Courtesy of the New York Public Library.*

Carry Nation, the Kansas Smasher, embarking on "a little hatchetation."
Courtesy of the New York Public Library.

Lillian Russell, before the lobster palaces took their toll. *Courtesy of the New York Public Library.*

commiserated with her and ordered a gold-plated, pearl-and-diamond-studded, ruby-and-sapphire-spoked bicycle constructed for pedaling through Central Park and shedding a pound or two along the way. On this eye-catching contraption, dressed in gabardine white as a seagull's wing and a matching ostrich-trimmed hat, Lillian could often be spied careening over metropolitan hill and dale, a pouter-phoenix rising out of two flaming Catherine wheels. After two years of dubiously Spartan living, she triumphantly returned to the boards as Lady Teazle in an operetta based on *School for Scandal,* and to Rector's on the arm of Diamond Jim.

Wilson Mizner, returning from the Yukon, gave the dining room of Rector's a cursory once-over and cracked that he recognized all of the same old faces—it was just that they were paired off differently. Rector's reputation was not only dazzling, it was becoming increasingly more *louche.* In 1907 Florenz Ziegfield, the "mephistophelian" husband of the piquant French music hall star Anna Held, made history by producing his *Follies* of 1907, an evening's tribute to flesh, feathers, and a few strategic strings of pearls. The hit of the production was a song entitled "If a Table at Rector's Could Talk." All this pleased the Rectors, and with lack of forsight they were also pleased when the toast of the town, Miss Frances Starr, portraying a fallen woman, declaimed histrionically as the curtain fell on *The Easiest Way,* "Get out my best dress and hat! I'm going to Rector's to make a hit, and to hell afterwards."

In 1911 the Rectors, father and son, opened a Hotel Rector. Its grand debut, a mortifying dud, indicated the inevitable fizzling-out and final collapse of their once inviolable position. The risqué jokes, the leering songs, the scandalizing curtain lines, and a persistent rumor that Papa R. paid an urchin ten cents a night to press his nose to the windowpane as an appetite-stimulant for the customers, while causing talk across the country, had in the long run done the family name irreparable damage.

But even while Times Square was at its peak of elegant prosperity, fashionable rumblings could be heard further uptown. A brother of Jacques Bustanoby opened the Domino Room at Columbus Circle and introduced midnight-till-dawn dancing in New York. Nearby, Reisenweber's, which could claim to have brought cabaret to America, had a rooftop dance hall with the added attractions of a marimba band and Sophie Tucker. For all intents and purposes, the lobster palace had given birth to the night club.

According to Larousse, in the first decade of this century "cabaret" still officially meant a "tavern, bar, or little inn; house where one sells drinks or where one gives also to eat." Rabelais haunted early cabarets with such names as The Sucking Calf, The Pine-Cone, and The Valley of Misery. Much later, in the era of Henri Murger's *La Vie de Bohème*, semi-private clubs called *cabarets artistiques*, where poets, singers, and song-writers gathered to perform their latest works, honey-combed the Latin Quarter. After a curiousity-seeking public had succeeded in destroying the atmosphere of these intimate little dens, the *cafés concerts*, with their music-hall turns, took over. Somehow the connection between "cabaret" and "entertainment" became fixed in American minds, and when, in the mid-Edwardian period, Reisenweber produced a floor show in the main dining room of his Columbus Circle complex, the word was telegraphed around that New York finally had acquired a "cabaret."

Dancing girls were the sole attraction of this first show, and within two weeks every lobster palace with a dance floor had a chorus line. Keeping well ahead of Churchill, Shanley, Murray, Martin, and other fledgling impresarios, Reisenweber hired Gus Edwards to stage a miniature revue. The stars of this production, an acrobatic team called the Five Jansleys, climaxed their act with a mouth-drying display of derring-do atop a fifty-foot ladder; the ceiling easily accommodated them, as the dining room, which seated seven hundred fifty, was vast in all directions. Reisenweber went still further: to his upstairs 400 Room he introduced a Southern group which was to kindle a musical revolution—the Dixieland Jazz Band. What is more, his Hawaiian Room was the first in New York to echo with the pitter-pat of turkey-trotting feet at the ungodly hour of four in the afternoon.

The women were something to see. This was the age of incense, beaded headache bands, and blazing boudoir fireplaces. In Paris Colette was appearing at parties in black silk leotards and cat make-up; American temptresses hesitated to carry the animal analogy to these extremes. Nevertheless, with their hair worn like inverted birds' nests over smudgy lioness eyes, their dead-white skin, feathered turbans, yapping Pomeranians, hobble skirts, bare backs, jade cigarette holders, and pointed nails, they did manage to suggest a composite sketch by Aubrey Beardsley, James Montgomery Flagg, and Beatrix Potter.

Cabaret singers went in heavily for rice powder, rouge, lipstick bright as a stoplight, and strident flirting with gullible old men at ringside tables, mussing their ties and ruffling their disappearing locks while intoning such lyrics as:

I can see that you are married
And you know I'm married, too,
And nobody knows that you know me
And nobody knows that I know you.

If you care to we'll have luncheon
Every day, here, just the same,
But swee–tart if you talk in your sleep
Don't you men–shun my name!

Maurice, who had inadvertently spawned a whole salmon leap of slithering, dipping, and sliding imitators, assumed a more impersonal approach. Equipped with flawless evening clothes, a synthetic French accent, eyes like La Belle Otero's, and swivel hips, he and his transient partners nipped across the most expensive dance floors in the city. Louis Martin first presented him to an audience already aware of Maurice's reputation for sizzling tangos at the Café de Paris in France. By this time he had added some eccentric variations to his routines, twirling gauzy young blondes around his waist and slickly groomed head like so many tulle scarves. With New York's leading society women literally at and on his feet (sometimes at one hundred dollars a go-round) Maurice moved on to Reisenweber's and helped make the tango and the gigolo the ultimate in wicked chic. He was followed in Martin's Paradise Room by the dance teams of Hawkworth and Durant, Moss and Fontana, and the Castles.

Irene and Vernon Castle put the fixative on the dance craze; within the batting of an eye and the twinkling of a toe all New York scrambled onto the floor between courses and, following their energetic example, shook a leg to beat the ragtime band. Overnight the Castles became a national institution, a social hit, and the proprietors of an exclusive dancing school, a supper club named the Sans Souci, and Castles in the Air, a smart cabaret where, like theatrical house physicians, they appeared briefly once a night.

The turn of the century had brought with it a rash of new hotels. The first of these, and the most formidable, was the Waldorf-Astoria. It was the Grand Central and the Hoffman House and

Delmonico's and Rector's all rolled into one, and something else besides.

The Waldorf-Astoria was conceived as the result of an Astor family contretemps. When tall, shy William Waldorf Astor decided to seek socially greener fields abroad, it was suggested to him by his aid, Abner Bartlett, that the site of his mansion, on the northwest corner of Fifth Avenue and Thirty-third, would be perfect for a hotel. Astor was particularly interested by the idea, as the new architectural extravagance would cause the acute discomfort of his aunt, Mrs. Astor, who had made his life miserable by retaining the title of *the* Mrs. Astor, even after he had become head of the family, with a wife of his own. Her residence lay next door on the southwest corner of Fifth and Thirty-fourth Street, and a hotel would ruin the view from her garden. To manage the new hotel, designed by Henry J. Hardenbergh, Astor hired George C. Boldt, director of the Hotel Bellevue in Philadelphia. Boldt, a short, bearded, dynamic man from the Baltic who had started his career as an oyster opener at a Sixth Avenue restaurant, was to become the managerial genius of the era.

The Waldorf Hotel opened on March 14, 1893, and by the year's end Mrs. Astor could bear the hubbub no longer and moved uptown. Her son, John Jacob IV, grudgingly constructed another hotel on the site of her house, taller and more magnificent than the Waldorf, but to be run in conjunction with it, hence the Waldorf-Astoria. He cunningly inserted a clause in the contract, however, whereby he could block every passage between the two buildings at any time he wished.

The Waldorf-Astoria accommodated a thousand guests with every sybaritic contrivance of the scientific age, including private baths and electric lights. Everything about it loomed larger than life. According to Oliver Herford, the Waldorf-Astoria with its Palm Garden, denser than the Amazon, fulfilled an American dream: it brought exclusivity to the masses.

Peacock Alley, that legendary three hundred feet of fashion runway, sideshow, and appraiser's gallery, teemed with wide-eyed life from breakfast time until the middle of the night, when both the fascinating foreigner in her Worth gown and the small-town housewife ruched to a fare-thee-well by the local seamstress collected their gloves and fans, their reticules and men, and promenaded off to bed. Preachers in every state of the Union were sure to include the Waldorf-Astoria and its spiritually pestilential Peacock Alley in their tirades against New York and its unholy flesh-pots, its palaces of material lust. Not in its wildest dreams could the local Chamber of Commerce have fomented tourist trade as effectively as did the clergy. And thanks to their bony-fingered invective, Peacock Alley emerged as the all-time corridor of worldly sin.

If the truth must be known, Peacock Alley would not measure up to any contemporary vision of grandeur. Lithographs of the period reveal it to have been distinctly narrow, its walls dadoed with stingy, shelflike, high-backed leather benches; an icy-looking tiled floor and clusters of gentlemen standing about in heavy overcoats further preclude any sense of warmth.

Peacock Alley may have had its limitations, but not the gentlemen's bar, an unalloyed joy by any standards. The massive superstructure of the bar itself encompassed a soaring, tiered, cleverly refrigerated buffet table surmounted by an avalanche of caviar, canapes, cheeses, cold meats, anchovies, olives, radishes, and salads. Dominating one end of the bar was an authoritative-looking bronze bear, at the other a rampaging bull; in between, a miniature lamb stood impersonally surveying a pool full of brook trout which lived in constant danger of being netted by an energetic luncher. Nude studies with ornate gold frames in the shape of keyholes, such as appeared elsewhere, were not for the Waldorf bar. With all deference to their unassailable masculinity, the regular patrons of the Waldorf bar and men's café would not have put up with it; J. P. Morgan, as we have seen, fancied himself something of a divine, and Henry Clay Frick preferred Old Masters. Only the occasional characters in wide-brimmed hats and cowboy boots who came clambering in from a gold killing at Cripple Creek or the Klondike, ordering champagne and paying with nuggets, might have felt the walls could have done with a little gingering up.

The Waldorf's most talented bartender arrived shortly after the Spanish-American War. His name was Johnnie Solon, and it was he who invented a classic drink named for a zoo. One day as he was oscillating a Duplex (equal parts of French and Italian sweet vermouth, shaken hard with squeezed orange peel or orange bitters), Traverson, the headwaiter of the Empire Room, challenged him to create a great new cocktail for a customer who had bet he couldn't. Johnnie's smile tightened. Reaching for the Gordon's bottle, a shot glass, and some fresh orange juice, he splashed out a two-to-one combination in favor of the gin. A dash each of

French and Italian vermouth and a muscular shaking completed the operation. Traverson took a tentative sip, then drained the glass in a gulp, ordered one for his customer, another for himself, and advised Solon to stock up on oranges. Until this historic moment, no more than a dozen oranges had been used daily by the bar. With the advent of the Bronx cocktail, a case had to be delivered every morning.

A staggering repertoire was still demanded of all the city's best bartenders. The Waldorf's claimed to have four hundred ninety-one different preludes, fugues, and codas at their fingertips and constantly memorized new ones: the "Santiago" for the Spanish American War, the "Coronation" for Edward VII, the "Dr. Cook" for polar exploits, the "Chocolate Soldier" for the hell of it. These were but a few. "Brain-dusters" or early morning pick-me-ups were an essential part of the daily concert. For a customer who suspected his hangover might be terminal, chilled absinthe drinks were commonly prescribed; for less advanced cases, Black Velvets and Shandigaff (equal quantities of porter and champagne, beer and gingerale, respectively). Disbelievers in the popular theory that a drink in the morning will do you no harm unless you let it die later in the day could safely cling to mineral water, a wide variety of which were available—Carl H. Schultz's Seltzers, Vichy, Lithia, Double Carlsbad, Marienbad, Kissingen, Emsswalbach, Pyrmont, Pullna, etc.

At the turn of the century, a martini was still a sweetish drink, dry Vermouth having not yet been introduced to New York. Rickey's, punches, and "cups," the last compounded of mineral water, light wines, lemon slices, mint, and pineapple and cucumber strips, survived as year-round favorites. All strata of drinkers clamored for milk punch; one bar on Eleventh Avenue did such a roaring 5 A.M. business with stevedores on their way to the docks that the owner frequently had to warn his bartenders to hold back on the milk.

Boldt's prize employee, Oscar Tschirky, however, lurked not behind the bar, but out front, behind the red velvet rope separating the Palm Garden, the most exclusive restaurant in New York, from the outside world. Oscar, who had made his way to Delmonico's and risen to become headwaiter for all the private dining rooms of the Twenty-sixth Street restaurant, coolly wrote himself a letter of recommendation on Delmonico's stationery and obtained the signature of every one of its patrons of note. This he sent along to Boldt, and was immediately hired as maître d'hôtel. Ever the perfectionist, Oscar promptly contacted his first employer Ned Stokes, and took a temporary job in a downtown restaurant Stokes then owned, so that he could bone up on Wall Street faces and fortunes before assuming his position at the Waldorf.

Oscar (he was always known simply as Oscar) was a punctilious gourmet and so hidebound a snob that he claimed he would rather see Mrs. Astor sipping hot water in the Palm Garden than a nameless nouveau-riche devouring a ten-course dinner. This did not prevent him, on occasion, from discreetly coaching the unsophisticated in the art of dining.

Oscar was the man who remembered that J. P. Morgan liked raw oysters and Pommard 1884, although "there were years when I knew he would want a bottle of Chateau Lafitte," and that William Howard Taft preferred a grilled lambchop, simply prepared vegetables, and a baked apple without sugar. But the guest who really drove Oscar to extraordinary exertion was Li Hung Chang, the Prime Minister of China, who stopped at the Waldorf in 1906. Li Hung Chang was quite outside Oscar's ken, and he recalled for readers of the Ladies' Home Journal the indelible impression made by him.

Always he appeared in his robes of heavy silk brocade; and when they were of yellow silk and he moved through Peacock Alley with an inscrutable gleam in his fathomless eyes, two-inch finger nails tucked in his loose sleeves, the red button of a mandarin as a kind of seal of authority on the lozenge hat, from the rear of which hung the root-like queue of his glossy black hair, you were staring at the personification of mysterious China.

The widely heralded visitor brought along his own cooks. If Oscar was offended, he didn't show it, and magnanimously assigned these attendants ("with eyes like shoe buttons" was his only remark) to ranges in the kitchen. Yet some slight exasperation creeps through Oscar's unruffled prose when describing the three-hundred-year-old eggs that Chang for some reason preferred to Oscar's own chafing-dish concoctions. "Personally, I felt that not in a thousand years could eggs become so terrible." He hastened to add that "it is with such confused points of view that a hotel man must be tolerant," an avowal which was shortly put to the test when Li Hung Chang requested that the bed be removed from his apartment so that he could sleep on the floor. Oscar rallied in a burst of self-abnegating rhetoric: "In the same spirit that had flags flying from the hotel, during his visit, embroidered with the Chinese dragons, I believe—if it would have made Li Hung Chang happier—then all of us would have taken down our beds and slept on the floor. Whether they are presidents or kings and queens, important personages are simplicity itself."

Oscar officiated as New York's high priest of the banquet mystique. Before the era of public relations tacticians, any organization or company, whether commercial, scientific, patriotic, political, artistic, or merely commemorative, was almost certain to launch a publicity campaign with a banquet at a prominent hotel; this being one of the few sure techniques for claiming public attention known to nineteenth-century man. Inventions were demonstrated at banquets; awards were given and received at banquets; political candidates, silently backed, were honored at banquets. And where, without disturbing democratic ideals, could pomp, and display, and public ostentation be as safely indulged, yet as fully covered by the press, as at a banquet?

When Oscar organized one, an unspoken policy giving him financial carte blanche was generally observed by the client. At one such feast, in celebration of a Tammany victory, the Myrtle Room was turned into an orchard complete with grass and warbling birds, the guests were given goldhandled grape scissors with which to cut the fruit, and blue raspberries were featured on the menu.

At the turn of the century, New York society, under the supervision of its self-appointed board of directors, settled down after several enforced stock splits to the routine business of daily social rounds and the strict observance of aristocratic order. To the thinking element, its excesses became irksome and its limitations a bore. As Ralph Pulitzer observed in *New York Society on Parade*:

The conversation which crackles through the rooms is at once animated and detached. Men and women address each other with the impersonal loquacity of barbers. Their attitude toward each other is much like their attitude toward the chauds-froids and the galantines which are set before them—familiarity with externals tempered by ignorance of contents . . . while in Europe the mutual entertainments of an inherently stable upper class create Society, in New York the constant contortions of Society are indispensable to create and maintain a precarious upper class. . . .

Never did more tentative novices do more for the sake of a social function or drive any society to greater contortions than the Bradley-Martins. The Bradley-Martins (history has bestowed on them the hyphen never entirely achieved in life) hailed from Troy, New York, and were hell-bent on making a profound impression on Manhattan society. This they did. In February of 1897 (a depression year, unfortunately) they staged a costume ball. The ballroom of the Waldorf was converted by fifty florists into a version of Versailles no Sun King would ever recognize, with upwards of six thousand orchids and a number of roses past counting. The guests were requested to appear in costume as courtiers of Louis XV and there were to be several orchestras and a midnight supper of roast suckling pig. The preparations, in great detail, were virtually cabled (well in advance) to the press and reprinted both here and abroad, not always to the advantage of the hosts.

On the appointed evening seven hundred of the twelve hundred invited members of "the Four Hundred" appeared—in any case, a full house. It had been made known that Johnson, the famous detective who recognized everyone in New York by sight, would be stationed at the door, a necessary deployment as the Bradley-Martins had received reams of beseeching letters from the uninvited. More protection was in order, however, as the ball roughly coincided with the Second International. Anarchism was in full swing and not all the letters the Bradley-Martins received beseeched the hosts for invitations; several colorful threats of violence lent a bloodcurdling note to the morning mail. Mrs. Theodore Roosevelt attended in costume, but her husband dutifully remained outside in the uniform of a Commissioner of Police, holding back the crowds.

The costumes worn by the Bradley-Martin guests would have given the original Louis XV (impersonated in this instance by Mr. Bradley-Martin) a nasty turn. Apart from the usual number of Fragonard reproductions in taffetas and laces, there appeared a seventeenth-century Dutch colonial, a knight in full armor (purchased from the Metropolitan Museum), an Algonquin chief carrying a war-pole complete with scalps, one Pocahontas (Miss Anne Morgan in daily life), and several George Washingtons. But it would have shaken the monarch the most to find himself mated to Mary Queen of Scots (Mrs. Bradley-Martin). As a hopeful touch of authenticity, however, the hostess wore a ruby necklace that had belonged to Marie Antoinette.

The final bill for the ball was estimated at over three hundred fifty thousand dollars, including the price of champagne shipped from London and opened at a corkage fee of one dollar and fifty cents per bottle. The Bradley-Martins appeared not so much dismayed by the bill as by a new tax assessment they were obliged to pay as the result of publicity surrounding the ball. This, along with the unpleasant reportage itself, led them to flee America forever and seek asylum in England. As a final salvo they threw a ten thousand dollar dinner for eighty-six of New York's richest citizens just before embarkation. It was held at the Waldorf, in a private dining room sentimentally decorated with smilax and pink roses. When a group of visitors put their heads through the door for a peek, Mrs. Bradley-Martin welcomed them with her legendary grace and bawled, "Who's paying for this room anyway?"

There were mixed reports as to the intent of the Bradley-Martin ball. The less charitable suggested that it was a method of social advancement, but the Bradley-Martins themselves always defended their expenditure as an act of welfare promoting business for cooks, florists, dressmakers, and such in hard times. If this was true, their kindness was most certainly misconstrued.

Any economist will admit, however, that the Bradley-Martins, the most extravagant hosts in the country, were better for the general economy than Hetty Green, the nation's least extravagant hostess. Hetty, one of the wealthiest women in the world, was an incurable miser. She lived for years with her children in a rented room in New Jersey, commuting daily to Wall Street, and her son Ned sported a game leg for life as a memento of her parsimony in the matter of consulting doctors. When Hetty lived at the Saint George Hotel in Brooklyn, there were sure to be no extra charges for room service on her bill. If her daughter was too ill to come down for breakfast, Hetty would wrap up a well-buttered hot roll, pinch a pitcher of coffee and cream, stuff the lot into her handbag, and carry it up to the room. The same scruffy bag doubled as a safe-deposit box and lunch pail during all the years she spent in her private bank vault sitting on the floor totting up her millions of dollars and riffling through stacks of securities. She died, finally, of apoplexy during an argument in which she extolled the virtues of skimmed milk.

Not even the penny restaurants that mushroomed in the shabbier parts of town, where one could get a square meal for five cents, met with Hetty's approval. Higher up the economic ladder were the fifteen-cent houses, providing for that sum a cut from a hot joint accompanied by bread, butter, potatoes, and pickles. For the genuinely destitute, the Saint Andrew's Society operated one-cent coffee stands, in reality the first American soup kitchens. That five such stands dished out one hundred fifteen thousand meals during the first three weeks of their existence in 1887 is a grim comment as to what was happening at the other end of the scale while Diamond Jim was wheezing through his gargantuan snacks. Yet cheap eating enjoyed improvements too.

In the nineties another development in frugal dining, the inspiration of Harry M. Stevens, an English-born food concessionaire at the Polo Grounds, permanently augmented the national menu. One chilly fall day, he decided to tuck sizzling sausages into lightly toasted buns and hawk them as "red hots." Stevens went on to fame and fortune, and the "red hots," represented in a cartoon as a dachshund between two slices of bread, came to be known as "hot dogs."

With the arrival of Childs on the New York scene, the basic concepts of restaurant service were drastically altered. The Childs brothers had worked for A. W. Dennet, owner of the first nation-wide chain of eating houses, and from him they borrowed the trademark of the pristine chef at work in the window, in this case juggling flapjacks. William and Samuel Childs, starting with only sixteen hundred dollars in 1889 and an entirely new format, were the success of the nineties. By 1898 they owned nine restaurants at which fifteen thousand to twenty thousand people were served a day. White-tiled floors and walls, white marble tabletops, waitresses in starched white, and oddly enough, crystal chandeliers, attracted a newly germ-conscious public. Then at their 130 Broadway branch, something totally new was attempted. According to *The Caterer's Monthly* of February 1898:

As you enter the restaurant you notice a large lunch counter on your left, filled with piles of sandwiches and . . . pastry. Opposite the counter, and about four feet from it, is a metal, ornamental open partition three feet high extending the full length of the counter. Before stepping into the passageway you notice a big pile of empty trays from which each guest takes one. Now as you pass along the lunch counter, you take off whatever you want and place it on your tray, not, however, forgetting to take a cup of coffee from the big urns. . . . Mr. Childs has introduced music (by a five-piece orchestra), and his verdict is that it pleases the majority of folks.

Germ-free breakfast at Child's. *Courtesy of the Museum of the City of New York.*

In other words, whether the majority of folks like it or not, the cafeteria was born; in 1903 its apparent stability gave courage to Messrs. Horn and Hardart of Philadelphia when they installed their first New York Automat on Broadway and Thirteenth Street.

A general craving for newness brought about the demolition of many of New York's older hotels. The U.S. Hotel was torn down in 1902, and the Fifth Avenue in 1907, although the Astor House, "commercial" by then, succeeded in lasting until 1920. The Waldorf-Astoria was not the only one to take their place. Shrewd hotel men built further uptown, and by 1900 the Plaza, the Netherlands, and the Savoy, all built on the Waldorf scheme of Palm Gardens and gentlemen's cafés, guarded the Fifty-ninth Street entrance to the park. In 1899 the Astors stuck their finger in the pie yet again with the building of the Hotel St. Regis. John Jacob Astor IV had purchased the plot on Fifth Avenue and Fifty-fifth Street to accommodate the new Astor mansion, but capriciously changed his mind and decided to build a hotel on the site. It was his first plan to call it the Astor, but this too was reversed and the new hotel was finally named after Lake St. Regis, near the Astor lodge in the Adirondacks. The St. Regis was to provide greater luxury than ever—the air would be filtered through cheesecloth, warmed, and pumped constantly throughout the house, and thermostatic regulators were to be installed in every room—but it was to be smaller, in a way simpler and more select, than the Waldorf-Astoria. Its much-bruited decor was largely planned by Ava Willing, then Astor's wife and later Lady Ribblesdale. Her plans were much erased when the hotel was purchased some years later by Mrs. Duke Biddle and Ava's voluptuous Salle Cathay (the main dining room) was converted into the Egyptian Room.

It was at this time that New York acquired its own Ritz Hotel. Swiss born César Ritz had worked his way up from a position as assistant wine waiter in a small Swiss town to owner of the most smoothly oiled establishments in Continental Europe. In 1889, at the behest of D'Oyly Carte, he took over the management of the Savoy Hotel in London, and the two masters of entertainment created a genuinely American revolution in European hotel-keeping. The Savoy was Europe's first grand hotel in the American tradition, with seventy bathrooms (its chief competition, the Dublin, could claim only fifteen), full electricity, and hydraulic lifts. Moreover Ritz, having hired the great French chef Escoffier to conduct his kitchen, and none other than Johann Strauss to play dinner music, managed to lure the upper-class English gentle-woman out of her house and into a restaurant for an evening's meal, something as yet unheard of. As the Savoy was situated in the heart of London's theater district, members of the aristocracy found themselves rubbing elbows with the stars of the very productions they had just attended, so that Ritz could claim he had single-handedly created a British "lobster palace society." Ritz next inaugurated a hotel bearing his own name in Paris, and despite his untimely collapse in 1902 under the pressure of preparations for the Coronation, hotels bearing his name, all the products of his organization, appeared as far afield as Madrid and Barcelona, Buenos Aires and Budapest.

In 1902 the real estate magnate Robert W. Goelet decided that New York was in need of a "Ritz" and organized an American venture on Forty-sixth Street and Madison Avenue. The new Ritz Carlton, joined the other hotels of the turn of the century in reverting from Queen Anne wainscoting to various persuasions of "Louis," in this case a delicate and restrained classical "Louis XV" that contrasted with the luscious and baroque "Louis XIV" of the new Plaza (the Plaza Hotel of 1900, confined in its construction, was rebuilt in 1907 at a cost of twelve million dollars). It has been said that were it not for the construction of the Ritz-Carlton, Madison Avenue, then so near the Grand Central trainyards, might have developed to resemble another Sixth or Amsterdam. The hotel brought elegance east of Fifth (where, incidentally, Mr. Goelet's properties were situated), and Vichyssoise to the world. It was the Ritz Carlton chef Louis Diat who created the original Vichyssoise, a velvety refinement of the traditional potato and leek soup beloved by Parisian concierges.

The opening of the new Plaza Hotel, attended by the obligatory host of Vanderbilts, Harrimans, and Goulds (and by Lillian Russell and Diamond Jim), was also accompanied by a sure sign of the decisive arrival of the automobile on the New York scene. Along the Fifth Avenue side of the hotel, guests at the opening found a string of green and red motor "cabs," their chauffeurs dressed like Hussars and assisted by tiny dials that registered the fee. Harry N. "Taxicab" Allen's brainstorm, however, led to the most infamous act of violence in a luxury hotel since the shooting of Jim Fisk. New York's hansom cab owners were at first only a little worried by the presence of what they took to be a gimmick. Within a year Allen was operating seven hundred taxis, neatly lined up not only at hotels, but at railroad stations and docks as well. He soon had a strike on his hands, and the horse cabmen of the drivers union as his deadly enemy. In the winter of 1909 he was lunching at a window table in the Plaza's park-side dining room when a shot rang out from the foliage across the street. It broke the window but narrowly missed Allen, who promptly sold his taxi business, despite the assailant's inability to arrest the march of progress.

The Times Square area had its own share of splendid hotels. The Hotel Knickerbocker made its debut on Forty-second Street and its bar became headquarters for the city's tippling hierarchy. Further up Broadway glittered the monumental Hotel Astor, constructed in 1904 as another excursion into New York hotel building by William Waldorf Astor. On a summer's night, the Astor roof shone like a beacon.

Roof gardens and canopied terraces began to multiply after the initial success in the 1880s of the modest Café Boulevard, an *al fresco* restaurant in the German quarter. Occupying a large plot behind an old Knickerbocker mansion built by the Remsen family, the café superficially recalled the early pleasure gardens; however, a full dinner menu and polished orchestra had replaced the guileless pound cake, lemonade, and oompah-oompahpah band of the past. Society turned thumbs down on it as being "Bohemian," but alert hotel owners sensed something big in the balmy breeze.

The instant success of the Casino in the Park, regardless of its indifferent, overpriced food and the seedy squabbles by city officials over the catering concessions (finally seized by Boss Richard Croker's brother-in-law), further convinced hoteliers of the beneficial aspects of outdoor dining. The old United States Hotel threw together a rooftop setting more rough than rustic and was immediately deluged by customers, and the Hoffman House and Waldorf-Astoria followed suit. The Waldorf's yearly florist's bill now regularly exceeded fifty thousand dollars. Cosmopolitan New Yorkers, reminded of Paris and Vienna, now flocked to the roof gardens like homing pigeons. When the Astor Hotel opened its aerie to the public in 1905, the best-dressed women in the world would have had a hard time standing out against the breathtaking background of flowers by the hundred thousand, fountains, trees, forty-foot parapets smothered in geraniums and fuchsias, Italianate promenades, mossy grottoes, and waterfalls tumbling into pools of lilies, frogs, and goldfish. The roof covered roughly the area of seventy city lots and could accommodate several thousand people. Countless gazebos dripped Virginia creeper, honeysuckle, and moonflower; a continuous white-columned, ivy-covered pergola ran the length of the building and from its beams swayed romantic wire baskets filled with flowers and ferns. Immense vases of rhododendrons, hydrangeas, and eucalyptus fenced off formal gravel paths; overhead explosions of tiny electric lights splintered the night like clouds of incandescent fireflies, lending the rooftop the appearance of an Adriatic brig that one could marvel at from any point in the city.

On June 25, 1906, the original Madison Square Garden opened its roof-cum-theater with a production of *Mamzelle Champagne*. Designed by Stanford White, the building incorporated two of his favorite materials, green granite and terra cotta, in a manner he vaguely attributed to the Early Christians. White was a paradox to his admirers. His visionary plans for a magnificent, monumental New York had been put forcefully into effect with the construction of the Metropolitan and Brooklyn museums, Pennsylvania Station, Washington Arch, and other public, commercial, and private structures too numerous to list; in this respect he was recognized as one of the city's presiding geniuses and leaders. On the other hand, there were a raft of lurid rumors about White's string of mirror-lined love nests stocked with nubile concubines. On the night of the twenty-fifth the two clearly separated lines of his double life fatally converged; the curtain fell on the last act of *Mamzelle Champagne* and White, having graciously accepted compliments all evening on the beauty of his roof, turned as a gentleman in white tie swiftly approached the table, his wife and a party of friends dogging his heels. The gentleman's salutary purpose was not to convey his best wishes but to pump three bullets into the architect's frizzy gray head. According to his murderer, Harry K. Thaw, White had forever dishonored Mrs. Thaw, née Evelyn Nesbit, prior to her marriage, by setting her up as a "love-slave" in one of his hideaways, drugging her, seducing her, and having her perform on a red velvet swing.

Evelyn, a twenty-two-year-old ex-Floradora and Gibson Girl, Harry, a weak-minded heir to millions, and "Stanny," the lamentably deceased, remained an inescapable topic of conversation for over a decade. Thaw's two trials and ultimate commitment to an asylum were plastered over every front page in the country, Evelyn's theatrical career was immeasurably, if temporarily, benefit-

A New York roof restaurant as painted for the *London Graphic* by Charles Hoffbauer. *Courtesy of the New York Public Library.*

ted, and White's private vices appeared rehashed at family tables with the monotony of left-over Thanksgiving turkey.

Society was hard hit. White had changed its tastes, built its mansions, and danced with its wives. Just what was going on? There was too much change in the air. An army of suffragettes had brazenly marched down Fifth Avenue. Women were painting their faces and smoking in public. In 1907, when Mrs. Patrick Campbell was requested to put out the cigarette she had lit in the tearoom of the Plaza Hotel, she replied, "I have been given to understand this is a free country. I propose to do nothing to alter its status." The press was outraged. Mrs. Clinton B. Fish, president of the Home Mission Society had the last word: "Smoking by women is definitely un-American." Elsie de Wolfe defected from the acceptable ranks of the gifted amateur and sent out professional cards, emblazoned with a little wolf bearing a flower in its paw, announcing her services as an interior designer. Shortly before his death, White hired her to decorate the premises of the newly formed Colony Club. Elsie established her trademark by strewing acres of chintz in every nook and cranny—the bedrooms, the hallways, and finally the reception rooms, in some of which alcoholic drinks were served by maids. When this last bit of information leaked out, the club was denounced from the pulpit, the pavement, and the tea table; the members merely shrugged.

What was going on? The Waldorf-Astoria changed its policy and welcomed unescorted women to all its dining rooms. Oscar followed this up with the depressing news that the traditional ten-course dinner had, by 1908, gone inexplicably out of vogue. In that year a demented and senile Mrs. Astor (*the* Mrs. Astor) was observed in Newport extending her hand in royal greeting to imaginary guests and chatting with distinguished phantoms. On June 10, 1912, the mayor gave a dinner at the Waldorf for the officers of a German squadron about to visit the city, but no one gave it a second thought. In 1913, an exhibition of works by obscure European "modernists" variously described as insane, color-blind, and otherwise depraved desecrated the Twenty-fifth Street Armory and caused a public furor.

On April 5, 1914, Josephus Daniels, a confirmed teetotaler and Secretary of the Navy, issued an order whereby no alcoholic beverages could be consumed by Navy personnel. His action prompted nothing more than a rash of cartoons and satirical editorials. By 1917, however, war hysteria had mounted beyond control and the prohibitionists leaped at their big chance; alcohol, they shrieked, was the major contributing factor in the rise of venereal disease, Bolshevism, and spies! Alcohol would deliver the world into the hands of the Hun! A sober America was a victorious America!

The people listened. The government listened, and a law was passed making it illegal to sell alcohol within a five-mile radius of any military base. What was going on? The drys were slowly but surely nailing a lid down on the wets—that's what was going on.

Chapter V

———— ••◦◦◦◦•• ————

1914-1940

Little Birds are hiding   *Blessed, I say, though beaten*
  Crimes in carpet bags   *For our friends are eaten*
Blessed by happy stags   *When the memory lags.*

The attitude of New Yorkers to the First World War was as equivocal as had been their attitude to the Civil War. Elsa Maxwell, a pudgy young woman from Keokuk, Iowa, returning to New York in 1915 after a prolonged tour abroad as a vaudeville accompanist, found the city totally changed: "The New York I remembered was a raucous, provincial town trying too eagerly and self-consciously to assume the air and attitude of a metropolis. The metamorphosis was almost complete in the summer of 1915. In seven short years New York had graduated from shirtsleeves to white piping on the vest."[3] But she was shocked to find her hotel flying the German flag and Count von Bernstorff himself, the German Ambassador, standing in the lobby. There was a surprising amount of pro-German sentiment, and New Yorkers seemed irrepressibly gay and indifferent to the war, or actually irritated by what seemed to the "social leaders" to be an inconvenience.

As in the case of the Civil War, actual hostilities and the sight of men marching down Fifth Avenue and off to the front led to a monumental change of heart. The Armistice was greeted with mass hysteria, the blasting of horns, and a rain of one hundred fifty-five tons of ticker tape, while effigies of the Kaiser were burned, hosed, and triumphantly carried in a coffin marked "resting in pieces." And when the boys came marching home they were greeted by a vast plaster triumphal arch at Madison Square, a Court of the Heroic Dead (in the Egyptian style) in front of the Public Library, and an "arch of jewels" further up the avenue.

The "psychology of war," however slow to be aroused, was also slow to die, and the wartime prohibition of alcohol remained in force, both legally and emotionally, in the minds and hearts of Americans finally excited to a moralistic fervor for curing what they took to be mankind's ills. The possibility of the spread of "Bolshevism" from Russia to these shores, however faint, was snuffed out by a series of vigorous "anti-Red" riots, fueled by the killing of thirty people by an anarchist bomb on Wall Street in 1917, an event which, had it taken place a half hour later, would have caused literally thousands of deaths. America was going to create a clean world. There was no room for the old ogre-drink.

"The Non-Drinkers had been organizing for fifty years and the Drinkers had no organization whatever. They had been too busy drinking." Thus the situation of 1920 was neatly condensed by journalist and author George Ade in his book-length epitaph, *The Old-Time Saloon*. Ten years later in the March 12 edition of the New York *Evening Sun*, H. Phillips neatly condensed American history to that date in eleven words: "Columbus, Washington, Lincoln, Volstead, Two flights up and ask for Gus." The social and political corruption of the corner saloon had instigated Prohibition which in turn fostered organized crime, and, in the cities, a rampant disrespect for the law by the citizenry. Abuse had bred abuse.

The very word "saloon" became anathema to a wide segment of the population through the callous machinations of owners and managers, ward heelers, brewers, distillers, bartenders, and graft-ridden police. No serious legislative precautions were taken to protect the public against the pitfalls of a saloon autocracy until the Raines law was passed toward the end of the nineteenth century. This law stipulated that no liquor could be served on Sundays except in such quarters as provided lodging facilities and food in conjunction with drink "regularly and properly set out on the table." Only the most meager ingenuity was demanded to elude this stricture, heavy fines and severe sentences for infractions notwithstanding; every saloon acquired a couple of sleazy cots, a prop hotel register, and a rickety dining table. The final transforming touch in the law-abiding metamorphosis from grubby barroom to cozy, family-style hotel appeared in the shape of a gray and lifeless sandwich, its edges curled in perpetual rigor mortis.

Still, offenses against the regulations were not always amusing.

The Mister-Barman-Have-You-Seen-My-Darling-Daddy-Dear type of tearjerking ballad was doubtless based on hard familiar reality; working-class families frequently starved as paternal paychecks were weekly confiscated by smiling bartenders who were only too glad to extend credit once again. "If drinking interferes with business, cut out business!" was a favorite saying displayed over the bar. If a wife angrily protested or beseeched the management to curtail her husband's drinking, she would be solicitously heard, gently escorted out the side door, and summarily forgotten. Sometimes it was not her husband but a high-school-age son who caused her distress by squandering family funds in a neighborhood saloon. And then again, if she too loved her beer and didn't feel like going to the corner, her small child might totter into a saloon with a pail or pitcher to be filled for home consumption. It is no wonder that the saloon fell into such inextricably disreputable straits and hid itself behind cluttered grille-work windows and louvered swinging doors. Small wonder the Prohibitionists patiently honed their axes in anticipation of the inevitable cataclysm.

With wartime Prohibition, the drys got their high-button shoe, their stove-pipe hat, their scruffy black umbrella, and their big blue nose in the door. Warring furiously after the Armistice, they brought pressure on government at all levels and the Eighteenth Amendment was conceived.

To enforce the amendment, an act of Congress had to be written and passed. Sponsored by Representative Andrew Volstead of Minnesota, the amended act with sixty-one articles was passed by the House 287 to 100. The bill was then sent to a Senate Judiciary Subcommittee and finally to its parent committee for approval. Woodrow Wilson perused the law, which was as full of loopholes as an antimacassar, and vetoed it without giving any specific reasons. The House and Senate, in a puritanical tantrum, reversed this decision and passed the Volstead Act over the President's head. On the calamitous morning of January 16, 1920, the Eighteenth Amendment, decreeing national prohibition, became federal law.

The world into which this great act of moral reformation was born just wasn't ready to receive it. At least not in New York. The current of reform had run into a tidal wave of revolution rolling in from the opposite direction.

Sinister intimations of change in social mores had been detected before the First World War. Women, wearing the suggestive hobble skirt, minced off to *thés dansants* at the more respectable hotels and danced with "gigolos" hired for that purpose by the management. Dances had undertaken a new form, and couples at arm's length since the invention of the waltz were suddenly cheek to cheek dancing the tango, as popularized by Maurice. Women went to the polls and threw off almost all their clothes at virtually the same moment. The flapper was not the wayward daughter of the Gibson Girl; she was frequently one and the same person. Stiff collars, corseted waists, bosoms, bustles, skirts, sleeves, hair—everything went. The female calf appeared for the first time since the dawn of the Christian era, and was soon followed by the knee. Some blamed it on Irene Castle, who shingled her hair and lifted her hems in order to be free to dance. Pomp and circumstance went the same way as the train and stays. Even the very wealthy dislodged themselves from their mansions and moved into apartments, where "cocktail parties" gained favor over other more constricting forms of entertainment.

The world was soon to be introduced to such twentieth-century commonplaces as women's suffrage, bare bathing suits (and bathing beauty contests), the crossword puzzle, the notions of Sigmund Freud, and greeting cards for all occasions (including hitting one's hostess with a bottle). One Frank Conrad of the Westinghouse Company was broadcasting speeches and music for owners of radio receiving sets from a barn near Pittsburgh, and in 1920 his success led the company's officials to found a permanent broadcasting station in order to encourage the sale of their equipment. Everyone was singing "Yes, We Have No Bananas."

If radio was about to draw the nation together, it would only be completing the task already begun by the motion picture. The cold winter of 1917–1918, with its rationing of coal and electricity, closed down the film center at Fort Lee, New Jersey, just across the river from New York, but early film makers were quickly learning the advantages of production in sunny California, where taxes were low, labor cheap, and every kind of scenery, from jungle to desert, near at hand. They were also discovering the advantages of the five-reel movie, the star system, and of course, publicity. The public was discovering a great deal more. It was learning how the other half lived. Fantasy sequences from the lives of the "rich," told in terms that even an immigrant could understand without speaking one word of English, led to impossible ideals (as well as a general, more sane notion of "the American

Way of Life"). There was a nationwide thrust toward self-improvement. Everyone turned to *Etiquette in Society, in Business, in Politics and at Home* written by Emily Post, a gentle Baltimore society divorcee, endeavoring to support her two children. The book soon outsold *The Life of Christ* and more copies were stolen from public libraries of this than any other volume, with the exception of the Bible. Meanwhile magazine advertisements preyed on the insecure, prescribing a "five foot shelf of books" for couples who were "one year married and all talked out." They also suggested "How to Make a Hit with Influential People!" and warned "How Little Social Errors Ruined Their Biggest Chance"—(the ordeal of a wife who says "pleased to meet you" and takes olives with a fork).

In April of 1914, the Strand Theater on Broadway, the first true "motion picture palace," opened in New York. The working class "nickelodeon" and small neighborhood theaters were no longer capable of containing the crowds, nor did they attract the sort of upper-middle-class audience the showmen wished to attract. The Strand had everything of which a legitimate theater could brag, and much more—gilt, marble, thick rugs, original works of art, crystal chandeliers, and a thirty-piece symphony orchestra as accompaniment to the feature. By 1915 the magnet of Hollywood had drawn from Broadway its greatest stars—De Wolf Hopper, Sir Herbert Beerbohm Tree, Billie Burke, Weber and Fields, Mrs. Leslie Carter—but they were coming back to "the great white way" on celluloid. Other legitimate theaters on Broadway itself, such as the Astor and the old Criterion, were built. At the Rivoli and the Rialto, stage shows presenting stars of the Metropolitan Opera and concert stage accompanied the feature.

Fifth Avenue was soon commandeered by expensive shops as far north as Fifty-ninth Street, and the sky-scraping Sherry-Netherlands and Savoy-Plaza replaced the senior Netherlands and Savoy (the Plaza had been rebuilt in 1907). The Woolworth Tower had reached sixty stories in 1912, and from then on the sky was quite literally the limit. As all buildings generally required "architectural detail," these were allocated to the top, resulting in a city of Tudor cottages, mosques, rococo pavilions, and Art Nouveau tabernacles suspended in space a thousand feet above the pavement.

F. Scott Fitzgerald was, even then, regarded as the spokesman for the post-war generation, and more than anyone, he and his wife Zelda were able to get the feel of the excitement of the new New York in the era of the post war "tea dances":

Twilights were wonderful just after the war. They hung above New York like indigo wash, forming themselves from asphalt dust and sooty shadows under the cornices and limp gusts of air exhaled from closing windows, to hang above the streets with all the mystery of white fog rising off a swamp. The far-away lights from buildings high in the sky burned hazily through the blue, like golden objects lost in deep grass, and the noise of hurrying streets took on that hushed quality of many footfalls in a huge stone square. Through the gloom people went to tea. On all the corners around the Plaza Hotel, girls in short squirrel coats and long flowing skirts and hats like babies' velvet bathtubs waited for the changing traffic to be suctioned up by the revolving doors of the fashionable grill. Under the scalloped portico of the Ritz, girls in short ermine coats and fluffy, swirling dresses and hats the size of manholes passed from the nickel glitter of traffic to the crystal glitter of the lobby.

In front of the Lorraine and the St. Regis, and swarming about the mad-hatter doorman under the warm orange lights of the Biltmore façade, were hundreds of girls with marcel waves, with colored shoes and orchids, girls with pretty faces, dangling powder boxes and bracelets and lank young men from their wrists—all on their way to tea. At that time, tea was a public levee. There were tea personalities—young leaders who, though having no claim to any particular social or artistic distinction, swung after them long strings of contemporary silhouettes like a game of crack-the-whip. Under the somber, ironic parrots of the Biltmore the halo of golden bobs absorbed the light from heavy chandeliers, dark heads lost themselves in corner shadows, leaving only the rim of young faces against the winter windows. . . .[4]

As New York spread almost to its present limits and took its position as economic capital of what had recently become the most powerful nation on earth, there was a cumbersome amount of money around. The stock market climbed, then climbed again. Everyone invested their savings and profited. It was hardly the *moment juste* to drive Delmonico's and Sherry's out of business, to padlock Martin's and Shanley's, to wreck Bustanoby's, Churchill's, and Reisenweber's. Rector's was spared the indignity, having expired a few years before, but Murray's Roman Garden was converted into a flea circus. On the last day bartenders stood with trembling lips behind bars, while their patrons openly wept into their drinks. A frustrated society was to take a terrible revenge.

At the onset of Prohibition, Louis Mouquin, whose excellent French restaurant was forced out of business by its inability to

The Salle Cathay of the St. Regis, 1929. *Courtesy of the New York Public Library*.

dispense the contents of its superb cellar, was surprisingly magnanimous: "Wines formerly drunk at table will be diluted by several parts of water, making beverages of not more than 3 per cent strength. We can also produce excellent 3 per cent wine in still form, and to be served as sparkling champagne at soda fountains. American ingenuity will certainly produce healthful, palatable and temperate beverages within these limits from fruit juices." Mr. Mouqin was certainly correct to surmise that American ingenuity would be able to cope with the restrictions of Prohibition, but very little time and effort was devoted to fruit juice.

Ring Lardner, writing about Prohibition in 1925, summed up the nation's reaction: "Well they was a lot of people in the U.S. that was in flavor of [Prohibition] and finely congress passed a law making the country dry and the law went into effect about the 20 of Jan. 1920 and the night before it went into effect everybody had a big party on acct. of it being the last chance to get boiled. As these wds. is written the party is just beginning to get good."[5]

Within days, possibly hours, of the fateful January 16, speakeasys shot up like crab grass. The wets were making up for lost time. At the peak of Prohibition there were thirty two thousand "speaks" in the city, actually twice the number of saloons that had been padlocked. A speakeasy might be anything from a hole in the wall where liquor alone was served, to an establishment providing food and drink on any scale, or a night club with dance floor and cabaret. The simpler formula was fairly uniform: a door with a Cyclopean peephole (possibly hidden behind a florist's showcase, an undertaker's coffin, or a telephone booth), several more shackled portals, and finally a well-carpeted back room and bar.

The quasi-madness of the institution appears when seen through the eyes of a foreigner, a visitor from a land of less eccentric laws and customs. Paul Morande, a French diplomat, passed through New York for the first time in 1925 on his way to Bangkok. In his subsequent book, *New York*, he viewed the speakeasy with anthropological dispassion:

The speakeasy . . . is a clandestine refreshment-bar selling spirits or wine. . . . The door is closed, and is only opened after you have been scrutinized through a door-catch or barred opening. At night an electric torch suddenly gleams through a pink silk curtain. There is a truly New York atmosphere of humbug in the whole thing. The interior is that of a criminal house; shutters are closed in full daylight, and one is caught in the smell of a cremation furnace. Italians with a too familiar manner, or plump, blue jowled pseudo-bullfighters, carrying bunches of monastic keys, guide you through the deserted rooms of the abandoned house. Facetious inscriptions grimace from the walls. There are a few very flushed diners. At one table some habitués are asleep, their heads sunk on their arms; behind a screen somebody is trying to restore a young woman who has had an attack of hysteria. . . . The food is almost always poor, the service deplorable. . . . The Sauterne is a sort of glycerine; it has to go with a partridge brought from the refrigerator of a French vessel; the champagne would not be touched at a Vincennes wedding-party.

A good deal of the hocus-pocus was probably for the sake of drama, as "police protection" accounted for a lion's share of all speakeasy overhead, and the force could generally be counted on to make only those raids absolutely necessary for the record. The agents of the Prohibition Bureau were another matter.

Izzy Einstein and his fat partner, Mo Smith, all agreed, ranked among the zaniest practical jokers of the era. A laugh a minute. Sometimes Izzy would stumble into a speakeasy dressed as a gravedigger tired out from a deep job and thirsty for a drink with his pal Mo. On other occasions, honest workingman Izzy, or football player Izzy, or iceman Izzy, or musician Izzy, or automobile cleaner Izzy, and friend, would be admitted. Their impersonations were perfect down to the dirt-smudged shirt, the shiny trousers, or the dog-eared black tie. Once they had reached the bar and been served, Izzy and Mo would raise their glasses and cordially announce, "This is a raid." Izzy and Mo were Feds. In fact they were the most effective agents ever to grace the Federal Prohibition Bureau. Unfortunately, they received too much publicity (about as much as the President and more than the Prince of Wales) and consequently were soon unemployed.

The Bureau itself was much maligned. When it employed women to escort its agents as part of their official charade, it was accused of corrupting the very sex it was meant to protect. It was also accused of being the best training school in the country for bootleggers, as so many of its agents sold their information to or actually went into business with the enemy. In fact, the Bureau was just poorly organized, ill-trained, and run on the always faulty spoils system. Finally, it had little money at its disposal, while huge sums were falling into the hands of bootleggers. The wets may have been caught off guard, but they were shortly more numerous, more powerful, and much better paid than the professional drys.

Speakeasies existed on all levels, from dank back rooms where a home-brew called "smoke" was served for only fifteen cents to the fashionable "clubs" on the ground floors of brownstones lining the side streets of the West Forties and Fifties. The former patrician owners of most of these houses had already moved uptown; those who had not soon regretted it. If they wished to get a decent night's sleep, it was necessary to post a large sign proclaiming, "This is a Private Residence" over the bell.

Unquestionably the street with the highest proof was West Fifty-second, between Fifth and Sixth avenues, where both Leon and Eddy's and Jack and Charlie's thrived. Jack Kriendler and Charlie Bern's "21" was to the average speakeasy what Delmonico's had been to the average saloon. Like Delmonico's, it occupied an entire mansion with two bars and several dining rooms, a dance floor and orchestra. Jack and Charley encouraged their speakeasy's confusion with a private club by designating rooms in which ping-pong, backgammon, and mahjong might be played. They also strove to preserve the remnants of haute cuisine in New York throughout Prohibition. Unlike Delmonico's the cellars of the "21" were hidden behind carefully camouflaged fortress-thick brick walls, which sprang open at the insertion of wire in an invisible chink. At its opening in the mid twenties (the owners had operated another establishment on West Forty-ninth Street), invitations went out reading: "Luncheon at twelve, Tea at Four and until closing."

More typical was Frank and Jack's, a jolly place where there were generally a hundred people jammed into a tiny kitchen barely large enough to hold three tables. Among those struggling for air and room to laugh might be Jimmy Durante, Pat Rooney, or Peggy Hopkins Joyce. It was Frank and Jack who perfected the gambit of getting rid of one drunk by asking him to assist another out the door.

The most obvious difference between the speakeasy and the old-time saloon or gentleman's bar was its clientele; the residue of Lobster Palace Society was augmented by members of High Society on the one hand and gangsters on the other. The official birth of what is called Café Society actually coincided with the dawn of Prohibition—February 1919. One evening of that month Maury Paul, the first Cholly Knickerbocker, espied an unlikely group seated together in the Ritz dining room—his secretary Eve Brown, Mrs. Allen Gouverneur Wellman, Joe Widener, Laura Corrigan, Whitney Warren, Jr., and an "assorted pair of Goelets." Although members of the "Four Hundred" enjoyed meeting artists and writers over tea at the home of Elsie de Wolf earlier in the century, at that date a more unlikely mixture of old and "nouveau" society, with a smattering of the arts, could not be imagined. At the lobster palaces women of the old guard and the music hall star might dine in the same room, but they did not sit down at the same table. This was a major break with tradition, and Maury Paul, Hearst's flabby, snobbish, and eternally genial gossip columnist, was astute enough to notice and label it. A firm believer in the reality that "the newcomers of today are the Old Guard of tomorrow," Paul recorded exactly who was a member of Café Society (the lines being no longer clean cut) and what they did.

But if Maury Paul identified Café Society, it was Condé Nast and Frank Crowninshield who nurtured and firmly established it. When Nast produced one of his exquisitely choreographed penthouse parties, he claimed that he could not fail to invite his good friend Mrs. Vanderbilt. Nor, on the other hand, could he leave out his old friend George Gershwin. The result was instant Café Society. When Nast, who already published *Vogue*, bought *Vanity Fair* magazine, he asked his friend, witty and chivalrous Frank Crowninshield (whose great grandfather was Secretary of the Navy under both Monroe and Madison), to edit it. It was Crowninshield's notion that a magazine should cover "the things people talk about—parties, the arts, sports, humor and so forth." *Vanity Fair* was an immediate success, and in its pages the nation was introduced to an altogether new aristocracy—the aristocracy of achievement (in the arts, films, sports, or even in "life style") rather than money—the members, for better or worse, of Café Society.

Unfortunately, Café Society no sooner had a name than it began to develop a shady one. Later in life Crowninshield felt that it had "drifted into barrooms and nightclubs and lost its chic. . . . What actually ruined the whole goddam thing was alcohol and noise." The fact is that the only "cafés" provided by America in the twenties were speakeasys, and Society flocked to these in such large numbers that presently, Lucius Beebe was later to remark, "New York society . . . found itself living frankly, unabashedly, and almost entirely in saloons." For Café Society going to a "speak" became the fashionable thing to do and getting drunk more fashionable still. Except for the very lowest on the scale, speakeasies were frequented by women of the most respectable order, those selfsame women who would have averted their eyes from a saloon and would only enter the Hoffman House to view the Bouguereau on a guided tour. There they were, drinking away, seated casually at the bar or collapsing onto the shoulder of their escort, if they had one.

Women exerted a certain influence on speakeasy decor. Gone were the days when a popular bar could merely comprise a highly polished wooden fixture with a mirror behind it and a tiled floor beneath. Now there were circular bars, revolving bars. Raymond Anthony Court, one of New York's most imaginative decorators, made a fortune designing them. The "Park Avenue" glittered like a black, gold, and red Follies set, and at his Aquarium, the bar itself was a mammoth tank of fish indirectly illuminated. There the specialty was a Goldfish Cocktail, compounded of equal parts of Goldwasser, gin, and French vermouth. Highly flavored mixed drinks grew in favor as the quality of liquor grew in anonymity; at Zani's the biggest seller was the Zani Zaza, a concoction of gin, apricot brandy, the white of an egg, lemon juice, and grenadine.

One could usually trust the gin at a first-class speakeasy. This was not always the case at less meticulous speaks. By the mid-twenties, an estimated fifty to sixty million gallons of industrial alcohol were annually diverted into the hands of bootleggers. However, under the Volstead Act and the persistence of the Prohibition Bureau, toxic denaturants of one sort or another would have been added to it. In the New York area, wood alcohol was most commonly used to render spirits impotable, and if it was not properly extracted by an expert at recovering drinking alcohol, the aftermath could be horrifying in incidents of death or blindness. Glycerine, water, and oil of juniper were its flavoring agents when masquerading as gin, and caramel, prune juice, or creosote when an order for Scotch had to be filled. In the instances of local moonshining from scratch, fuel oil and rotten rat meat gave an extra kick to the finished product. Brand names for moonshine varied with vicinity of origin, but all implied an electrifying jolt: Jersey Lightning, Jackass Brandy, Panther Whiskey, Yack Yack Bourbon, or Straightsville Stuff. One was much better off knowing a bootlegger who could lay his Black Hand on some of the million gallons of trustworthy spirits officially exported, if unofficially imported, into the country by Canadian sources every year.

The dispenser of Prohibition alcohol, the bootlegger, carved himself an interesting niche in history. Prior to Prohibition, the

individual gangster had been a private entrepreneur on a modest
scale. The toughs at Billy McGlory's were as expert at "rolling" a
drunk as their successors, but their bank accounts were negligible.
With the advent of bootlegging, a number of slouch-hatted, chalk-
striped public enemies found themselves millionaires overnight,
with business obligations that could best be handled by professional strong-arm men. The result was what is termed "organized"
crime. Never had so many hoodlums had so much money at their
disposal at one time. Like all tycoons on the crest of a wave, they
"diversified," some investing in other illegal enterprises and some
in legal undertakings. Many of the most unsavory nursed secret
yearnings to become the partners or imitators of the quasi-legiti-
mate businessmen with whom they dealt, such as speakeasy and
night club owners. The new gang lords had a weakness for the
glamorous world of entertainment, and by the end of Prohibition
many speakeasys had fallen into the hands of the mobs. Their
determined association with the world of catering could lead to
grisly results, even under the nose of the emancipated debutante
getting fashionably oiled in a nifty speak.

The carnage known to history as the Hotsy-Totsy Murders is a
case in point. On the inauspicious night of Friday the thirteenth
of July 1929, the brightest rhinestone in the Broadway tiara, the
Hotsy Totsy Club, was blazing with simulated glory. Its large dining room and smaller bar were packed as usual with newsmen,
entertainers, sportsmen, and brilliantined delegates of the underworld. Its legitimate proprietor, Hymie Cohen, was one of the
best-loved figures around Times Square. Described by Mark Hellinger in the *Daily News,* Hymie "was a small, roly poly sort of guy.
He was short and fat and had a face like a moon. When I knew him
he ran a speakeasy. That made him a criminal, I suppose, but I
never figured him as such. . . . He was too good natured to be an
out and out crook. His sense of humor was too large. He was too
jovial and too big-hearted ever to be pointed out as a man who had
no respect for law and order." One day he told Hellinger, "Some
day Hymie Cohen is goin' to disappear. Fast like. And that'll be the
last you'll hear of him. See if I'm not right." But back to Friday the
thirteenth.

It was after 3:00 A.M. when an altercation broke out over the
comparative merits of Ruby Goldstein, prizefighter. Gunshots
were heard, twenty-four in all. When the smoke cleared and the
police arrived, two bullet-riddled bodies, one belonging to a

"Sammy" Walker, ex-convict, and another to a William "Red" Cassidy lay splayed out on the floor. Red's brother, Peter Cassidy, was taken to the hospital with a blackjack wound. It was soon established that the participants in the argument, aside from the victims, had been Hymie Cohen's silent partner Jack "Legs" Diamond, an infamous bootlegger known as "the clay pigeon" for his ability to absorb bullets, and his lieutenant, Charlie Green, alias Entratta. That these two had disappeared did not disturb the police: they would be apprehended, and there were hundreds of witnesses.

Soon there followed an overall depopulation of the Hotsy Totsy Club. The cashier disappeared. The bar waiter was murdered. The bartender was shot. The hatcheck girl disappeared. Hymie Cohen disappeared, "fast like." Three assorted witnesses were killed.

Green presently gave himself up, and Diamond followed. Both were arraigned and brought to trial. But for some reason, the police could not produce a single witness. If the customers who had been present at the Hotsy Totsy during the disruptive interlude were now to be believed, each had been so drunk or engaged in such earnest conversation that he had failed to notice who in particular had fired the shots. Peter Cassidy had amnesia. The defendants were acquitted.

Despite their perils, speakeasies remained popular as did the "night clubs," those tiny-tabled playrooms featuring dance bands and "floor shows," which took the place of the padlocked Martin's and Reisenweber's.

"Never give a sucker an even break," drawled Wilson Mizner in 1905. Twenty years later Mary Louise Cecilia Guinan was still wringing laughs out of this big city commandment as, perched atop a piano, she rattled a noise-maker, finger-crimped her brassy marcel, and greeted her customers with a raucous "Hello, sucker! Have all the laughs you can, because they'll be on the bill!"

"Texas" Guinan, the queen of night club "hostesses," had come a long way since her early days as a movie cowgirl and vaudeville performer; from 1923 when she opened the El Fey Club in partnership with Larry Fay, an underworld taxi czar, until the end of the decade when she mysteriously fell out of grace with the mob, she reigned as the country's number one party-girl and indefatigable mistress of ceremonies. In night club parlance a heavy spender was known as a "live one" until Texas, with her gift for coining blunt and colorful phrases, asked a customer who was tossing fifty-dollar bills about like so much rice at a wedding exactly what his profitable line of business was. "Dairies," he allowed. "Well, let's have a hand for the big butter and egg man," she crowed, nailing down a type once and for all. "Give the little girl a great big hand," became a more familiar directive, and punctuated Texas' all-night harangues like invocations at a tribal rite. She herself was not an entertainer in the strict definition of the word. She knew how to start a party and keep it going, how to pick pretty chorus girls (at one time she had seventy-eight on the payroll), and, most important, how far she could go as a "character" and still keep an international audience coming back for more of her seemingly endearing insults and practical jokes. When asked where she would be without Prohibition she answered "nowhere," and to the accusation that her only interest was in fleecing the customers for all they were worth, she patiently explained that there were no charity wards in her club, the boys came to spend, and she was not about to disappoint them.

Texas' turn-over in clubs was considerable, as was the case with the great night club team of pianist Jimmy Durante, singer Eddie Jackson, and hoofer Lou Clayton. Durante started playing in Bowery and Chinatown dives when he was a mere boy; at one time he worked at Carry Walsh's "joint" in Coney Island along with a pint-size singing waiter, or "nickle kicker," whose piping voice and look of perpetual glassy-eyed astonishment would later become famous. The nickles to be kicked were called "throw money" and indicated the audience's favorable reception of an act; the nickle kicker, artfully skimming and skittering this tribute into a small pile near the piano while simultaneously taking beer orders and improvising lyrics to unfamiliar requests, was Eddie Cantor. By the twenties, however, Durante and company were riding the crest of the wave at such clubs as the Silver Slipper, where Ruby Keeler tapped her first sweet staccato steps into the hearts of millions, and the small, expensive Les Ambassadeurs.

In fact, the amount of talent available in New York night clubs was impressive by anyone's standards. Bea Lilly could be seen chatting and croaking out songs at the Sutton Club on Fifty-seventh Street, while Libby Holman, the Society blues singer, entertained at the Lido. Clifton Webb was dancing with Mary Hay at Ciro's, and Fred Astaire and his sister Adele punished the parquet at the Trocadero. Meanwhile, on top of the New Amsterdam Theater, Ziegfeld's *Follies* girls starred in his Midnight Frolic. For the more serious-minded the Mirador provided Michael Fokine and his Russian ballet.

Many night clubs priced themselves out of existence. The "cover charge" in New York dates back to Reisenweber's and the introduction of the floor show; however, some time elapsed before customers began to complain that they were receiving less for their money. But the twenties were nothing if not roaring, and the elite of suckerdom would have felt somehow unfulfilled without their nightly trip to the cleaners. It was the modest New York job-holder, the out-of-town buyer, and the college boy with limited funds who felt the pinch and bitterly blamed the cover charge.

After the sound and fury began to die down in the midtown area, the place to hustle for the late show was Harlem. Harlem had come into existence as New York's Negro quarter during the First World War, before which only a small group of blacks had taken up residence along Lenox Avenue in the vicinity of One Hundred and Twenty-fifth Street. At this time the majority lived cheek by jowl with Irish immigrants in the Pennsylvania Station area of Hell's Kitchen, a section of the city that was bearable only because it was mercifully near Broadway. After a series of ugly interracial skirmishes, the blacks, many of whom could afford a better place to live, looked desperately for homes elsewhere. Soon an enterprising black realtor, Phillip A. Payton, discovered that, in the rush of New York northward, speculators had overbuilt, and there were apartment houses standing empty north of Central Park. Offering rents higher than the whites would pay, the blacks moved in, at first to the area between One Hundred and Thirtieth and One Hundred and Fortieth streets from Fifth to Eighth avenues. As the new tenants were obliged to double up and rent out rooms in order to pay the exhorbitant rates, black Harlem was overcrowded from the very beginning. When the war brought waves of southern Negroes to find employment in the cities, Harlem became still more congested and spread further.

The new settlers brought with them not only their downtown churches, but also their cabarets. And their white patrons followed. For night-owl New Yorkers, Harlem in those days had a surface atmosphere all its own—bittersweet, clamorous, bursting with life; to the casual outsider it was a three-ring circus of jazz, jubilation, and high-kicking Charlestons; to the initiate, it meant hard times and the blues as well.

Even before the apocalyptic arrival of the original Dixieland

Jazz Band, an outfit calling itself the Memphis Students played Proctor's Twenty-third Street with effective success. The banjo remained lead instrument and pulse of their "sound." The Dixie-land group radically changed this early jazz style by dispensing with strings and limiting their instruments to cornet, clarinet, trombone, piano, and drums. As their music caught on, similar bands were hired for private parties, and by 1912 a jazz concert drew a standing-room-only turnout at Carnegie Hall. But it was not until the twenties that jazz became what might be justifiably called a national rage. And not surprisingly, when everything else had broken its moorings, music had to cut loose as well. The new era simply refused to move in three-quarter time.

The Broadway production of *Shuffle Along* with an all-black cast set cash registers ringing not only at the box office, but in Harlem's more successful night clubs. The star of the show, Gertrude Saunders, failed to appear for a performance one evening and a tiny singer named Florence Mills stepped into her place. In 1926 Florence died prematurely and tragically in France but not before she and two other young Harlem favorites, Bricktop and Josephine Baker, had made themselves the toasts of Paris. It was artists like these, and Bessie Smith, Ethel Waters, and Bill "Bojangles" Robinson, who made New Yorkers aware of first-rate jazz and its availability at almost any hour, and who triggered the nightly migrations north.

The "swankiest" places to go were the Cotton Club with its sardonic "ol' plantation" decor and Duke Ellington's band, and Connie's Inn on Seventh Avenue, where a red canopy underlined its reputation as Harlem's most expensive "hot spot." Connie's was dark, subterranean, and exotic—with a waist-high barrier around the dance floor on which "Snake Hips" Tucker performed to the blare of Fletcher Henderson's group. Racially mixed couples were denied entrance at both of these clubs. This restriction was unknown at Jeff Blount's Lenox Club, where the Monday morning breakfast dance blew the last undamaged weekend gaskets, or at Ed Small's Small's Paradise. Sunday was, for some reason, Harlem's most frenetic night; as a result, there was an inevitable sprinkling of empty seats in Wall Street Monday morning.

There were many other Harlem night spots to explore; some are nostalgically remembered to this day, others scarcely recalled at all. The main thoroughfare attracting "ofays" was "Jungle Alley," covering the One Hundred and Thirty-third Street area from

Lenox to Seventh Avenue; among its major attractions were Til-
lie's, The Clam House, The Nest, The Catagonia Club, and a clutch
of others. There were also quieter places where the only performer might be a blues shouter, a girl dancer exhibiting an alarm-
ing and acrobatic elasticity when it came to scooping coins off a
table, or a down-and-out pianist shrouded in sweet marijuana
smoke.

Harlem's own Café Society, its own intellectual and in a way
bohemian elite, lived on Sugar Hill, where, at rents unrealistic
even for Harlem, they could command a view of the Bronx and
try to create a world of their own. It was here, in small, comfortable and brightly lit speakeasies that artists' artists played for an
audience of cabaret and theater performers. These were quiet
places with no extravagant floor shows. For "hot" dancing the
"Sugar Hillies" were obliged to descend to Seventh or Lenox
avenues, where they could join the nightly blow-outs at the Renaissance or Savoy ballrooms. Neither were owned by Negroes, as
in fact none of the Harlem night clubs were, with the exception
of Small's and just a few others.

In the twenties, laying hands on decent liquor proved as much
a problem in Harlem as anywhere. Gin was Harlem's favorite
drink, followed by bourbon, yet the forms could vary. Something
called "Chicken Cock Whiskey" came unpromisingly in cans, and
at Mexico's the house brand was called "Old 99." According to
Duke Ellington, everyone took turns making it, and Mexico felt
they never got it more than 99 per cent right.

According to Duke Ellington's son, Mercer, it was Mexico's that
molded his father's philosophy:

Whiskey came in two classifications in those days—No. 1 and No. 2. If you
ordered a drink, you got No. 1. But if you came in and had more money
to spend, you were allowed the oral privilege of ordering No. 2, thus
showing everybody you were a big shot. Of course No. 1 and No. 2 were
the same whiskey.

So Ellington . . . began to learn how to convince people. Not necessarily
how to convince them, *but how to make them want what you've got to
give them.* . . . He learned there what so few people know—so few leaders
know—he learned *presentation,* one of his great assets all through his
career. . . . Even now, though Mexico is gone, you can still hear the echoes
of his idea of No. 1 and No. 2 in Ellington. . . . He never scolded Ellington
or told him he was wrong. He'd simply say, "A Wise man would do so and
so," and nothing was ever *wrong.* Mexico knew all the people that could

lead him down the wrong road and Mexico used to steer him and give him knowledge as to who the real good influences were in New York and who the bad influences were. During those crazy twenties, a guy who was just a musician could sit in a place in Harlem and get a fifty dollar tip for playing just one tune. . . .[6]

Downtowners went to Harlem for "Yardbird" (chicken), "strings" (spaghetti), or a melange of brains and scrambled eggs. If they went to Craig's on St. Nicholas Avenue, they could try the very best in "soul food"—collard greens and hog jowl, pigs' tails and neck bones, grits and red-eye gravy, corn pone, sweet potatoes, black-eyed peas, and spare ribs.

If a sightseer was looking for drugs, sex, or trouble, he could easily find it. But on the whole, the cabaret world of Harlem in the twenties was as safe as anywhere else on the island, that is until the gangsters muscled their way in, as was their lethal habit, and the muggings and the shootings began. Musicians and other artists were among the first to give up and move elsewhere; wealthier Harlemites followed them, the ofays stopped commuting, and one of the greatest shows of the Prohibition era ground to a halt. Harlem's magic powers, like those of the "Tree of Hope" outside Connie's Inn (giving this scrawny elm's tattered bark an affectionate rub was meant to bring good luck), had inevitably worn out.

According to Oscar Tschirky, more than magic powers were needed downtown to restore the pre-Prohibition standards of New York's great kitchens, and he delivered an ultimatum: "In the interests of good eating, speakeasy dining must be eliminated." Prohibition had not succeeded in eliminating drinking, but it had very nearly put an end to a century's development in haute cuisine. The depressing truth was that New Yorkers could not be enticed into a dry restaurant, no matter how superior the food; on the other hand, they seemed perfectly content to eat sawdust as long as it came with a drink. Delmonico's, already dying on its feet, gave up the ghost, and Louis Sherry went into the commercial confectionery business. The site of the Horseback Dinner became a bank. The best hotels served a captive audience (they remained the scene of all official entertainment), and maintained the standards of haute cuisine as far as they were able. Oscar, it is true, had managed to stay in business.

But he was not pleased. With or without Prohibition, changing tastes had buried the era of twelve-course dinners. Like pompadours and large houses, they were just too much trouble to keep up. Women wanted to be thin and flat, men wanted to be dapper. Extra courses went the way of the extra curves they engendered. The menu for the dinner given in honor of Thomas Alva Edison at the Astor in May of 1928 may be regarded as transitional:

Coup of Grapefruit Fraisette
Petite Marmite aux Nouillettes
Celery   salted nuts   Olives
Supreme of Striped Bass Veronique
Roast Baby Lamb Renaissance
Vegetables Bouquetière
Boned Long Island Duckling Rossini
Fruit salad Astor
Pavé glacé Monte Christo
Petits Fours
Friandises
Moka
Cigarettes

At least this Astor menu included both a roast and a game course, as well as soup and fish. It is with complete horror that Oscar records the dinner given for the Crown Prince of Sweden at the Waldorf-Astoria:

Leaf of Lettuce, Swedish Style

Egg Princess with Asparagus

English lamb chop, Waldorf

String beans, Mornay

Génoise Glacé

Coffee

As if in a dream, the great feasts had vanished. Oscar noted the changes, detail by chilling detail. People no longer wanted their fish swimming in sauces. Heavy lobster or chicken salads were now considered a meal in themselves rather than an accompaniment to dinner.

As with drink, the law stepped in. The serving or selling of freshly killed game was forbidden in New York State, and all game had to be sent frozen from Europe or South America. Proprietors had to be more careful about what they served. Barrington vs. the Hotel Astor (184 App. Div; 317,171 N.Y. Supp., 840) was a case in point: The plaintiff walked into the dining room of the hotel, sat down, and ordered kidney sauté. He found a mouse in it. The facts were expressed succinctly by Mr. Justice Dowling in his summation: "As soon as the plaintiff discovered the unexpected addition to his order, he became violently sick and remained so for some weeks, and suffered illness and other discomforts as the result thereof, including a pronounced loss of appetite." The Justice concluded: "A guest of a hotel who orders a portion of kidney sauté has the right to expect, and the hotel keep impliedly warrants, that such dish will contain no ingredients beyond those ordinarily placed therein . . . the defendant does not seek to justify the inclusion of the mouse in the dish as any proper part of its menu."

Although haute cuisine had reached a point of retrenchment, or stagnation at best, something very exciting and wholly unique was happening to New York dining. The children of the bathtub gin era were willing to try anything, from couscous to Pekin duck. Nowhere else in the history of the world had so many disparate ethnic groups mushroomed as prolifically as they had in New York by the first decades of the twentieth century. To their "typical" restaurants, eating houses, and cabarets they welcomed not only their own, but anyone else who cared to join them. Gone were the days when a New Yorker might screw up his courage to set foot in the Atlantic Garden but still shrink in dread from the "Opium Dens" on Pell Street. Now everyone went down to Chinatown for chop suey. In fact they wanted their chop suey served to them right on Columbus Circle.

For many Chinese restaurateurs, the adventurous Jazz Age came as an answer to a prayer from Jok-Quon, the kitchen god; chop suey, chow mein, and jazz bands brought hordes of Occidental dinner dancers to brand new second-floor palaces, from The Oriental, located at 4 and 6 Pell Street in Chinatown, to the Palais d'Or in the heart of Broadway. At the Palais d'Or the ceiling was draped with loops of lace, conveying nobody knows what dynastic period, and boxes ringed the dance floor, where a cast of thirty flounced through a revue at seven and half past eleven.

In Chinatown, only a second- or, better still, third-floor establishment was considered acceptable by the cautious; first-floor and basement restaurants which were cheaper, and catered to Chinese workingmen, bachelors in particular, were dismissed as "too foreign."

The first Chinese to attract attention in New York sailed into the harbor aboard the junk *Keying* on July 5, 1847. There were thirty-five of them in all, and it had taken them two hundred and twelve days to make the trip from Canton. New Yorkers were fascinated by the extraordinary paintings of birds and beasts decorating the junk, and while it remained in port, fifty thousand curiosity seekers paid twenty-five cents apiece to investigate its interior and a ten-armed idol hung with flowers, trinkets, and beads.

No appreciable number of Chinese took up residence in the city until about twenty-five years later. Cantonese immigrants started pouring into California at the peak of the Gold Rush, but they stayed there. Coming to America (Mei Kowk, or "Beautiful Land") not as immigrants but as sojourners to Gum-San (the informal name for America, literally meaning "Land of the Golden Mountains"), they planned to prospect for gold and take it directly home. Approximately half of them did just that. Despite their background as farmers, others stayed and took jobs as laundrymen, or hired themselves out as track-layers for the transcontinental railroad. It was from this last group that the founders of New York's Chinatown came, settling on the old Doyers farm and tavern property hard by the Bowery in 1869. To this day, Chinatown encompasses no more than five square blocks bounded by Canal and Mulberry streets and the Bowery.

The Tongs, business or fraternal organizations, came into being basically to fill the emotional needs of lonely Chinese. And restaurants serving simple "home-style" dishes such as salted eggs, cabbage, and plain rice hastily opened to feed the growing influx of expatriots, most of them bachelors. It was during the 1890s that newsmen from Park Row began frequenting the Chinese eating houses. Their more trusting readers followed to sample "chop suey" (invented on the spur of the moment by a non-English-speaking Cantonese cook in the Far West and meaning "chopped finely" or "hash") or to catch a performance at the dark, pagoda-like Chinese opera house recently built at 12 Pell Street. This building was bought in 1904 by "Nigger Mike Saulter," who converted it into a Bowery-style dance hall and solicited the trade of the quarter's most awesome toughs. But the crowd at Nigger

Mike's didn't stint when it came to "throw money" for the entertainers. Their favorite performer was a singing waiter named Izzy Balin who one night chased an ambitious young hoofer out of the place and into the street with a few smart cracks of a broom handle. The hoofer, who later became known to audiences as George White of the *Scandals* fame, had had the audacity to break into someone else's territory and the singing waiter, who later became known as Irving Berlin, was having none of it.

The Tongs had come to be credited with both the opium and the white slave trades in Chinatown, however, and the area was "cleaned up" in 1910 following the discovery of the dismembered body of Elsie Sigel in a trunk on Doyers Street. By then Jimmy Kelly owned the Mandarin, which stood across Doyers from the Chatham Club; Callahan's, O'Rourke's, and "Big Aggie" Stanton's 7 and 8 Club stood nearby. "Big Aggie," a celebrated "blues-shouter," repeatedly turned down offers from Broadway producers and uptown cabaret owners so that she could stay in her beloved Chinatown.

The Chinese Delmonico's at 24 Pell Street offered relative peace and quiet in mother-of-pearl, rice-paper, and red silk-fringed surroundings. Others, such as The Pekin, The Port Arthur, and Lum Fong's, decked themselves out in dragon tapestries, gongs, colored lanterns, and incense burners. The Chinese Delmonico's clientele was mixed, and its menu indicated a growing interest in dishes of greater authenticity than the ubiquitous chop suey. With Prohibition and the emergence of the rococo Oriental jazz palace, New Yorkers began to order egg rolls, sweet and sour pork, spare ribs, roast duck, lobster Cantonese, and various other adventurous items. The family plan was initiated with its "two from Group A" and "three from Group B," etc. This inspiration made it possible for a family, large or small, to dine out interestingly, economically, and well.

In the 1880s and 90s many thousands of Jews, fleeing the pogroms of central and eastern Europe, had taken ship for America. They came from Russia, Poland, Austria, Hungary, and Roumania, but they spoke a common language, Yiddish, and settled near each other on the Lower East Side between the Bowery and the river, and later as far north as Fourteenth Street, east of Third Avenue. The immigration was attended by overcrowding, poverty, and pushcarts, and the newly arrived worked exhausting hours, often in the notorious sweatshops of the garment industry. But from the point of view of daily diet the picture was far from depressing. Their "Jewish style" of cooking was that generally common to eastern Europe, and was built on a solid foundation of borscht, sour cream, boiled beef, chicken soups, gefiltefish, herring, horseradish sauce, buck wheat groats, poppy seed pastries, and strudel. Jewish housewives who were known as good cooks often converted their front parlors into diminutive dining rooms where they were able to cater to friends from their own part of the world who lacked such satisfactory domestic arrangements.

At Bohemian cafés along the avenues, Yiddish-speaking artists, actors, musicians, and literary figures dined and drank coffee in the open air rather as if they were in Vienna or Budapest. From this vantage point, the free thinker in floppy hat and flowing cravat could most easily fall under the disapproving glance of his more religious compatriot in beard and black coat. By the eve of World War I, the Yiddish theater was regularly visited by uptown audiences even though they couldn't understand a single word, and probably missed the hastily improvised dialogue when Joseph Adler and his wife pursued a domestic battle between the lines of *The Merchant of Venice*. The dramatic peformances, private as well as public, would afterward be dissected by the cognoscenti over chopped goose liver and stuffed peppers at the Royal Café, the Charlie Pfaff's of the Lower East Side.

Emil's on Essex Street was the home of political argument, and specialized in Roumanian food (the Royal Café was basically Hungarian), as did Haimowitz's, which encouraged bombastic pinochle after dinner. Undoubtably the most famous restaurant was Moscowitz and Lupowitz, where the food was unsurpassed and the atmosphere harmonious. Mr. Moscowitz himself had spent his childhood playing gypsy tunes on a dulcimer aboard a little pleasure craft that plied the Danube.

All these Lower East Side restaurants were kosher, and as dairy products could not accompany meals at which meat was eaten, or indeed be consumed with the same dishes and cutlery, specifically "dairy" restaurants came into being. (At least one restaurant guide of the period, obviously ignorant of this fact, mistakenly identified the "dairy" restaurants as vegetarian retreats founded by disciples of Count Tolstoy.) At Rapoport's dairy restaurant the choice of first courses is fairly typical to this day of Lower East Side cooking. Among the entrees listed were chopped salmon salad, chopped spinach and eggs, chopped sturgeon salad, pickled herring, schmaltz herring, and chopped eggplant. The soups included cold borscht with cream, potato soup, cold fruit soup, brown cabbage soup, mushroom barley soup, and cold chav (a deliciously sour sorrel soup) with cream. It is when one arrives at the unexpected "Choice of Roast" that impressive ingenuity is shown: These include vegetable cutlet and mushroom roast, as well as the usual blinzes, kashe (buckwheat groats), and pirogen (a stuffed puff pastry). Two New World specialties, spaghetti and chow mein, were borrowed from Rapoport's Italian and Chinese neighbors.

Other central and eastern European groups clung to the same neighborhood. Many Germans still lived there, although a large number had moved up to Yorkville. On Houston Street stood Little Hungary, not actually an area, but a restaurant frequently patronized by Theodore Roosevelt when he was Police Commissioner. To the management's increasing jubilation, he returned as New York's Governor, and again as the nation's President. The Little Hungary was closed by Prohibition, but by then its owner, Max Schwartz, had made enough to retire—to a castle in Hungary.

The Revolution of 1917 sent a whole new flock of Russian immigrants to the Lower East Side. They were neither all Jewish, nor in sympathy with the freethinkers of the Royal Café. But they settled down not too uncomfortably with the rest, and opened their own restaurants featuring rye pancakes with sour cream, caviar, and gypsy music, which was not necessarily ersatz. The gypsies had somehow crossed the ocean too and pitched camp behind bright red curtains in empty stores on Houston Street.

West of the ghetto and south of Greenwich Village lay Little Italy, still milling with new arrivals as the old, or their descendants, moved on. The supposed myth of Little Italy's armies of organ grinders has yet to be exploded. Organ companies charged extortionate rates to rent their instruments, knowing full well that few opportunities presented themselves to a man who spoke no English. Other Italian street merchants sold sausages, peppers, and zeppele (deep fried pastries dusted with sugar), and their clam stands lurked in every other doorway. Most of the residents of Little Italy came from Naples and points south, which is why Italian cuisine is frequently and erroneously associated with the immigrants' southern predilection for tomatoes, garlic, olive oil, and farinaceous dishes, which were as cheap in America as they were in the old country.

But the most distinctive culinary aspect of the twenties and early thirties was an outbreak of small foreign restaurants all over Anglo-Saxon New York. The "talkies" helped build a romantic legend about the Darling-Hole-in-the-Wall presided over by a heavily accented, lugubriously sentimental European, and where, in the final reel, the small-town heroine was expected to clutch the picturesque checked tablecloth as Jim or Rod or Bill, a newspaper man or the boss's son, would lean perilously over a candle-stoppered Chianti bottle to Pop the Question. Moreover, Prohibition, with its lackluster speakeasy cuisine, had led many diners to investigate small foreign restaurants; and as their prices remained within reason, the popularity of these little restaurants was cemented permanently by the Depression.

Although Monetas persisted on Mulberry Street, Del Pezzo's moved up to the neighborhood of the Metropolitan Opera House where several other Italian restaurants, among them Barbetta's and Guffante's, dispensed pasta to Verdi and Puccini lovers as well as to Enrico Caruso, who appears to have eaten in more Italian restaurants than Washington slept in colonial beds.

German cuisine had its musical following, too; Fritz Kreisler favored the Blue Ribbon in the Broadway district, a Teutonic combination of pewter-lidded steins, checked tablecloths, murky vistas peopled by Wagnerian heroes and Rhine maidens, and *Pommersche Gansebrust*. The exodus to Yorkville brought the beer hall to East Eighty-sixth Street. At Otto Boege's Zum Drei Maederl Haus, German nursery rhymes were sung under the disciplined direction of the bandleader as a huge scroll imprinted with the lyrics unfurled behind him, and at the same time the diner might acquaint himself with sauerbraten, schnitzel, cheese cake, or Viennese pastry. (Directly across the street stood the offices of the Bartenders' Union. Regardless of Prohibition, the Union was pleased to report that it kept active the year around recruiting able craftsmen for any employer who could provide the raw materials. The old breweries on East Eighty-ninth Street kept open by making 1.5 per cent near beer, a pallid and insipid imitation of lager that stayed within legal limits.)

But there was no question that the small French restaurant had captured New York's doting attention more than any other. Although French haute cuisine had been familiar for almost a century, these small and unpretentious restaurants, with their limited menus, were a far cry from Delmonico's. Probably the Divan Parisienne on Forty-fifth Street was best known. Organized by members of the defunct Delmonico's staff, it claimed to possess a machine for the extraction of acid from strawberries, and contributed "Chicken Divan" to the national menu (chicken slices on broccoli, the whole luxuriating in *sauce Mornay*, sprinkled with Parmesan cheese and glazed under the grill). Also much touted was Mme. Giraud's, guarded by a stuffed bear, on Fulton Street in the financial district. There the cooking was supervised by Monsieur Giraud, sometime chef to Queen Victoria herself, and whose specialty was frogs' legs. In the sidestreets a horde of Henri's and Pierre's and Robert's could be observed sweeping through the city's streets like expeditionary forces.

Further waves of foreign invaders staked out conquered territories in all parts of Manhattan. Turkey and Armenia commandeered the east side area from Twenty-sixth to Thirty-fourth streets, and before anybody knew it, lamb was being cooked in forty different ways on Fourth Avenue. The Greeks held Ninth Avenue. Little Spain occupied Cherry Street, although the American Hotel on Fourteenth Street reportedly served the best Spanish food in New York. An Egyptian Garden bloomed on Washington Street, a Swedish Ratskeller entrenched itself on East Fifty-second, and the steamed fish with rice at the Miyako Japanese restaurant on Fifty-eighth soon became as famous as the curries of the Ceylon-India Inn. The vogue for foreign food was such that an authentic Mexican restaurant was conducted by two spinsters from New England who didn't know a word of Spanish between them.

The Russian formula of candlelight, chintz, balalaikas, and primary-colored walls went over particularly well uptown, where society was only just recuperating from the thrills of Diaghelev. The most popular, the Russian Bear, with both downtown and uptown quarters, claimed to have served Leon Trotsky during his brief stay in New York. Trotsky would have called himself Leon Bronstein at the time. It was not until reaching Halifax that he realized a pseudonym would be expeditious, and seized upon the very first name to come to mind—which happened to be that of Herman Trotsky, the manager of the Broadway Central Hotel he had just left. We have no record of what he thought of the fare of the Russian Bear, but his brother, who called himself Trotsky out of fraternal pride, later opened a kosher restaurant on West Thirty-fifty Street, highly praised for its chicken and noodle soup.

It is not to be thought that the influx of flappers and their "dates" drove the Bohemians from their habitual haunts. Far from it. But by the twenties, Bohemia had evolved a "local habitation and a name"—Greenwich Village. The area around Washington Square, site of New York University, enjoyed a long guest list of literary residents; Henry James, Brander Matthews, and Mark Twain each had lived there in their time. Walt Whitman and his crew at Pfaff's had been drawn to Union Square and never tempted further uptown. They were charmed by the small, winding streets, picturesque old houses, and low rents of the section taking its meandering form from the colonial village of "Greenwich," so rudely engulfed by the city.

Toward the turn of the century, scrimping journalists and hopeful artists who earned precarious livings etching pictures of the latest catastrophies for the New York papers, were joined by liberals from the Lower East Side, recognized writers such as John Masefield and Willa Cather, and others, like Stephen Crane and Alan Seeger, who, for one reason or another, found themselves arbitrarily lumped together under the genus "Bohemian." The ranks of this category had swollen greatly since the days of Whitman and Adah Isaacs Menken. The new Bohemians were united by a fierce desire for personal freedom and political rights for all, although critics interpreted these goals as simply meaning Free Love and Socialism.

According to Emily Hahn, who traces the course of Bohemianism in America in *Romantic Rebels*, the heart of the Village before World War I could be located at The Liberal Club and Polly's Restaurant, which occupied one and the same structure, the club operating upstairs and the restaurant below. Talk was the Bohemians' common addiction, an opiate recklessly enjoyed in both places. The club staged exhibitions of cubist art and speeches by Margaret Sanger, as well as dances called "Pagan Routs." Prior to her catering venture, Polly, born Paula Holladay in Evanston, Illinois, had taken up anarchism, and acquired one of the many ex-lovers of Emma Goldman, the celebrated Communist coquette. Her paramour, Hippolyte Havel, expended his energies equally as anarchist and Polly's chef, and has been described as swarthy, broad-browed, and obscene of speech.

Village distractions at this time consisted of reading the *Masses* (to which Max Eastman, John Sloan, and Floyd Dell all contributed), or listening to Bobby Edwards accompany himself on a cigar box ukulele at his Crazy Cat Club:

Way down South in Greenwich Village
Where the spinsters come to thrillage,
Where the ladies of the Square
All wear smocks and bob their hair.

The club itself, called the Circolo Gatti Matti degli Stati Uniti, was actually a Tuesday-night distraction at Enrico and Paglieri's Italian restaurant. Then there was Mori's for more Italian food, or the Pepper Pot, or Romany Marie's. Emma Goldman's one-time aid, Romany Marie, with her red bandanna wound rakishly around her head, became so much a Village fixture that she trouped off yearly with the Bohemians for their summer holidays at Provincetown. Her restaurant's trademarks were once described as being "batik and paper napkins," but the Roumanian cooking was good.

The Black Cat, first opened by a Signor Mazzini in 1890, came closest to the public's notion of a "typical Greenwich Village dive" with its white-washed brick walls adorned by stenciled black cats, raftered ceiling, permanent aura of smoke, and low prices. It is hard to say which is a better summation of the wicked innocence of the Village before and during the Great War, the entertainment at the Black Cat or Anna Alice Chapin's description of it in her *Greenwich Village* (1917):

There is a pianist at the Black Cat—a real pianist, not just a person who plays the piano. She is a striking figure in a quaint, tunic-like dress, greying hair and a keen face, and a personal friend of half the frequenters. She has an uncanny instinct for the psychology of the moment. She knows just when "Columbia" will be the proper thing to play, and when the crowd demands the newest rag-time, she will feel an atmospheric change as unswervingly as any barometer, and switch in a minute from "Goodbye, girls, Goodbye" to the love duet from Faust. She can play Chopin just as well as she can play Sousa, and she will tactfully strike up "It's Always Fair Weather" when she sees a crowd of young fellows sit down at a table; "There'll be a Hot Time in the Old Town Tonight," to welcome a lad in khaki; and the very latest fox trot for the party of girls and young men from uptown, who look as though they were dying to dance. She plays the "Marseillaise" for Frenchmen, and "Dixie" for visiting Southerners, and "Mississippi." And, then, at the right moment, her skilled fingers will drift suddenly into something different, some exquisite, inspired melody—the soul child of some high immortal—and under the spell the noisy crowd grows still for a moment. For even at the Black Cat they have not forgotten how to dream.

When the Villagers yearned for clear air, clean linen, or just a square meal, there was always hope of being invited to the salon of Mabel Dodge at 23 Fifth Avenue. Mabel had come to town from Buffalo, via Florence, and by 1913 established herself as the patroness of the advance guard. She was a friend of Gertrude Stein, she kept about her Carl Van Vechten and Jo Davidson, and she was involved in the Armory Exhibition. In her apartment, a vacuum of white in the school of Elsie de Wolf, she gathered together, in the words of Miss Hahn, "Socialists, trade-unionists, anarchist, suffragists, poets, lawyers, murderers, old friends, relations, psychoanalysts, IWW's, single taxers, newspapermen, artists, club-women, clergymen, and a few out-and-out nuts." It was Mabel Dodge who, during one anarchist scare, invited Emma Goldman to address the guests at her Saturday night salon on the subject.

Mabel soon acquired a virile young lover, John Reed, a wealthy boy down from Harvard with a thirst for Village life and a desire to embrace the cause of the workingman. Together they staged what must be called the most elaborate political extravaganza of all time. It was a pageant to benefit the silk weavers of Paterson, New Jersey, then striking for an eight-hour day. Reed and Mabel Dodge rented Madison Square Garden, and staged the pageant with a cast of two thousand strikers and their families for an audience of fifteen thousand people. The sets were colossal, with the aisles of the Garden decorated to represent the streets of Paterson. When Mabel discovered that for all her effort the pageant had yielded a deficit, she simply gave up. With her lover on her arm she sailed for her villa in Italy.

It was a blow for the socialist movement in the Village, and another was yet to come. The *Masses,* apart from its Marxist credo, also had pacifist leanings, and at the outbreak of World War I its editors were officially charged with conspiracy to obstruct recruiting. But the trial was turned into a farce when the sheet's cartoonist, the good-natured Art Young, was overheard snoring in the defendants' box. There followed an acquittal. Still and all the city was in no mood for an anti-war magazine, and the *Masses* folded.

The Villagers did not see its likes again, nor did it matter. By 1920 the Village resounded with the voice of Edna St. Vincent Millay and a chorus of thousands. Young people with talent and without it, with money and more often without it, herded to the Village in search of excitement, fulfillment, spaghetti, or love.

It was an easy situation to exploit while the thirst for "romance" endured. The same slumming parties that went to pay their respects to the black singer uptown or the "heathen Chinee" downtown came to survey the starving poets and have a good dinner. The ubiquitous jazz bands moved in. Restaurants became "villagey." Electric lights were turned off, candles were lit, and everyone donned his beret. Decorative attempts at the artistically quaint proved more Coney Island than Montmartre. The Country Fair invited diners to eat in grandstand boxes, dance inside a picket fence, and race each other in kiddy cars. At the Pirates' Den (an inspiration of the same entrepreneur, one Don Dickerman) guests were greeted by a doorman in kneeboots with a parrot on his shoulder. The interior was littered with encrusted chains, cutlasses, muskets and maps, the waiters and musicians were dressed as buccaneers, and the specialty of the house was Black Skull Punch. As if for deliberate color, an insane Baroness haunted the streets with her shaven head painted a bright red.

With Prohibition, New Yorkers, about to break with the rules of society for the first time in their lives, turned very naturally to the Village, where in their opinion all the rules had already been broken. They were perfectly correct. Barney Gallant, a favorite Village restaurateur and bar owner, thought the Volstead Act so ridiculous he simply ignored it. Barney, a Latvian immigrant, onetime roommate of Eugene O'Neill, and member of the Liberal Club, enjoyed his moment of public protest when he went briefly to jail as a bootlegger. From then on, he served liquor under the counter, and did so well that he eventually employed a band and cabaret acts to attract the white-tie bracket. His was just about the only place in the Village to do so.

When the Villagers themselves had a few dollars in their pockets, they looked, as ever, through Stanford White's Washington Arch to the grandeur of lower Fifth Avenue. There the grass grew greenest under the doormats of the Brevoort and Lafayette hotels.

Exploring the Village during the twenties could be infinitely more rewarding than any similar expedition today, particularly if one went to either of these two beguiling landmarks, which then stood back to back—the Brevoort at Fifth Avenue and Eighth Street, and the Lafayette on the corner of Ninth and University Place. Both were operated by Raymond Orteig, a man of many distinctions.

The Brevoort's history extends back into the middle of the last century, when it was owned by a Bostonian, a Mr. Albert Clark. Clark was so devoted to undemocratic principles that when Lincoln was elected, he had a time capsule containing a Brevoort menu buried in a brick wall behind his office so that future generations would know that such refinement had once existed.

The Lafayette, a small, white, and cosy pension, was bought in 1883 by Jean Baptiste Martin of the Café Martin. His first venture in the Western Hemisphere was as an innkeeper in Panama where diggings had been initiated by the French. For a while he called his New York house the Panama Hotel, but the name soon became locally associated with fever and Spaniards. In 1886 it was changed to Hotel Martin, thriving as a French retreat under that name until 1899, when the owner moved to Fifth Avenue, and the premises were leased to the dining room's headwaiter, Raymond Orteig, who renamed it the Lafayette.

Throughout his career, Orteig maintained the deceptively simple continental style first projected by Martin. His food was impeccable; the latest Parisian dishes appeared on his menu and were gleaned weekly from the pages of the *Brasserie Universelle* as soon as it arrived from Paris. The corner café with its plain tile floor, racks of foreign newspapers, and bare little tables and chairs, which were set outside in fair weather, remained a morning to midnight clubhouse for the French colony and more affluent Bohemian circles for decades. Checkers and chess, cards and dominoes usually occupied its regulars as they sipped Amer Picon and Grenadine or brandy.

With the continued success of the Lafayette, Orteig took over the Brevoort and made the joint kitchens of the two hotels among the most celebrated in the city. *Escargots* and frogs' legs *provençal* first became popular there. Omelets, onion soup, pastry, and wines were exceptionally fine at the Brevoort as was the delicious breakfast of iced melon, shirred eggs in blistering hot ramekins, fresh croissants, and coffee. The Lafayette especially prided itself on its *hors d'oeuvres variés*, squab *en casserole*, filet mignon *à la Bearnaise*, and pears flamed in a secret blend of liqueurs.

Greenwich Villagers loved both hotel dining rooms, but rarely could afford a complete dinner. The Brevoort's two-roomed basement café, which stayed pleasantly cool even in the dog days, was more within their means. No one made an appearance until 10:30 at night, but by midnight the café would be jammed with Bohemians and their chain-smoking girls wearing wild "pre-Raphaelite" and Dutch boy hair-dos, "artistic" jewelry of bone, wood, or simulated jade, and bilious chartreuse, henna, puce, or magenta smocks. After gorging themselves upstairs, sightseers always made it a point to drop in for an instructive taste of the aesthetic life.

Orteig's passion for aviation came second only to his devotion to food and his native France. In 1919, he posted a Raymond Orteig Prize of twenty-five thousand dollars to be awarded to the first pilot who could successfully nagivate the Atlantic. The prize, which he hoped would foment activity in French aeronautics, was to be honored for five years. Nineteen twenty-four passed and still no adventurer had flown either tri-color over the ocean. The prize was reoffered in 1925 and the rest is history, although not a particularly gracious one in all details. Orteig was in Paris when Lindbergh victoriously flew into Orly, and most naturally was invited by the government to take part in the official celebration there. The spoils were not to be handed over, however, until all parties returned to New York. Whether this delay triggered the intangible hostilities sensed at the presentation ceremony in New York or not is a moot point; suffice it to say that no great warmth of affection was felt in the public rooms of the Brevoort on that momentous afternoon. Afternoon, not evening, for Lindbergh canceled the celebration dinner which Orteig had dreamed of for years at the last moment, the hero's representative pleading lack of time on the agenda. A reception-tea on June 16, 1927 was made to do, and even that was graced but briefly by the guest of honor. The trustees and advisory committee for Orteig's prize, among whom was Orville Wright, forgathered at the appointed hour, as did a mob of dignitaries and gentlemen of the press, to hear a presentation speech of unrelieved effusiveness to which the recipient tersely replied "I have often been asked what it was that first directed my attention to the flight from New York to Paris. I believe Mr. Orteig's offer directed the attention of most aviators to that flight. The offer was nothing more or less than a challenge to aviators and engineers in aeronautics to see whether they could build and fly a plane from New York to Paris. I do not believe that any such challenge, within reason, will ever go unanswered." The chairman then turned to the other guests and, visibly upset, announced, "It is a source of great regret to us that Colonel Lindbergh is obliged to leave almost immediately to fulfill another engagement. You are all invited to participate in the tea and refreshments now to be served." Money in hand, Lindbergh left.

Lindbergh had been directed by the White House to put in an official appearance in Washington before submitting to the hoopla of New York's distinctive welcoming ceremony, the ticker-tape parade. Under the ebullient administration of spiffy, knife-pleated Mayor James. J. Walker the parades had turned into an issue of fierce local pride on the one hand, and diplomatic concern on the other.

A year prior to the historic flight of *The Spirit of St. Louis* an Amsterdam Avenue sausage-maker, Henry "Pop" Ederle, drove exultingly down to the bay with forty relatives in tow to welcome home his sturdy Nereid daughter, Gertrude, from the icy currents of the English Channel. She had traversed it in less time than had any of the five men before her, and her home town was in a frenzy of excitement over their "Trudy's" accomplishment. Grover Whelan, New York's official greeter, sometimes known as "The Black Hussar," sailed out to meet her along with a convoy of hysterically whistling craft. Once ashore, she had to be protected from an estimated fifteen thousand admirers, each intent on wresting a personal memento from his idol's toilette. The parade itself put all previous ones in the shade, as ticker tape, telephone books, and scrap paper rained heavily down on Miss Ederle. Only Lindbergh's accolade was to surpass its density of flying wood pulp— seventy-five thousand pounds of it. Gertrude, rather than feeling slighted by the fickleness of the crowd that hailed Lindbergh, modestly declared her feat not comparable to his, as it had been merely a sporting proposition while his effected a "help to humanity through the advancement of commercial aviation."

In 1926, Queen Marie of Roumania was more troubled by the explosion of over-filled flash guns than by hurtling telephone directories; in 1928, the Governor of Rome, Prince Potenziani, presented Jimmy Walker with a statue of Romulus and Remus breakfasting lupinely and received a stimulating tome entitled *Abroad with Mayor Walker* for his pains; but in 1931 the Italian Foreign Minister, Dino Grandi, was rerouted to Washington for his first reception as a federal sign of disapproval of New York.

Meanwhile, New York was treated to another form of commemorative parade, namely the spectacular gangster's funeral. Dominick Scocozza, the east coast's number one undertaker to the underworld, usually acted as general overseer and was considered by those who knew to be unexcelled in the arts of flower-arranging and post-mortem cosmetics; nobody could camouflage bullet holes

with such deft expertise as he, and when it came to fifteen-foot
Heavenly Gates constructed of roses, lilies, and carnations, he was
the master. The glorious standards for gangland funerals had been
earlier set at the lavish interment of a Chicago hood by the name
of Dion O'Bannion. New York did not see such a full-scale spectacle until the sudden demise in July 1927 of Frankie Uale; his last
rites introduced three basics of necrophilic elegance—the silver or
bronze casket, the endless motorcade comprising at least thirty-
five cars to carry floral tributes alone, and the prominent display
of a mammoth clock or pocket watch made out of violets, roses,
lilacs, and other sentimental posies, the hands indicating the precise moment of a victim's hastened departure. Nineteen thirty-
one proved an outstandingly good year for Scocozza, and apprentices were kept busy around the herbacious clock recording the
correct hour at which "Joe the Baker" Catania, "Joe the Boss"
Masseria, or Daniel J. Iamascia went to their rewards.

Titanic arrangements of laurel and dogwood were used by
Frank Case to herald spring at the Algonquin Hotel rather than
the abrupt leave-taking of a dear lamented thug. He and his wife
had great boughs of them carted into the lobby each year as a
morale booster for the Algonquin's legendary clientele, city-white
and winter-drawn after a season in town. This was just one of the
innumerable extra touches that made Case a phenomenon of devotion to his pampered patrons.

The Algonquin opened on Forty-fourth Street between Fifth
and Sixth avenues on November 22, 1902, and soon attracted a
certain number of theatrical personages, among them the polo-
playing, prodigiously profiled John Drew. His catatonic reserve
and stagey manners were such that one critic claimed that if Drew
were to essay the role of Simon Legree, the character would be
brought off as a misunderstood aristocrat of unsullied breeding
and noble purpose. The New York *Herald*, in a more charitable
frame of mind, came limply to his defense, insisting that while he
"never smiles except when he is acting, and he is as serious as a
new policeman and quite as uncommunicative as his own valet
. . . he is a gentleman, but he is taciturn. That is the John Drew
of East Hampton and the Hotel Algonquin."

Not until 1919, however, was the Algonquin permanently put
on the map by a tubby, beaky, bespectacled newspaper reporter
named Alexander Woollcott. He had been assigned to write an
article for a Sunday edition about the Algonquin, which he passed

over as "a little, unpretentious hotel, tucked away on a side street," sporadically frequented by Evangeline Booth, Raymond Hitchcock, and William T. Tilden II. But he loved the apple pie, and that, more than any other contributing factor, launched the hotel's international reputation. Woollcott made a habit of lunching with two colorful colleagues, Heywood Broun and Franklin Pierce Adams ("F.P.A.") every Saturday, and suggested that they meet at the Algonquin. Harold Ross, Robert Benchley, and Brock Pemberton began to join them, forming the nucleus of "The Round Table," "The Vicious Circle," "The Sophisticates," or the less endearing soubriquets suggested from time to time by such square outsiders as O. O. McIntyre and George M. Cohan (but then, as George S. Kaufman was to point out at a future gathering of the clan, "one man's Mede is another man's Persian").

The hotel's manager, Frank Case, gave the unholy alliance a dining corner of its own, and to their ranks they systematically gathered a bunch of crackling dry wits and raconteurs, self-styled prophets and glittering sybils, good eggs and sour apples. The all-star cast reads like the rousingest party list every compiled in the twenties and infant thirties: Marc Connelly, Russel Crouse, Donald Ogden Stuart, Robert E. Sherwood, Dorothy Parker, Kaufman and his wife Beatrice, Ring Lardner, Peggy Wood, Irving Berlin, Edna Ferber, Neysa McMein, Raoul Fleischmann, Alice Duer Miller, Deems Taylor, Tallulah Bankhead, Ruth Gordon, the Gillmore sisters, and more and effervescent more. Fringe groups occasionally blossomed and bore indistinct benefits; Dorothy Parker and Irvin Cobb organized a Chowder and Marching Club, its one cardinal rule being "that there be no chowder eating, or marching, otherwise the sky's the limit."

The Round Table as a whole was indifferent to drink; if anything, poker was its major vice, next to scathing literary debates. Case provided the upstairs quarters, and every Saturday night after dinner the Thanatopsis Literary and Inside Straight Club sat down with a vengeance to sweat it out until late Sunday afternoon. One hundred dollars was about as astronomic a sum as any of the Fourth Estate could afford to lose, so the atmosphere grew more than electric with the arrival of bigger-time bettors like Jerome Kern, Herbert Bayard Swope, Prince Antoine Bibesco, and Harpo Marx. Woollcott frequently whimpered that his doctor warned him of damaging his nerves by losing money, a diagnosis which somehow did not restrain him from twenty-four-hour stretches at the gaming table. Frank Case patiently put up with the club's weekly *valpurgisnacht,* and when the gamblers started sending out around the corner for food, went no further than to tack up a sign in their rent-free suite proclaiming BASKET PARTIES WELCOME.

The publicity brought to the Algonquin was both invaluable and unique. Not only did the most agile minds of the twenties and thirties busy themselves making witty remarks, but the selfsame group, many of whom were columnists, including Franklin Pierce Adams (F.P.A.—who wrote simultaneously for the *Tribune, World,* and *Post*), Heywood Broun, Dorothy Parker (who, along with Robert Benchley and Robert Sherwood, wrote for *Vanity Fair*), and Alexander Woollcott—reported these remarks to the waiting world. The Algonquinites' specialty was the pun, like F.P.A.'s "She suffered from fallen archness," or "If you take care of your peonies, the dahlias will look after themselves," and the quip, "Don't look now, Tallulah, but your show is slipping" (Heywood Broun), or "Tell us your phobias, and we will tell you what you are afraid of " (Robert Benchley). The rapier insult was another specialty: "This must be a gift book, which is to say, a book you wouldn't take on any other terms" (Dorothy Parker), or "He should be gently but firmly shot at sunrise" (Alexander Woollcott). Still, many of the Algonquin witticisms will probably rank as twentieth-century proverbs: "Nothing is more responsible for the good old days than a bad memory" (F.P.A.); "A liberal is a man who leaves the room when the fight starts" (Heywood Broun); "Being an old maid is like death by drowning, a really delightful sensation when you stop struggling" (Edna Ferber). In any case, it was all great repartee, which, according to Heywood Broun, was "what you wished you'd said."

Business at the Algonquin boomed. Case bought the entire hotel outright in 1927. A native of Buffalo, he had worked there for years, first as clerk and eventually as lessee. He honestly loved his patrons and showed it by prying from them their preferences and dislikes. His enthusiasm for food was one of an ambitious participant and, along with Rex Stout, Wilfred Funk, and Walter Slezak, he religiously attended meetings of the Society of Amateur Chefs. Parsnips, he discovered, took the prize as celebrities' pet aversion. The spectacle of a fish on a platter with head and tail intact distressed Mary Pickford. Lillian Gish liked Algonquin apple pie, garlic salad, ice cream, champagne, and oysters. Eddie Dowling preferred caviar, lobsters, corned beef, and a pot of oolong with two bags in it. And Frank "Bring-'Em-Back-Alive" Buck inevitably ordered Indian curry. Joan Crawford, on the debit side, disclosed to Case that, while she adored eating, she could not abide coy women or drunkenness; Clifton Webb felt the same way about oysters out of season and "Mr. Lindbergh's speeches."

Ben Hecht's great addiction, target practice, gave Case only a moment's pause. Armed with an air rifle, Hecht liked to lounge around his suite taking pot-shots at a paper bull's eye propped over the dressing table mirror. Case claimed that Hecht favored minute steak with garlic sauce because he smoked so much he needed the garlic to tell where the meat was. The recipe for this sauce, which involves nothing more than adding a mite of minced garlic to the juices of a small steak pan-broiled in butter, was revealed to the public in *Feeding the Lions,* a cookbook composed by Case and published in 1942, the day before rationing went into effect. More interestingly, the formula for the fabled apple pie was disclosed, along with a few general pointers on the making of a perfect pastry crust.

The Algonquin's Famous Apple Pie

8 medium apples, peeled and sliced thin	1 teaspoon cinnamon
2/3 cup granulated sugar	1/8 teaspoon nutmeg
grated rind of 1 lemon	2 tablespoons butter

Pastry Crust

2 cups flour	1 teaspoon salt
1/2 cup butter	3 teaspoons sugar
1/2 cup shortening	1/2 cup cold water

Mix flour, salt and sugar together in bowl, adding small pieces of butter and shortening until flour is all absorbed. Add the water slowly and mix, working it as little as possible. Put in refrigerator for half an hour to chill. Then roll out half the dough and put in a 9-inch pie plate, leaving about half an inch of pastry around edge. Fill this with the prepared apples and grate on them the lemon peel.

Mix sugar, cinnamon and nutmeg and sprinkle over the apples. Put small lumps of butter here and there.

Roll out balance of dough and perforate with fork in several places. Lay on top of apple filling and press around edge of pie with fork, to hold under and upper crusts together. Then trim off around edge of tin with scissors. Dot top of crust with small bits of butter, and sprinkle just a dash of sugar and flour on top crust, to get an attractive uneven brown surface. Bake in hot oven (450° F.) 15 minutes, then reduce heat to moderate (350° F.) and finish baking for 30 minutes.

Serve warm or cold.

But by the 1930s not everything was apple pie. Most literati were forced to forgo the pleasures of the Algonquin's table and many went to Hollywood; in Greenwich Village the artists were starving in earnest. Manuel Muino, in later years an owner of El Quixote, the Spanish restaurant in the hotel Chelsea, liked to recall his youth when he worked as a busboy in the restaurant of the New York Stock Exchange. One day in October of 1929 a small, kindly-looking man entered the dining room. He was immaculately dressed in spats, pin stripes, morning coat, dove-colored vest, pince-nez, and bowler hat. Carefully drawing off his gloves, he ordered a modest meal and ate quietly. He lingered over his coffee and pie. When the last crumb was finished, the little man quietly took a dainty pearl-handled revolver out of his pocket, aimed it at his head, and blew his brains out. "That," concluded Manuel, "was the Depression."

It was above all, a *psychological* depression. The problem was not merely that stock values that had created overnight millionaires were now creating overnight paupers. Thousands on every level of society, from the candy store clerk and local hairdresser to the hopeful young broker, had invested their savings and gambled in the market. Now the crash was taken by many as an almost personal blow, for which they felt themselves vaguely responsible, a cataclysmic revenge of God on the reapers of easy profits. There seems to have been an overall embarrassment, a desire to retrench and retreat into moral austerity. The first sure sign was the retreat of the skirt—back across the knee to mid calf. The long dress, trailing the floor, was the "new" fashion for evening, and the thirties saw a spectacular return to white tie and tails for men, after a decade of the casual dinner jacket. On the other hand, stiff collars, vests, spats, and gaiters became a thing of the past. According to Cecil Beaton, fashions reflected the disjointed quality of life: "Women's fashions in the name of practicability comprised street suits of indeterminate shape and length, formal pajamas, tea gowns with horse halters around the neck, and the creations of . . . Schiaparelli. . . . She often put women in unfeminine clothes, going as far as to introduce bus-conductor outfits."[7]

But Americans did not so much deny the newly won social freedoms as lose interest in them. In the *New Yorker,* Robert Benchley announced: "Sex, as a theatrical property, is as tiresome as the Old Mortgage." The freedom to drink again was won almost without notice. In a time of national economic disaster, the very idea that Prohibition had for decades been a burning issue seemed ridiculous. And when drinkers could drink freely again, they may actually have drunk less.

It was one of history's nicer ironies that Americans were allowed to openly purchase liquor when many no longer had a nickel for a cup of coffee. Roosevelt brought in the New Deal and with it Repeal in 1933, when the total wages paid to laborers in New York were just under half what they had been in 1929 and over thirteen million unemployed languished throughout the country. In Chicago, police broke up a pitched battle in which fifty men fought over the contents of a restaurant garbage can. National events were fatal to the digestion: over a decade of wood alcohol was followed by another decade of soup kitchens.

Many of the smaller speakeasies collapsed between October of twenty-nine and the dimmed jubilance of December of thirty-three. Now their operators faced the gray dawn of legitimate business in the Depression. Those who were gangsters had several other tricks up their sleeves. Many simply remained bootleggers —untaxed, and therefore cheap, alcohol was still in demand. Those who were not gangsters opened restaurants, bars, and night clubs, and some succeeded surprisingly well. The reason for this was that "Café Society" had become, once and for all, "night club society." Hotels in large cities attempted their own night clubs, complete with floor shows. Jack and Charlie just unlocked their front door.

Many night clubs found themselves in fairly desperate straits. Some which depended on bootleg prices folded abruptly. There were few who could afford a three-hundred-dollar evening, and many of the perennial playboys with fortunes so huge they were unshaken by the Depression were actually dying off of old age. To bring in the crowds, night club operators fell back on their self-appointed, underpaid, and resourceful ally, the "press agent."

The press agent had come into existence in response to a crying need for information on the part of the Broadway columnist. And the Broadway columnist had come into existence in response to the need of the newspaper editor for a safe place to print the distinctly untrustworthy but edition-selling stories they were constantly handed about stars of the stage and screen who felt a splashy milk bath or a suicide attempt would put them on the front page.

It all started with Walter Winchell. Winchell began his career in journalism as a vaudeville trade paper's advertising salesman. From that post he moved on to a tabloid called the *Graphic*. For the *Graphic* Winchell wrote his first column, "Your Broadway and Mine," and the reaction was immediate. Soon other columnists surfaced in other papers, Mark Hellinger and Louis Sobol among them. After 1929 gossip was gobbled up by people who were obliged to live vicariously, and there was not always that much on which to live vicariously. Any mention of a night club or restaurant in a column brought in customers, and soon proprietors were hiring "press agents," armed with typewriters, their "items," and precious little else, who were kept scurrying back and forth between two implacable enemies—the proprietor who simply must be mentioned in the columns, and the columnist who simply did not have space or who did not trust the story at hand.

A press agent's imagination could frequently get the better of him. The personal histories of the dance team of Gomez and Weinberg provides a perfect clinical example. Jack Tirman, a struggling young agent, was faced with obtaining the necessary mention for his employer, the owner of a night club with no floor show and within the precincts of which no celebrity had recently bloodied the nose of another celebrity, or been seen with another celebrity's wife. Tirman hit on a scheme. He sent out a release to the effect that the dance team of Gomez and Weinberg were appearing at the night club in question. Gomez and Weinberg were taken up with alacrity by the columns. Tirman couldn't let a good thing go. Soon Gomez and Weinberg were being held over. Then Gomez was pregnant. Then fate struck a blow—the child was lost. This kind of trauma led to the usual complications. Gomez and Weinberg split up, and each began to be seen with various of the jaded mashers and concubines so frequently named in the columns. But virtue triumphed in the end, and they were reunited to appear at the aforementioned night club. Then, just when all was going smoothly for Gomez and Weinberg, a villain by the name of Dick Manson, then night club editor of the *New York Post*, decided that for figments of the imagination they had received entirely too much publicity. He finally devised a subtly

When asked to revivify the faded Rainbow Room, architect Hugh Hardy recognized a need to intensify sensations of light and color for new generations desensitized by color films and television. (Photos courtesy of the Rainbow Room.)

1934

1986

At its lowest ebb, the rear façade of the 42nd Street Public Library looked upon a drug polluted Bryant Park; Hugh Hardy's affixing to it of a new restaurant pavilion and umbrellaed café propelled the physical reclamation of the disreputable Time Square sprawl (*see opposite*). (Photos © Paul Warchol, courtesy of Hardy Holzman Pfeiffer Associates.)

The speakeasy heard around the world: Jack and Charlie's "21" Club presently owned by the Orient Express Company, is champing to enter the 21st Century with iron jockeys, bright as ever, sporting favored patrons' racing colors. (Photo courtesy of the Orient Express Company.)

In the 1980s and '90s, weddings took on the free-spending lavishness of the Gilded Age. The Plaza Hotel, a prime venue, fomented its own bestselling guide by Lawrence D. Harvey, executive director of catering. (Photos courtesy of The Plaza Hotel.)

In 1997, the repeatedly voiced belief that change is needed to revitalize a restaurant after twenty years led Sirio Maccioni, an amalgam of Henri Soulé, Oscar Tschirky, and Flo Ziegfeld, to resettle his fashionable Le Cirque in the baronial vastness of the Villard Mansion, where he posed, mid-move, with his sons and dynastic heirs apparent Mario, Marco, and Mauro, and executive chef Sottha Khunn. (Photo by Courtney Grant Winston.)

The Hoffman House era of ebulliently elegant bar life was translated into electric futurism by Le Cirque 2000 designer Adam Tihany in the 19th Century Villard Mansion, a landmarked appendage to the modern Palace Hotel; in respect to landmark commission laws, all of Tihany's decor and furnishings had to be freestanding. (Photo by Courtney Grant Winston.)

Restaurant kitchens of the 1990s often deliberately commandeered patrons' rapt attention, and cost small fortunes to install. At The Typhoon Brewery (*above*) diners could follow the flurried maneuvers of Thai-cuisine artists while sipping house-crafted beers.

At Le Cirque 2000 (*right*), magnificent "cooking suites," custom made in France, recalled an earlier age of Delmonican formal grandeur. (Photos by Courtney Grant Winston.)

Since 1906

BARBETTA
1906 ~ 1996
90th Anniversary

Opened in 1906 by Sebastiano Maioglio and now owned by his daughter Laura Maioglio, Barbetta is
the oldest restaurant in New York that is still owned by its founding family.

Barbetta's 90th Anniversary cocktail: Barbetta 1906 7.00

LUNCH

Specialties of Piemonte in Red

This menu features vintage Barbettta dishes from the restaurant's 90-year history.
Beside each dish is noted the year it was introduced at Barbetta.

~

FIRST COURSES
Fonduta Valdostana with Polenta (1962) 8.95   Roasted Fresh Peppers alla Bagna Cauda (1962) 7.00
Prosciutto imported from Parma with Persimmons or Melon 9.00
Our own home~smoked Atlantic Salmon (1992) 9.00
Minestrone Giardiniera (1906) 5.75   The Soup of the Day 5.75

SALADS
(For salad as a Main Course, please double the price)
Rollatine of Piemontese Robiola in Grilled Zucchini (1995) 8.00
Roasted Red Beets, with Lola Rosa (1993) 7.00
Field Salad Piemontese (1962) 8.00   Lingua Piccante (1906) 8.00
Fresh Mushrooms alla Schobert ~ *compositore di musica e piatti ultimi* (1962) 7.00
Mesclun Greens (1990) 6.00   Finocchio and Parmigiano Salad (1970) 8.50
Salad of Mâche, Roast Peppers, Smoked Salmon and Mushrooms (1992) 8.50

GAME
varying daily:
Charcoal~grilled Squab with cranberry beans, foie gras, and a red beet~olive oil sauce (1991) 26.00
Lepre (wild Hare) in Civet (1906) 26.00
Roast Pheasant in a white wine~juniper berries sauce with chestnuts (1989) 26.00
Rack of Venison with Hudson River Valley apple (1992) 26.00
Quail stuffed with a mushroom medley, served with Polenta   (1988) 25.00

MAIN COURSES
varying daily:
"Tajarin" with salsa di campagna (1906) 18.00   Agnolotti ~ made by hand ~ (1906) 19.00
Cannelloni alla Savoiarda (1962) 16.00   Lingue di Passero al Paté d'Olive (1975) 19.00
Linguine al Pesto (1914) 19.00
Risotto alla Piemontese with wild Porcini mushrooms (1906) 19.00
Risotto di Stagione 18.00   Risotto al Rosé Champagne (1973) 18.00
Risotto con le Seppioline, served only when the seppie are running very small (1971) 20.00
Risotto di Gamberetti, served only when tiny fresh Gamberi are available (1971) 20.00
~
The Fish Specialties of the Day 21.00
Charcoal~grilled Swordfish over a bed of lentils with a warm tomato vinaigrette (1991) 21.00
Carne Cruda, hand~chopped raw filet of Veal, alla Piemontese (1962) 22.00 Pollo al Babj (1962) 20.00
Roasted organic and herb~fed Rabbit with Savoy Cabbage (1925) 21.00
Bue al Barolo, beef braised in red wine with Polenta (1962) 20.00
Free~range organic Chicken, the day's preparation 18.00
Bistecca alla Fiorentina, aged Angus steak in the Florentine style for two persons, (1972) 26.00 per person
Veal Scaloppine al Prezzemolo (1906) 22.00   Calf's Liver Veneziana (1962) 21.00   Veal Kidneys Trifolati (1906) 18.00
Our Vegetarian Dish, a sampler of seasonal Vegetables cooked in different Italian regional styles (1996) 19.00

DESSERTS
(all 6.00)
Seckel Pears "Martin Sec" in red wine (1962)   Monte Bianco (1962)   Torta di Nocciole (1992)
Panna Cotta (1984)   Gianduja Chocolate Cake (1996)
Crème Brulée (1962)   Mousse of Bittersweet Chocolate (1962)   Zuppa Inglese (1906)
Baked Hudson River Valley Apples or Pears   St. Honoré (1962)   Almond Cake (1992)
Red Fruits Soup (1991)   Fruit Tarts (1980) of the Day
Apples, Pears, Raspberries, Strawberries, Grapes ~ Fresh Fruits in season

~

Decaffeinated Espresso (1962) 2.25   Vin Brulé (1962) 7.00   Caffé Espresso (1911) 2.25
Decaffeinated Coffee 2.00   Coffee 2.00   Teas 2.00   Decaffeinated Tea 2.50
Franck Marelica Croatian Tea (1996) 2.50
Coffees and Teas served with Piccola Pasticceria Piemontese

To the joy of amateur food historians, Barbetta, the oldest restaurant in New
York still owned by its founding family, has, since its 90th anniversary in 1996,
added the dates on its menu of the debuts of the house's signature dishes.

Reborn SoHo and TriBeCa owe much of their enduring artistic allure to talented chefs like David Waltuck and his wife Karen, pioneer downtown Food Revolution restaurateurs whose inspired seasonal menus such as this one dated January 19–February 14, 1998 continue to set a fresh culinary pace at Chanterelle. The powerful menu cover art (*above*) is by Elizabeth Murray.

January 17 ~ February 17
1998

Menu.

The soup of the Day
or

Cassolette of Prince Edward Island Mussels & Spinach

Salmon sauté with Cloves of Garlic & Verjus
or

Beef Fillet with Star Anise
or

The Entrée of the Day

An Assortment of Sherbets
or

Hazelnut Creme Brulée with Apple Hazelnut Strudel

Coffee or Tea

$35.00

Appetizers

Medium Plate $7.50

The Soup of the Day $7.50

Grilled Seafood Squares $13.50

An Assortment of Fish, Japanese Style $13.50

Cassolette of Prince Edward Island Mussels & Spinach $8.00

Entrées

Salmon sauté with Cloves of garlic & Verjus $20.00

Striped Bass with Fresh Sage & Red Wine $19.00

Line caught Maine Sea Scallops with Sesame Seeds & Gingerado $19.00

Roast Venison with Red Wine and Prunes $20.00

Rack of Lamb with Mint & Vinegar $22.00

Beef Fillet with Star Anise $23.00

The Entrée of the Day $8.50

An Assortment of Cheese $10.00

Desserts

Ice Creams & Sherbets $8.00

Hazelnut Creme Brulée with Apple Hazelnut Strudel $9.00

An Assortment of Chocolate Desserts $6.00

Tropical Fruit Soup with Spiced Ginger Ice Cream $8.50

Crisp White Chocolate Soufflé Cake with Local Oranges and Cointreau Ice Cream $10.00

Infusion - Tea $3.00

Coffee Espresso $3.00

Four New York covers of *Food Arts* magazine, founded by the authors in 1988 to give the Food Revolution's brave new world of restaurateurs, hoteliers, and chefs a periodical of their own, mirror some milestone phenomena and trends.

Right
Maguy Le Coze, who, with her late brother Gilbert (*in background*), elevated the "fish house" to unprecedentedly luxurious heights, chicly embodied the new-found power of women in the restaurant industry. (Photo by Dan Wynn.)

Left
Greenwich Village coffeehouses of New York Bohemia's halcyon days, such as the venerable Café Borgia, served as spiritual paradigms for a sudden citywide surge of cappuccino-scented cafés. (Photo by Brooks Walsh.)

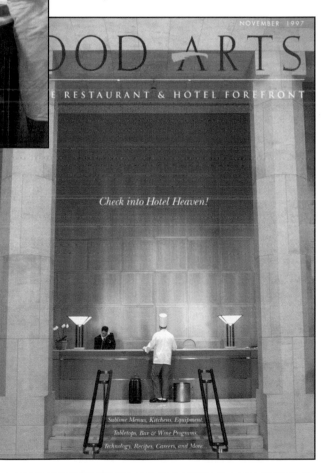

Left
The stage-like open kitchen, particularly in larger restaurants, was to become almost obligatory; at Restaurant Associates' Café Centro, adjacent to Grand Central Station, even the pastry chef was given a stage on which to perform. (Photo by Courtney Grant Winston.)

Right
After decades of decline into atrophied "continental" style cookery, New York hotel kitchens vied to recapture the culinary excitement of the 19th Century grand hotels; this vignette was staged at architect I. M. Pei's Four Seasons Hotel. (Photo by Courtney Grant Winston.)

Herb Wilson

Michael Schenk

Marcus Samuelsson

Nobu Matsuhisa

Brian Whitmer

Gray Kunz

Red Pepper Blinis with Smoked Salmon, Coconut Cream and Osetra Caviar

Patrick's Famous BBQ Spareribs and Pork Beans

First Course
Sauvignon Blanc, Meridian 1997

Steamed Wild Mushroom Crusted Bass with Black Truffle Broth

Lobster Sashimi Salad

Second Course
Chardonnay, "Estate," Monticello Vineyard 1994

Maine Sea Scallop on Crispy Farro Crouton with Artichoke Vinaigrette

Velouté of Provence Truffles and Frog Legs with Chervil

players. The pantheon of chefs shown in these pages collaborated on April 20, 1998 to produce a landmark banquet at Tavern on the Green that simultaneously served as a memorial to one of their own and a brilliant

Hors D'oeuvre
Mumm, Cuvée Napa, Brut N.V.
"Bianco," Francis Ford Coppola 1997

Cream of Ramp Soup

Gold Truffle Duck Confit Sandwiches with Sweet/Sour Rhubarb

Grilled Wellfleet Little Necks with Osetra caviar

Crab Cakes

Foie Gras Morel Poppers

Golden Osetra Beggar's Purse

Shrimp & Shiitake Strudel

Smoked Swordfish with Herb Potato on Flat Bread

Roasted Pepper Loin of Tuna with Chili Papaya Relish and Wakame Salad

Jimmy Sneed

Jody Adams

Donnie Masterson

Jean Georges Vongerichten

Danielle Reed

Michael Lomonaco

In the era of late 20th Century celebrity worship, the ascension of chefs to the status of box office stars saw them hotly pursued by charity planners, themselves now a professional group with their own roster of recog-

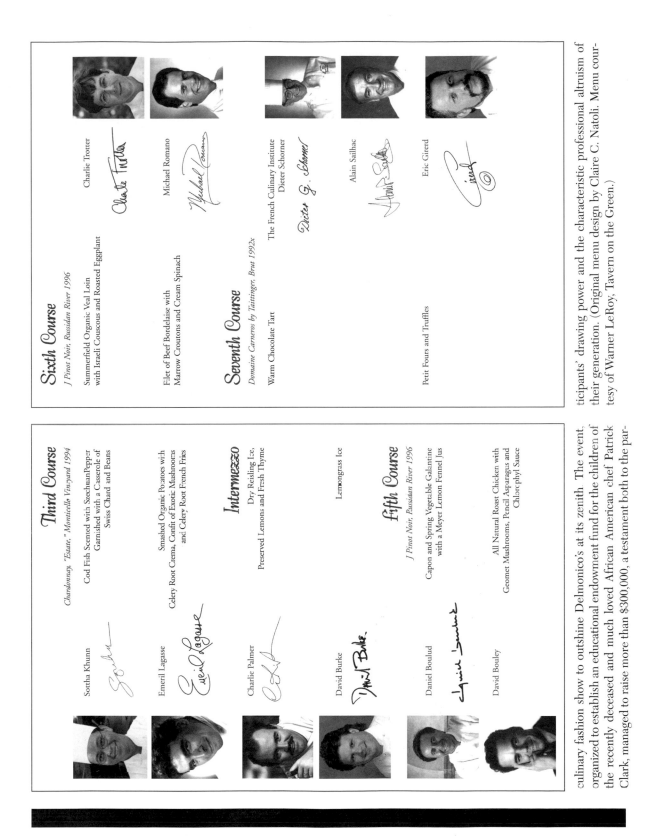

Third Course

Chardonnay, "Estate," Monticello Vineyard 1994

Cod Fish Scented with SzechuanPepper
Garnished with a Casserole of
Swiss Chard and Beans

Smashed Organic Potatoes with
Celery Root Crema, Confit of Exotic Mushrooms
and Celery Root French Fries

Sortha Khunn

Emeril Lagasse

Intermezzo

*Dry Reisling Ice,
Preserved Lemons and Fresh Thyme*

Charlie Palmer

Lemongrass Ice

David Burke

Fifth Course

J Pinot Noir, Russian River 1996

Capon and Spring Vegetable Galantine
with a Meyer Lemon Fennel Jus

Daniel Boulud

All Natural Roast Chicken with
Geomet Mushrooms, Pencil Asparagus and
Chlorophyl Sauce

David Bouley

Sixth Course

J Pinot Noir, Russian River 1996

Summerfield Organic Veal Loin
with Israeli Couscous and Roasted Eggplant

Charlie Trotter

Filet of Beef Bordelaise with
Marrow Croutons and Cream Spinach

Michael Romano

Seventh Course

Domaine Carneros by Taittinger, Brut 1992x

Warm Chocolate Tart

The French Culinary Institute
Dieter Schorner

Petit Fours and Truffles

Alain Sailhac

Eric Girerd

culinary fashion show to outshine Delmonico's at its zenith. The event, organized to establish an educational endowment fund for the children of the recently deceased and much loved African American chef Patrick Clark, managed to raise more than $300,000, a testament both to the par- ticipants' drawing power and the characteristic professional altruism of their generation. (Original menu design by Claire C. Natoli. Menu cour- tesy of Warner LeRoy, Tavern on the Green.)

French patissier François Payard's Payard Patisserie & Bistro, an Upper Eastside pastry, ice cream, chocolate, and catering shop-cum-bistro and café, opened in 1997 in partnership with restaurateur/ chef Daniel Boulud and designed by David Rockwell, harks back two centuries to the Bowery pleasure garden blandishments of Corré and Delacroix. (Photo by Courtney Grant Winston.)

cunning way to bring about their ruin. But he was not a poor sport, he never contemplated unmasking them. He *reviewed* them.

But if Gomez and Weinberg were totally the invention of the press, so was a good deal more of the 1930s' Social World than one might suspect. Night club owners were anxious to build a reputation for entertaining "high society," a group that had by the 1930s only the very blurriest outlines. The solution was simple, rather in the spirit of Pirandello's *Right You Are If You Think You Are:* if you say that your patrons are celebrities long enough and hard enough in all the papers and magazines that will print the fact, they become celebrities, *ipso facto.* Nothing more was needed for this venture than a group of people of some distinction who could be counted on to become regular patrons. The answer to this requirement was found, some say, by Sherman Billingsley, who was just then promoting his new Stork Club. Billingsley bought advertising space in the Harvard and Yale students' magazines, and paid with chits to be honored at his club. The young men who came to cash them in were attractive, and frequently members of distinguished families. The Stork Club invited the college men and their dates to come often and drink hearty on the house. These young blades were invariably reported as "scions of wealthy families," or "sons of Wall Street magnates," and with a few photos cleverly placed by a press agent, the Stork Club had a roster of ready-made celebrities willing to come back, free of charge, again and again. Other night clubs followed suit, notably John Perona's El Morocco.

But the Stork Club and El Morocco were not the only establishments to paper the house. According to Lucius Beebe, any presentable young man or woman who could afford a decent dressmaker or tailor, and boasted good manners and a refined appearance, could live in a style suitable to J. P. Morgan for next to nothing.

I for one, rode it [the Depression] out tearing at a pheasant *en plumage* like Henry VIII lunching with Cardinal Wolsey, and cuffing magnums of Bollinger '26, a particularly delectable vintage, from the Plaza in Fifty-ninth Street to the Bermudiana Hotel in Hamilton. I was repeatedly nominated for lists of the best-dressed and most versatile men about New York. My photo in glad evening attire glared at less fortunate passersby in Radio City, and Carino, the maître d'hôtel at El Morocco and the snootiest waiter captain in town, received me with princely honors. . . . The secret . . . was that young and reasonably attractive people of both sexes were in greater demand as shills in the luxury spots of town than may be even faintly imaginable to the current generation. For peanuts I occupied a five-hundred-dollar-a-month suite at the Madison, the only condition being that a half dozen times a day I walk through the lobby looking prosperous, and dine, on the cuff, in the Madison restaurant, then under the management of the now legendary Theodore Titze of the Ritz. I was supposed to look carefree and bountifully stoked with foie gras and forty year old Hennessy, which, in fact, I was.[8]

The degree to which proprietors would go to create in their establishments a Potemkin Village of mimed prosperity and gaity strike us today as almost sinister. Gene Cavallero offered Beebe and others a 60 per cent discount at the Colony, and car rental services offered their limousines free, provided that they be driven to the right places. Even the all important tailor did not charge if his suit appeared and was photographed at important gatherings. Henry Sell, later editor of *Town and Country,* actually ran a casting agency that sent forty couples a night to the Waldorf's Starlight Roof. There the poor souls in their gratis (or handed-down) finery were instructed to dine on a cloying diet consisting exclusively of flaming desserts and wine in double bottles. "The place doesn't look half as wonderful now as it did when it was losing seven thousand dollars a day," said Mr. Sell. Only Jack and Charlie's "21" charged full rates.

The Starlight Roof had to look well—it was posing for its picture. The new "flash bulb" indoor photography had in its own way revolutionized New York night life. Never has the power of the press meant so much and what the press needed, what everybody needed, were celebrities. Celebrities sold papers, and celebrities, as seen in the papers, brought clients into night clubs, restaurants, and hotels. There was a grand conspiracy to create celebrities. A flagpole sitter, aviator, member of the original "400," decorator, restaurateur, or of course a millionaire might qualify. An attractive appearance and a face that could be recognized at a glance were requisite. Movie stars were natural celebrities, but their snob appeal was not the strongest. The conspiracy then turned naturally on the young. Since 1870, the first time a debutante was presented outside her home—at Delmonico's as it happened—the debutante had become more and more a public figure, and the thirties became the era of the "Poor Little Rich Girl" and the "Glamor Deb." Unfortunately the year 1930 itself saw the debut of two girls who might have qualified for either title, and who also happened to be

Oscar greeting guests in the Sert Room of the new Waldorf. *Courtesy of White Studio.*

the two richest girls in the country. Doris Duke, lean, fair-haired and large-jawed, married fairly quietly and suitably four years after her showy debut, but Barbara Hutton seemed to have a knack for rubbing the public the wrong way. Barbara, who inherited thirty million dollars when still a child, came out at a sixty-thousand-dollar supper dance for which eucalyptus and silver birch trees were imported from California. Four orchestras played, Rudy Vallee sang, and the public somehow sensed that this was real and not sham wealth in their midst. Barbara's constantly photographed appearances in night clubs in Europe and America, and her marriage to Prince Alexis Mdivani, a young man of exotic Georgian antecedents and transparent intentions, further irritated the public, but her twenty-first birthday party in New York was the last straw. Barbara had just come into another thirty million. At the party, guests were entertained by Chinese dancers and acrobats, balalaikas and Cossack sword dancers. All this style would have been applauded in the twenties. Now, those lucky enough to be employed as Woolworth salesgirls had to subsist on ten dollars a week. The news of the party was greeted with hoots of derision and the plebeian rotten tomato.

Elsa Maxwell's famous "barnyard" party fared rather better in the press. During the Depression years, sham giddiness was in the very best escapist fashion. Elsa Maxwell based her career on an innate human love of sheer silliness. Her credo was that inside everyone (and especially the rich) was a voice crying out "make me a child again, just for tonight." She delighted Society with costume parties, cooking parties, painting parties, treasure hunts, and parlor games. In a way she was a kind of warm, smiling nanny figure, and the escape into childhood she offered seemed to be much in demand during the thirties.

As a child herself in San Francisco, Elsa claimed that she suffered her bitterest disappointment when she learned that she had not been invited to a party given by a "Vanderbilt" neighbor. Her family was too poor, her mother explained. Little Elsa vowed she would give many parties, and invite few Vanderbilts. Her career as nickelodeon player and vaudeville accompanist led eventually to a job as columnist, and this last put her in a perfect situation to pursue her career as party giver. By the thirties, with the power of the press behind her, and publicity-starved hotel and restaurant owners before her, she could not have been in a better position to exercise her fertile imagination.

The barnyard party took place in the Jade Room of the Waldorf. Girlishly, Elsa set about converting the ballroom into a barnyard ("my passion for Paradox"). She seemed little aware that fifty years earlier Delmonico's had undergone similar, and frequent, transformations. There were apple trees (fake, with apples sewn on), a well (yielding beer), hayricks draped with red flannel underwear (genuine), ten huge hogs (all genuine) and three cows, of which two were flesh and blood, and gave milk. The third was papier-mâché and gave champagne. There was also a hog caller (imported from Ohio) to create a stampede at the height of the festivities. The two-footed guests were the farmers and milkmaids of a thoroughly homogenized Café Society: Mrs. Vincent Astor, Bea Lillie, the Fredric Marches, Winthrop Aldrich, Cornelius Vanderbilt Whitney, George Jean Nathan, and others. One Lady Iya Abdy came in red tights with a crow's head, Prince Serge Obolensky came as a Russian peasant, and so did Mrs. James A. Corrigan (according to one tale, an ex-switchboard operator from Cleveland). Only Mrs. Ogden S. Mills wore a tiara with her denim overalls. Elsie de Wolf, perhaps more attuned to the times, entertained at the automat.

At Jack White's Club 18, the jokes were all on the guests, and nobody knew what would happen next. Waiters dumped soup in matrons' laps, and gentlemen making their way discreetly to the lavatory found themselves impaled to the top-hatted door by a traveling spotlight. If it sounds like "Hellzapoppin' " with dinner, it is hardly surprising Olsen and Johnson dropped in to do a little friendly stooging as often as they could. In retrospect, much of the humor of the era was all quite as depressing as the daily stock market averages. When the Chinese Ambassador came as an honored guest, he was handed a bag of dirty laundry.

Again, it is Lucius Beebe who best describes "high life" in the 1930s:

It was this amiably demented whirl of scrammy entertainments, Fifty-second Street Morris dancing, whoopsing, screaming, and clogging it to Eddie Duchin music at the Persian Room of the Plaza, making pretty faces for the cameramen at Gilbert Miller first nights, bicycling through Central Park to charity carnivals, keeping luncheon trysts at the Vendome in Hollywood and being at the old desk next morning, gossiping by the hour on the London phone, and living in a white tie till six of a morning before brushing the teeth in a light Moselle and retiring to bed, which constituted the life of Manhattan's café society.[8]

On the other end of the scale, life in New York during the earlier years of the Depression attained an unfamiliar grimness. Anyone who had the heart to notice would see that many shopfronts stood empty, and the sight of the possessions of the evicted, some interior life thrown out into the open, was not uncommon. Façades in a run-down condition stayed that way. Men and women could be seen sleeping on park benches and in doorways, and the stroller was stopped by beggars more frequently than before, or tempted by rather more pushcarts than he had seen in some decades. Men and women were waiting—waiting on lines for bread, or employment. And for the first time in a century, New York was again surrounded by a shantytown—the homes of the unemployed who had drifted to the city, and made lodgings out of whatever scrap they could find. Public relief for the needy did not yet exist but was soon to come.

Everyone had less money to spend on food, or to squander in restaurants. A man who might have paid a dollar fifty for a restaurant lunch now paid thirty cents at a lunch counter. The stenographer who might have bought a sandwich at the lunch counter now brought one from home. In response to the new demand for something decent to eat at a low price, cheap restaurants proliferated. The basic concept of the People's House, for example, was that all the profits were to be reinvested in the restaurant; in this way it became possible to serve tea for seven cents, vegetables for eight cents, and ice cream for twelve cents. Even at that rate the People's House (modeled upon similar experiments abroad) could scarcely meet the competition of the California Kitchen on Sixty-first Street, where meat and vegetables, salad, cherry pie, and coffee cost fifty cents.

The tearoom type of restaurant was considered suitable for women to own, operate, and frequent. When the Depression left many spinsters in reduced circumstances, and the wives of the unemployed attempted to go into business on their own, the results were predictable. At Aunt Clemmy's you could find any variety of fruit pie, including pumpkin and mince. The thirst for romance led to some unexpected nomenclature. Marie Antoinette's Tea Room specialized in Southern cooking, with waffles, creole soup, braised chicken, and ham and eggs. At the Samovar, on the other hand, the cuisine was old New England, with baked beans, baked ham, and the ubiquitous pumpkin pie.

New Yorkers fought the Depression in every way they could.

They fought back by laughing—at the new magazine *Ballyhoo,* which satirized and ridiculed everything and everyone on the industrial, financial, or political scene, and at the plays of George S. Kaufman and Moss Hart— *"You Can't Take It with You,"* or *"I'd Rather Be Right."* The effects of the Depression and the competition of movies may have cut the number of theaters and productions in half, but the quality of what was produced was higher than ever, and audiences who could muster a dollar fifty for a seat could escape dreary reality with the likes of Maxwell Anderson's *Winterset,* Thornton Wilder's *Our Town,* and S. N. Behrman's *No Time For Comedy.* George Gershwin's *Porgy and Bess* was produced in 1935, the season after Cole Porter's *Anything Goes,* with Van Johnson and Gene Kelly dancing in the chorus. The cacophony of lights and electrical signs that now marked the crossing of Broadway and Forty-second Street burned more brightly than ever. And those who could not afford the legitimate theater could escape, along with eighty five million other people throughout the country, *weekly,* to the movies—to a world in which hundreds of Busby Berkeley choreens clattered up mammoth keyboards, in which Fred Astaire tapped his debonair way into the heart of Ginger Rogers, in which Shirley Temple dimpled through her tears, and the Seven Dwarfs, never out of a job, whistled Hi Ho. Some people escaped by playing bingo and pinning their hopes to chain letters.

Music had always been a favorite escape hatch and this was the great era of "Swing." The jazz musicians of the twenties had largely improvised their performances, a few playing together in combination. It was first Fletcher Henderson and then the white musician Benny Goodman who adapted and arranged jazz so that it could be played by large bands. The success of the new technique was astounding. It caught on with the young, with college students, and jitterbugging bobbysoxers replaced the flapper of an earlier decade. By 1937 Benny Goodman's opening at the Paramount could command a crowd of six thousand screaming teenagers, and thoroughly commercialized jazz blared from radios and phonographs from coast to coast. But the real thing could best be heard on Fifty-second Street. In the mid thirties, the brownstones along Fifty-second Street, having outlived their glamor as speakeasies, were saved from indigence by Jack Colt, who opened the Famous Door in 1935 as a place where his friends could listen to sophisticated jazz over a drink. "It was about the size of a large

closet and, once inside, patrons found themselves sitting under the
guns. . . . You either loved it or hated it—there was no middle
ground."⁹ This was a place where a musician could "blow it out
clean"—his own invention, his own pattern. Overnight the block
became "Swing Street," with a jazz club in almost every door.
Unfortunately, Swing Street followed a pattern we have seen
before: first the show business world came to hear Count Basie,
Eddie Condon, Nat "King" Cole, Billy Daniels, Billie Holliday, or
Hazel Scott, and then Society followed. But inevitably the gangsters moved in, and gave the street the kiss of death.

New Yorkers also fought the Depression by direct political action. Paradoxically, the desire to escape was balanced by a tremendous crusading spirit. The thirties saw the greatest era of social and
economic reform in the nation's history. The very day on which
Roosevelt took office every bank in the country closed its doors in
final collapse. In the following month, while a congressional committee investigated an appalling number of frauds, swindles, and
sharp practices that had served to undermine the financial structure of the country, Roosevelt and his advisers took advantage of
the *tabula rasa* to pass through Congress a seemingly endless
series of laws reconstructing the system from its foundations, and
providing a minimal opportunity for corruption. But while Wall
Street's problems were set straight in Washington, New York itself
was undergoing a long-needed housecleaning. In this case the
object of public furor was, at long last, Tammany Hall. As the
Hofstadter Committee to investigate corruption in New York
City, with the implacable Judge Seabury as its counsel, revealed
the bribery of Tammany politicians by members of organized
crime and the misappropriation of funds for building and construction, right at the very height of the Depression, New Yorkers
finally rebelled. In 1932 Jimmy Walker was charged by Governor
Roosevelt with misconduct, and allowed to resign. The next year
Fiorello La Guardia, the "little flower," was swept into office by
the broom of reform. La Guardia, with young Thomas E. Dewey
as special prosecutor appointed by the governor, declared a war
on crime, gangsterism, and the rackets that has been described as
"one of the most extraordinary performances in the history of
criminal detection and prosecution." In short order over seventy
of New York's leading racketeers were convicted despite the fact
that witnesses were constantly threatened with extermination,
and the myth of the G-man was born. Among other triumphs stood

the liberation of two hundred forty restaurants that had been forced to regularly pay protection to gangsters.

As a savior, La Guardia presented an altogether unlikely figure. A greater contrast with the dapper tailor-shop dummy, Jimmy Walker, could not be imagined. Fiorello was tough, pug ugly, fast talking, and funny, a man with the incorruptability of Sir Galahad, and the physiognomy of a Damon Runyon bookie. He fought the Tammany politicians with craft, and with fisticuffs if necessary. And he was everywhere. If a fire broke out, he was on the scene. He read the funnies to children over the radio. He conducted the Police and Sanitation Departments' bands at Carnegie Hall ("just treat me like Toscanini"). And his wit was explosive. Above all, the underdog never had a lustier champion. When told that an alderman had proof positive that relief money was misdirected because prostitutes were on the rolls, La Guardia howled, "I thought that question was settled two thousand years ago, but I see I was wrong. Mr. Sergeant-at-Arms! Clear the room! Clear the room!—so this big bum can throw the first stone."

It was with La Guardia's championing of the underdog that the waiters' union first became a force in New York dining. New York waiters struck for more reasonable hours, wages, and conditions, but the hotels and larger restaurants continued to function without them. La Guardia stepped in with his usual sense of drama. He chose the most active moment of the dinner hour to stage health inspections in the city's major establishments—and of course found any number of scab waiters suffering from communicable diseases. Restaurant and hotel owners got the idea very quickly, the strike was settled, and the waiters' union under La Guardia became a force to be reckoned with.

While political reforms were enacted to the cheers of an avid audience, the social conscience of the nation at large, and of New York in particular, was prodded by the outraged pleas of the most prominent writers of the decade. When Scott Fitzgerald's masterpiece, *Tender Is the Night*, reached publication in 1936, it aroused scarcely any notice. His characters, totally involved in private emotions, were tortured figures of the twenties' revolt for personal freedom—a crusade which now appeared less important than the reorganization of society and the proper distribution of wealth. In works such as John Dos Passos' *U.S.A.*, Thomas Wolf's *Look Homeward, Angel*, James T. Farrell's *Studs Lonigan*, Erskine Caldwell's *Tobacco Road*, and John Steinbeck's *Grapes of*

Wrath, the protagonists challenge social injustice and blind fate. Meanwhile, the Group Theater, founded by Harold Clurman, Cheryl Crawford, and Lee Strassberg, produced realistic dramas of social comment such as Clifford Odets' *Waiting for Lefty* and *Awake and Sing.* "We all thought of what we wanted to express and what had to be said rather than about success." These are the words of Irwin Shaw, but they are echoed by Harold Clurman, and the Group Theater was joined by a list of well-established actors who could easily have earned more elsewhere—Franchot Tone, Stella and Luther Adler, Mary Morris, Alexander Kirkland, and many others. At the same time, a government-subsidized Federal Theater enjoyed an enthusiastic opening in Harlem, and dance groups featured social messages in their choreography.

If New York's anatomy possessed a social conscience, it was located on Union Square. Albert Halper recalls "14th Street . . . was thick with people and traffic. Along the crowded sidewalks hawkers were selling fifty-cent watches, fake 'blind' beggars led by mangy dogs were chanting ballads, and 'pullers' standing before the doors of stores were enticing passing women shoppers, whispering or calling, 'Psst, we got some terrific bargains in underwear and cotton goods inside, come in and feel the material. . . .' "[10] Changed from the days of Steinway Hall, Union Square was the scene of weekly left-wing parades and demonstrations (often ending with a clubbing by the police) and screaming soap-box attacks on capitalism, although an expanding Klein's displaced the headquarters of the Communist Party. The Academy of Music now exhibited a movie and "six great vaudeville acts," and in front of Luchows, still a popular and fairly smart restaurant, a newsboy shouted "Buy the *Daily Worker,* comrades, the only paper that prints the truth about the struggles of the laboring masses!" At Webster Hall, in which meetings and rallies were generally held (and the *New Masses'* annual ball), hillbilly music was performed by backwoodsmen sent by the committee for the striking Kentucky mines. At tiny nearby restaurants, young intellectuals and unpublished authors could celebrate with a good Italian dinner and a glass of wine for one dollar inclusive. Otherwise, there was always the automat. One veteran of the thirties recalled, "There used to be a rose on every cafeteria table, during the Depression, when coffee was five cents."

Meanwhile, Roosevelt's W.P.A. (the Works Progress Administration) set about giving employment to the jobless, in their own field in so far as that was possible. The Federal Theater provided a stage for unemployed actors, and the results were admirable, with talents like Orson Welles at work, and the production of plays like T.S. Eliot's *Murder in the Cathedral*. Unemployed artists were set the task of covering the walls of post offices, libraries, train stations, and other public buildings. These decorations generally featured heroic figures and allegorically represented the assault on one of the nation's many as yet unrealized objectives, brawny embodiments of Agriculture, Commerce, and Industry. But probably of most importance to the national economy were actual construction projects. New York itself underwent the largest single building project it had ever known, instigated, not by the government, but by John D. Rockefeller, Jr.

In 1928 Rockefeller lead a civic drive to build what it seemed to him at the time New York needed, a new opera house to replace the shambling yellow brick of the Met. For this purpose he leased from Columbia University a tract of land in the upper Forties and lower Fifties, west of Fifth Avenue, part of which was to be used for the opera house, and part for commercial development. By 1929, however, the picture had changed, and the opera company no longer cared to invest in a new house. It withdrew from the enterprise leaving Rockefeller holding a long-term lease on the land. He could abandon the project, or set out to develop the property on his own. Fortunately, for Depression-pressed New York, he decided to build.

Twelve acres were hastily cleared of brownstones, speakeasies, and shops. Some two hundred and twenty-eight buildings were demolished and four thousand tenants relocated. Most important, a grand total of almost a quarter of a million people were put to work, in one way or another, for the construction of no less than fourteen office buildings offering five million square feet of rentable space. Rockefeller Center, as it came to be called, ranks as one of the biggest projects ever undertaken by private enterprise.

New Yorkers observed an almost terrifying city within a city rising in their midst—something altogether new in urban design. They saw "modern" principles freely applied, with skyscrapers planned in relationship to one another, with open spaces, gardens, and plazas. Rockefeller Center embodied new ideas for achieving human comfort—an extra street to ease traffic circulation, small landscaped areas as "oases of beauty," underground passages, and a subterranean concourse area with shops, restaurants, and, of course, a subway. Moreover, it presented the public with the greatest movie palace of them all, Radio City Music Hall, the world's largest indoor theater, seating sixty-two hundred people who were treated daily to performances of a full ballet company, a symphony orchestra, and the largest chorus line in history (Rockettes, in deference to John D.), as well as a movie. And with the "Lower Plaza" at its heart, Rockefeller Center gave New Yorkers a place to dine in the open air in summer, and ice skate in the winter, while winning the attention of throngs gazing down from street-level esplanades. Diners in the lower-level restaurants, who enjoyed the cuisine of the Promenade Cafés flanking the rink, could watch through plate glass windows while their children fell on (but *not through*) the ice. And those looking up rather than down were treated to an uplifting prospect of vertical lines, symbolic of eternal optimism. The whole was decorated with murals, sculptures, and constructions by some of the century's greatest names in art—Paul Manship, Isamu Noguchi, José Maria Sert, Giacomo Manzu, Joseph Albers, and Diego Rivera. All, of course, interpreted a specified theme—New Frontiers. Rivera's painting, however, caused trouble. It was going along well enough, when one day it was noticed that one of Rivera's giant figures heartily embracing a representation of Abraham Lincoln, was beginning to bear a distinct resemblance to Lenin. The resulting outrage on the part of the Rockefellers can well be imagined. Rivera's indubitably excellent work was speedily painted out and covered over. As a final insult, the artist himself was summoned to a communist meeting at Webster Hall and drummed out of the party as "a bootlicker to the capitalists" for ever having accepted the commission in the first place.

If Rockefeller Center gave work to millions, it also gave New York's Café Society a new playground from which to observe the flickering lights of their kingdom, the Rainbow Room.

The Rainbow Room was first opened by the Rockefellers on October 3, 1934, and O. O. McIntyre, apparently unapprised of a last-minute change in name, had this to say in the *New York American:* "The Stratosphere's swanky night club on the 65th floor of a Radio Center building, may puff a breath of life into the

Al fresco dining and dancing in the Promenade Café, summer 1939.

dying night clubs. But not many believe the resuscitation will be permanent. At least fifty clubs that opened last winter will not brave this season. It has become a passé entertainment largely on account of the Depression but also because patrons are fed up. And in lieu of some diversion to supplant it, people are going home." But the management had taken every precaution to side-step the vulgar pitfalls of a run of the mill cabaret; *Variety*, in a state of alarm, feared they might go so far as to ban gossip columnists entirely in their quest for pursuit of dignity and equilibrium.

This reserved approach was enthusiastically endorsed by a flock of Rockefellers, Astors, Blisses, and Harrimans who winged in for the gala opening dinner and paid fifteen dollars a place to hear Lucienne Boyer trill "Parlez-moi d'Amour," then prance across a two-speed revolving dance floor.

Mlle. Boyer, the first in a long series of French *chanteuses* to find popularity in New York, managed to squeeze in a nightly theater performance in "Continental Varieties" between her eight o'clock and midnight appearances at the Rainbow Room and was adored by audiences and press alike; when she arrived on the Ile de France for her engagement, it was breathlessly reported that she had brought with her a manager, a pianist, a violinist, three maids, twenty-three trunks, fifty pieces of hand luggage, a twelth-century Buddha, and a stateroom full of identical blue-velvet Lanvin gowns. *Women's Wear* noted every detail of her wardrobe as minutely as it did the toilettes of the celebrities at ringside: for the opening "the choice of Miss Miriam Hopkins" was "old rose velvet combined with aubergine . . . with skirt full at the bottom, a treatment emphasized by the inside ruffles of taffeta which swished melodiously as she walked."

The Rainbow Room paid its artists twenty-five hundred dollars a week in the day of the bread line; Bea Lillie, Vivienne Segal, and Odette Myrtil replaced each other in effervescent succession and were spelled by the dance bands of Jolly Coburn and Ray Noble, and the terpsichorean fantasies of Alice Dudley and Jack Cole, Rosita and Fontana (the last having disengaged himself from Miss Moss of Louis Martin's Paradise Room days). They, in turn, rested during Richard Liebart's resonant recitals on a one hundred thousand dollar pipe organ. Two organs, in fact, were installed—the second to manipulate the changing colored lights on the circular ceiling.

By the time that the building of Rockefeller Center got underway, two other landmarks of the thirties had already risen over New York. In May of 1929 the palmy old Waldorf was abandoned. At the very moment of the crash, the Empire State Building was constructed on its site, rising to one hundred and two stories as a tombstone to the twenties' prosperity as much as a portent of things to come. And in 1931 the Waldorf itself moved into a formidable structure that towered over Park Avenue at Fifty-first Street.

The new Waldorf, like so many of its predecessors, boasted comfort and size never before provided to the weary traveler in the history of civilization. In general appearance, it was a sister to the Empire State and the soon to be constructed Rockefeller Center edifices. It stood forty-seven stories high, with numerous public dining rooms, a "Peacock Alley" for old times' sake, now a café restaurant, at least two night clubs, and twenty private banquet rooms. It could comfortably hold over ten thousand people, twenty-five hundred in its ballroom alone, and its wine cellar on the fifth floor was equipped with stabilizers to prevent the bottles from being joggled by the vibrations of the New York Central Line beneath the street. The hotel had its own railroad siding in Grand Central Station, its own driveway, bisecting the building, and its own medical staff. It also had the largest radio receiving system yet built, and over this system President Hoover himself addressed several thousand people on the eve of the hotel's opening. (Hoover later became a resident of the Waldorf Towers and a member of the hotel's board of directors). And above all presided Oscar himself, compiling his memoirs, nostalgic over the old Waldorf and secretly loathing the new.

The new hotel had no shortage of admirers, however, and there was one in particular, the down and out operator of a modest hotel chain in Texas. Conrad Hilton was no longer very young when the crash wiped out his small and over-extended business. The settlement of a half million dollar debt had left him in such a state of penury that he literally did not have the price of carfare. He "hid under a ten gallon hat" until one day, as an act of pity, an ex-bellboy pressed three hundred eighty dollars into his hand—his life's savings. Hilton was in such a state of desperation that he accepted the loan. His mother suggested other assistance. "Some men jump out of windows, and some go to church," she said.

"Pray, Connie. Its the best investment you'll ever make."[11] When the Waldorf opened, Hilton cut its picture out of the local paper and wrote across it "The greatest of them all." He kept the clipping.

It was not until the last years of the thirties, however, that there was a distinct return to economic health, and now, without the bugaboo of Prohibition, hotels burgeoned and chic restaurants began to genuinely prosper. Park Avenue and its adjacent side streets in the Fifties, the very knee of the Silk Stocking district, was the area of preference for the city's smartest restaurants, which were generally French. On East Fifty-eighth Street alone stood Le Merliton and La Rue, both "intimate," a new word that had gained a certain currency, but still meant small and ruinously expensive. La Rue had a canopied garden, and Urban Associates, responsible for the decors of innumerable restaurants, had treated its interior to their characteristic red, black and beige color scheme. One was expected to dance between servings of Boula Gratiné (a combination of green turtle and pea soups) and guinea hen Smitane (dressed in sherry, cream sauce, raisins, and mushrooms). At Louis and Armand's on Fifty-second Street, where the specialties of the house were red leg partridge *cocotte* with sauerkraut and broiled beef tenderloin "Henry IV." Victor, the maître, took a childlike delight in preparing *crêpes Suzette, cerises flambé*, or strawberries Cincinnati (also *flambé*).

The restaurant furthest downtown on Park Avenue was the Marguery, situated in the immense apartment block of that name. It was essentially formal, and prided itself on its hors d'oeuvres, salads, and garden for summer dining. Further up the Avenue were L'Aiglon and the Passy, known for its "ultra modern" decor, its potted palms, its mushroom soup, and its duckling with cherries.

Somewhat out of the way, but definitely among the elite, was the Café Chambord on Third Avenue near Fiftieth Street. Founded by Roger Chauveron, the Chambord indulged in the reverse snobbery of the genuinely French temples of gastronomy —the severe simplicity of its decor matched the baroque character of its prices. There was even a glass wall through which the homely vision of copper pots and chefs in white toques could be regarded. It is interesting to note that guide books of the period suggested black tie as the only proper form of dress for dinner in any of these establishments.

In the twenties and thirties a certain number of restaurants took root in New York and survived, despite boom, bust, and aftermath, to take their places as durable members of a kind of Hall of Fame, attaining the character of Institutions. Some are still in operation and, although they vary in character from aggrandized delicatessens to latter-day Delmonico's, they had one thing in common—each revolved around one or two obsessed men, or a family.

Le Voisin was the oldest of New York's restaurants to achieve true Institution status in the twenties and thirties. It was at the original Voisin in Paris, from which it received its name and tradition, that the inhabitants of the zoological gardens were served for Christmas dinner during the seige of Paris. The menu for that memorable feast included, among the appetizers and soups, butter, radishes, stuffed Donkey head, sardines, cream of red beans, and clear elephant broth, to be accompanied by Xérès and Latour Blanche 1861. Under entrees listed were camel, kangaroo stew, and rack of bear, and the roasts included leg of wolf and game sauce, cat garnished with rats, and truffled antelope pie, complimented by Mouton Rothschild 1846 and Romanée Conti 1858. A copy of this menu, along with a withered, framed piece of 1870 bread, hung for many years as a grim reminder of war, the lengths to which the French public will go for a decent meal, and the ingenuity of earlier Voisin chefs.

It was in 1912 that Messrs. Muller and Janderck, two Viennese friends of the owners of the Paris restaurant, signed an agreement with the latter and opened their own Voisin at 375 Park Avenue. Their enterprise gained in reputation during the Great War, and by 1924, when the parent Voisin folded in Paris, they had established their preeminence in New York. Throughout the thirties and forties Le Voisin reigned as the most august of Park Avenue restaurants, outshining the Marguery, L'Aiglon, and Passy with its soufflé Grand Marnier and heroic echoes of Sedan.

The life story of the Colony's owner and guiding light, Gene Cavallero, reads like a rollicking synopsis for a Hollywood operetta of the golden age. Cavallero first saw the light of day in his father's humble trattoria. The year was 1889 and the place Villimpenta, near Mantua, and the baby was named, prophetically enough, after Italy's first socialist deputy, Oswaldo Gnocchi. By the time he reached his teens young Oswaldo was working at a local *buffet de la gare,* or railroad station restaurant, where his activities seem to have been limited to leaping onto incoming trains with a rattling tray of espresso and brioches on his shoulder, distributing the steamy breakfast, collecting handfuls of lire, and nimbly hurling himself back onto the platform as the train panted out of the station. One morning, as the last car clattered out of sight, young Cavallero discovered amidst spoons and empty coffee cups a diamond bracelet, dropped accidentally, it transpired, by a beautiful principessa as she helped herself *molto allegro* to the sugar cubes. With the reward money, he proceeded hastily to London, the Olympus of Italian waiters. There he worked at the Hotel d'Italie, the Marguery, and Romano's, before finally achieving, at the age of twenty, the youngest captaincy in the history of the Savoy. Moving on to the even more voluptuous and elegant ambiance of the new Esplanade Hotel in Berlin, he acquitted himself with perfect grace until one hideous evening a young patrician, later to become King Paul of Greece, called for writing materials with which to compose a letter at table. In his rush to comply, Cavallero splashed a whole bottle of ink over the Prince's braid-festooned white tunic. He was fired on the spot.

1912?

Nineteen twenty-two saw Cavallero on the pier at Boston with fifty dollars in his pocket, no friends, and no glimmering of the future. The Copley-Plaza took him on, and when he eventually joined the Biltmore in New York, "Gene," as he now came to be called, found himself working once again with "Ernest," Esterino Umberto Guiseppe Antonino Cerutti, who had also served as a captain at the Savoy. These two were to establish the Colony as it later came to be known.

But we have not quite reached that climactic moment. Cerutti and Cavallero left the Biltmore in favor of the old Knickerbocker Hotel. As captain from 1917 through 1919, Gene served, without visible signs of exasperation, a monumental nuisance named Masury, a paint manufacturer from Brooklyn, whose finicky demands were backed up by a sizable fortune. But Gene received not a penny of it, despite Masury's reputation as a generous tipper. Suffering in silence, he officiated flawlessly over the chafing dish and anticipated every quirky wish the old ogre might express until his second Christmas Eve at the Knickerbocker, when his tormentor presented him with a plain white envelope, best wishes, and the observation that all headwaiters and captains, with the exception of Cavallero, were lazy bums. Optimistically hoping that a fifty dollar bill might be reposing inside, Gene tore into the hallway, then into the envelope, and finally into the the office to announce his resignation. He had been rewarded with two one thousand dollar bills.

His windfall was invested in a small New Jersey farm which failed and a Westchester hot-dog stand was purchased in its place. This succeeded magnificently and was promptly sold for a twenty-five hundred dollar profit. At Cerutti's suggestion, the next step was taken into a down-and-out bistro on the corner of Sixty-first Street and Madison Avenue, saved from bankruptcy only by the occasional patronage of furtive sugar-daddies and their one-night-a-week kept women, gamblers craving scrambled eggs and onion soup between hands at an illegal club on the second floor, or bibulous expectant fathers whose wives were undergoing labor in a private maternity hospital at the top of the building. It called itself the Colony.

Cerutti had previously been taken on as headwaiter, but quit when the owner threatened to hire a jazz band, a couple of good acts, a couple of bad acts, and turn the place into a night club. Cavallero stayed on, having secretly made a pact with Cerutti and Hartmann, the Alsatian chef, to buy out the owner, Joe Pani. With the notorious gangster Arnold Rothstein, the building's "real estate man," acting as a go-between, the conspirators made a deal with the landlord, raised eighteen thousand dollars, and took over the Colony once and for all.

The doors were closed to the public for ten days, and a staff of headwaiters, captains, waiters, chefs, salad men, vegetable girls, dishwashers, a *dame de lavabo,* a men's room attendant, and two mousing calico cats was assembled. In a daze of expectancy and a blaze of high hopes, the Colony reopened in March 1922. Nothing happened. Even the sugar-daddies had sniffed out new lairs during the brief limbo of renovations.

Two months passed. The kitchen was conducted as if playing to a full audience, but morale was severely shaken, and the new owners cursed themselves for having appropriated a reputed "cat house." Illicit couples began trickling back, but not enough to help, only enough to sully the disinfected atmosphere. And then one night what should Gene murmur casually to his few tawdry regulars but "Mrs. W. K. Vanderbilt is dining here with another lady this evening." This was no final leave-taking of a distraught restaurateur's senses. Mrs. Vanderbilt, albeit unaccountably, *was* dining with another lady, and liking it. Mr. Vanderbilt, a peripatetic gourmet and extremely difficult to please, came around for lunch the next day upon his wife's recommendation and thereafter returned almost every day of his life, a matter of another twenty years.

Joseph E. Widener, the racing-stable owner, followed Vanderbilt's example as did Reginald Vanderbilt, for whom the Colony's stinger became a favorite apéritif as well as an after-dinner drink. "Reggie" inaugurated the practice of bringing one's children to the Colony for a dish of the estimable ice cream; it was on one such occasion that he met a friend of his sixteen-year-old daughter, a Miss Morgan, whom he later married.

The Colony grossed seventy-five thousand dollars that first year. By the 1927–28 season, it was clearing half a million annually. The oldest and most famous names of the era were to be found scrawled at the bottom of the Colony's bills; 85 per cent of the business was handled by charge account. Everyone who was anyone signed, and if they absent-mindedly strolled out without doing so, Cavallero would considerately forge their signatures for them: names like Bernard Baruch, Gloria Swanson, Otto H. Kahn, Prince Youssupoff, the assassin of Rasputin, Louis B. Mayer, Archduke Otto, Queen Zita, Preston Sturges, the Duc d'Orleans, Tommy Hitchcock, Helen Hayes, Lords Northcliff and Rothermere, the King of Siam, Mrs. Franklin D. Roosevelt, and Jules S. Bache, the investment banker. During the Depression Cavallero learned the wisdom and discretion of credit. Old customers might take literally years to pay their bills. Cavallero had the taste not to nag.

Cavallero and his partners coddled their clients in other ways as well. They kept a card index of each of their regulars' tastes in wine and food. The piquant *"crêpes* Colony," stuffed with seafood, chicken Marengo, *homard à la Cardinal,* grilled lamb steak, boysenberry sorbet, and *poire belle hélène* with a soft light chocolate sauce are recalled by afficionados twenty and thirty years after the moment of tasting. And those who knew Madison Avenue in the thirties and forties remember that between expensive jewelers, dress shops, and art galleries, and dominating the corner of Sixty-first, stood Shaffer's meat market, its tall windows exhibiting to the public the choicest cuts of meat, displayed like haute couture with paper frills and fleecy white curls of sculptured fat. Shaffer's was among the finest, and most expensive, butchers in the city, and it was here, conveniently enough, that the Colony bought much of its meat. Cavallero even held a key to the premises so that, should the need arise, he could enter the store after hours and take whatever he required, leaving a chit at the cashier's desk. On one occasion, in dead of night, no less than a hundred partridges were removed from the refrigerators of Shaffer's to please Clarence Mackay who wanted to give a large and sudden dinner party.

It was in the thirties that the preparation of *crêpes Suzette* became a major dining spectacle. At the Colony it was a particular tour de force with a team of heavy-breathing technicians in crisp black and white hovering, whispering, peeling, and sprinkling, lighting matches, clinking bottles, juggling dishes, and bumping into each other with the urgency of brain surgeons. The recipe itself however, seems deceptively simple:

The Colony's Crêpes Suzettes

Put into a bowl six tablespoonfuls of flour, two whole raw eggs, and two egg yolks, two cups of cold milk (added by degrees), and a very little salt. Mix very well together, and strain through a fine sieve. Now fashion small round pancakes out of the batter; the best method for this is to pour the batter into a skillet, round and small and slightly buttered. Fry the crêpes on both sides, and keep them warm, but do not stack them on top of each other.

In a chafing dish rub five lumps of sugar to the rind of a lemon and the rind of an orange, until the lumps absorb the rind's flavor. Light a moderate flame under the chafing dish. Crush the lumps of sugar with your fork, and add a little orange juice. Let this simmer for a minute. Now add three tablespoons of sweet butter, and cook 'til the butter-sugar-orange juice becomes creamy. Now add your crêpes, and turn them frequently so that they will absorb the sauce. Mix in a cup some Curaçao liqueur, with a little rum, Grand Marnier, and brandy and pour over the crêpes. Heat, and finally ignite the crêpes and top them in the flaming sauce. Be sure to serve on warm plates. (Serves four persons.)

The fuss that was made during the thirties about precisely where one sat in a smart restaurant may be traced to the Colony. The blue and white striped Bar became a coveted spot for dining as well as drinking one day when the Duke of Windsor capriciously turned to Cavallero upon being told his table for six was in readiness and airily replied, "We'll dine here, Gene, the Bar has such a gay atmosphere." Overnight, *Vogue* was moaning that a table in the Colony bar was more difficult to secure than a membership in the Junior League. Moreover, for reasons totally unknown, the back of the red and crystal main dining room gained a reputation for being "unchic." Clients increasingly requested that they be placed in the front part of the room, and one society columnist spoke of the nether regions as the "doghouse," a name that stuck, although the impression was unintentional. But such was the cachet of the Colony that Jimmy Walker had the direction of traffic on Sixty-first Street between Madison and Park shifted from east-west to west-east to assist his Rolls in a speedier passage to the spot, and an aspirant could become a celebrity merely by habitually occupying the same "good" table. Placement in Sherman Billingsley's Stork Club was more deliberate. When Billingsley wanted to indicate to his maître that a guest was of no consequence, he put his finger to his nose.

The antecedents of Sherman Billingsley were less auspicious than those of a Soulé or a Cavallero. He was born in Enid, Oklahoma, in the year 1900, and claimed exactly four years of formal education. There are conflicting reports of his first occupation. Some say that at the age of seven he was busily pulling his little toy wagon loaded with alcoholic "soda pop," to the nearby Cherokee reservation. Others say he merely sold discarded whiskey bottles to bootleggers, Oklahoma then being a dry state. It is within the realm of possibility that he did both. At twelve he went to work in the family business of cigar and drug stores, and Prohibition found him the proprietor of a chain of pharmacies in the Bronx where medicine of a high proof could be secured "on prescription." In 1928 he went into the speakeasy business on West Fifty-eighth Street with two friends from Oklahoma. By the time repeal came along he was partnerless, and in a position to open a club at his final location, 3 East Fifty-third Street.

The basic appearance of the Stork Club conformed to certain ideals of the thirties and forties: black, white, or one primary color and mirrors added up to the aesthetic of the moment, an application of modern decor that originated in speakeasies and night clubs and eventually became fashionable in New York apartments. Those fortunate enough to be allowed past the solid gold chain at the entrance were met by a walk-around bar—long and generous in its proportions. From this room led the entrances to two others: the main night club, its mirrored Moderne lines softened by a bouncy ceiling full of balloons, where two dance bands alternately set Cole Porter and Tito Puente moods, and the Cub Room, solely for dining, where paintings of *Cosmopolitan* cover girls winked and pouted from the walls. Representations of a stork, top-hatted and beak askew, appeared everywhere. On the second floor were to be found rooms for private parties; the six floors above contained everything from Stork Club offices and giant deep-freezers to Billingsley's own apartment, in which he kept over twenty conservative suits, drawers full of conservative shirts, and a hundred conservative ties.

Three East Fifty-third Street was a hive of activity, a *castello* run along oddly feudal lines for the twentieth century. The Stork Club was picketed for years on end by the kitchen workers' union, among others, but it made no more of a dent in Mr. Billingsley's adminstration than a few archers against the Krak des Chevaliers. Although outwardly nervous, shy, and soft-spoken, he ruled with an iron hand, sending vitriolic notes to any employee who displeased him. When a hatcheck girl was seen shifting from foot to foot in an effort to keep warm at the entrance, she was bluntly advised "this place is not a dance hall."

A prime example of Billingsley's managerial behavior was the ordeal of the barmen caught exceeding the line on their whiskey shot glasses. When this practice went unabated, Billingsley instigated a series of fines against the undisciplined employees, but still they ignored the edict and continued to dribble a fraction over the line; after all, they reasoned, they could not be kept under surveillance around the clock. In this presumption they erred. One day, several weeks later, the management called them together. Their offenses, every possible one, were enumerated and proper fines administered. But how did Billingsley know? The answer was in character. One night after closing hour, or rather one morning, workmen had entered the club and substituted a one-way mirror for the more traditional type that stood behind the bar. On the other side, "Sherm" stationed an eagle-eyed hireling to scrutinize and record every drop from the whiskey bottle.

Precisely why the Stork Club was so very successful is a matter that defies analysis more deftly than the crash of 'twenty-nine. The fact is that everyone went to the Stork Club who was allowed through the door from Eugene O'Neill to Madam Chiang Kai-Shek. Everyone came who could, that is everyone whom Billingsley did not bar for misbehavior, like Elliott Roosevelt or Humphrey Bogart, or on principle, like the Maharajah of Jaipur ("I don't want none of those colored men in here").

The Stork was particularly popular for lunch and late supper. Its essentially "American" menu embraced everything from baked clams, "ministrone soup," and chow mein to "Noisette of Lamb sauté Lakmé," and "coupe Venus." But it was not the food that was responsible for the Stork's popularity. Nor was it the gifts. Billingsley, along with Morton Downey and Arthur Godfrey, owned a perfume company which manufactured "Sortilège" and various cosmetics. Guests were conscientiously plied with these and such other favors as neckties, cigars, champagne, and cameo soaps emblazoned with waxen profiles of Messrs. Billingsley, Downey, and Godfrey. Perhaps it was the external Café Society image of the Stork Club that prolonged its nirvana-like reputation or perhaps it was just Walter Winchell, Café Society's Boswell and Billingsley's best friend, who sat night after night in the Cub Room, reporting everyone, everything behind the solid gold chain. In any case, its name became as synonymous with New York as garrulous taxi drivers and Radio City Music Hall.

John Perona, the founder of El Morocco, was, like Cavallero, an Italian who received his early training at that international finishing school for waiters, London. He made his way to New York, and like so many of his contemporaries with a bent for catering, went into the speakeasy world. Having operated several "speaks" on the West side, Perona decided in 1931 that the less heavily populated East side (from the speakeasy point of view) would be more successful and safer. As Federal agents took somewhat the same attitude toward barroom furniture as Carry Nation, he felt that a sparsely decorated "desert"setting would not only be suggestive of the drinkers' dilemma, but would also present less opportunity for hatchetation. A site on East Fifty-fourth Street was chosen, and while Vernon MacFarlane was contriving blue and white zebra-striped banquettes, Perona pondered possible names such as the Sahara or The Sands. A penchant for North Africa, plus the success of another night club, El Patio, where Emil Coleman's society band was playing, influenced the final decision to call the club El Morocco.

El Morocco, more fondly known as "Elmo's," opened its doors in 1931 to several decades of unrelieved tablehopping. Mr. Perona, dashingly Italianate and speed-mad, with his North African oasis and Latin band, quickened the pulses of a new set that glided back and forth across the Atlantic on palatial liners. He welcomed Stork rejects, like the Maharajah of Jaipur, with deep salaams and brisk hand claps for prancing attendants. But Elmo's had its own snobbish house rules. Those unknown to the management were shown to tables on the far side of the dance floor, so distant a wasteland that it became known as "Siberia." The headwaiter, Carino, was a genius at "dressing a room" and could literally spread one celebrity over the entire club. Billingsley's clean-cut American was not Perona's ideal, but the fact that El Morocco did not achieve nearly the approval of the Stork on the part of college parents troubled Perona not the slightest. Nor did Jimmy Donahue hiding out in the covered roast beef wagon, nor Michael Farmer, Gloria Swanson's ex-husband, shaving at his table, nor Max Baer giving his fellow diners a hot foot. It was a dizzy age, and El Morocco, more than any place, was the home of Café Society —its members jammed the bar, crowded the dance floor with their tables, and appeared so regularly that a week's absence signified death. And if one were to believe the tabloid papers, a generation of the most powerful, creative, and talked-of individuals in America spent its glad existence against a background of blue and white zebra stripes. Neither El Morocco nor the Stork provided more than dance music. By the 1930s the night club with a long and lavishly costumed floor show had somewhat lost its status. Certain hotel night clubs, like the Empire Room at the Waldorf, or the Persian Room at the Plaza, booked the most "sophisticated" entertainers, but never quite achieved the cachet of the Stork Club or El Morocco, where dining and dancing celebrities were never obliged to share the spotlight.

For the professional world of the stage, however, the caricature-plastered walls of Vincent Sardi's theatrical institution have merrily, if mercilessly, mirrored the *gloria mundi* in transit. The restaurant itself pre-dated the Depression, and was founded in May of 1921 on West Forty-fourth Street. Sardi had originally worked for Rector's; he and his wife, Eugenia, were next employed at the Bartholdi Inn, a hotel with a theatrical following where they became acquainted with the hard core of their future clientele. No actor or actress, no producer, director, or playwright, no set, lighting, or costume designer, no stage manager or press agent worth his salt, Scotch, lasagna, or diet dessert has failed to appear at Sardi's sometime or other while in New York.

An opening night party at Sardi's has been considered as appropriate as a marriage at St. Bartholomew's or a funeral at Frank E. Campbell's for several decades of this century. The scene would vary only its starring cast, the supporting players being frequently one and the same, a sort of resident company of professional friends, flacks, and father confessors, soothsayers, goblins, and witches. If the producer had a run of luck, if one of the "angels" felt uncommonly expansive, or if the star's wattage was powerful enough, a large second floor, christened "Club Sardi," would be reserved. More intimate death watches might be held in the appealing Belasco Room, bearing a faint resemblance to an expanded old-fashioned private railway car, with paneled walls, a tiny bar, a buffet, and a scattering of tables and chairs. In both rooms the opening night party menu would follow a fairly set pattern of vegetable and chicken salads, lasagna, seafood Newburgh, cold sliced meats, chow mein, macédoine of fruit, cheese, French pastries and layer cakes. The atmosphere of such an event, however, was rarely conducive to digestion, particularly if there were any doubt as to the critics' verdict. Traditionally, when a star made his or her entrance through the downstairs front door on opening night, spontaneous applause erupted in the public dining rooms and bar; this would but temporarily assure peace of mind. Once upstairs the nerve-shredding process of smiling through commenced until the morning reviews came in. The press agent was generally obliged to break the news, his agony spent in a closet-like room down the hall with a dark brown highball in one hand and a damp telephone receiver in the other. When the presses began to roll, the review would be read to him in this gray and smoky cubicle. Only after he hung up did the producer and director finally get the irrevocable word. It cannot be said that a critical rave usually stimulated more drinking and hence more profits for Sardi's than an unqualified pan; many have been the times that the death penalty provoked a wake of Joycean proportions.

Whether or not a play failed or succeeded, an established theatrical personality could take solace in the knowledge that his face would still appear before the public at Sardi's, if not in the foyer where the reigning stars of the season are traditionally displayed, at least somewhere on the acres of wall. As of 1929, when Vincent Sardi gave Alex Gard two square meals a day in exchange for his caricatures of Broadway's best-known citizens, the latest crop of celebrities have been sardonically immortalized for the institutional rogues' gallery. Upon Gard's death, he was succeeded by Don Bevan, author of *Stalag 17*.

For years, hungry theatergoers more intent on simple, solid food than overt staring repaired to Reuben's after the performance. Arnold Reuben once boasted that his restaurant represented a development from a "sandwich into an Institution," and no one would gainsay the claim. Reuben's started in 1917 as a sandwich stand in Atlantic City, and arrived at its Fifty-eighth Street address between Fifth and Madison, in 1928. It laid no claim to being more than a delicatessen, and always had a sandwich counter at its entrance. But the mystique of Reuben's was that nothing ever changed—the stuffed-fish-and-sail-boat non-decor was changeless, the grouchy waiters grew old but never died, the basic menu, its duckling and red cabbage, its apple pancake and its Reuben's chow mein remained unaltered. Only its sandwich menu changed, very slowly, very hesitantly, until the list of its celebrities, whose favorites might be ordered, had all the historic interest of a geological cross-section, with everyone from Walter Winchell, Hildegarde (tongue, Swiss cheese, tomatoes, and Russian dressing on rye) and Judy Garland through Ginger Rogers (Nova-Scotia, cream cheese, and french-fried onions) and Mimi Benzell right up to Jackie Gleason and ZaZa Gabor (Turkey, tongue, cole slaw, whole wheat toast, and *no butter*). Final proof of Reuben's institutional character is the fact that one New Yorker, when traveling to France just after World War II, was requested by his Paris hostess to bring her two Reuben's cheesecakes, rather than the usual nylons and cigarettes.

It was on Arnold Reuben that gangster Arnold Rothstein played one of the telephonic practical jokes for which he was so well known, but not so very well liked. Rothstein was never away from a telephone for five minutes if he could help it. In this instance he was awaiting a call at Reuben's, and when the phone rang, he picked it up. It was Lillian Lorraine of the Ziegfeld *Follies*, who lived not very far away, and who wanted a chicken sandwich and a glass of milk sent up to her apartment immediately. Rothstein replied that the order would be right there. Then he slipped out of the restaurant, went to a pay phone, rang up Reuben's, and in what must have been a pretty extraordinary imitation of Miss Lorraine's voice, requested six dozen club sandwiches, a can of caviar, a gallon of pickles, and three gallons of milk. The results were predictable, but Arnold Rothstein's widow, Carolyn, in her book *Now I'll Tell,* tries to be conciliatory—"When this order reached Miss Lorraine, there was a great excitement. Miss Lorraine, as an artist, had a temperament."

It was the telephone that got Arnold Rothstein in the end. He was waiting for a call at Lindy's, the Broadway shrine to cheesecake, show business, and Damon Runyon, and this time the call came through. He arranged a rendezvous with parties unknown and was subsequently rubbed out.

Schrafft's, on the other hand, was an institution of a very different kind. The Schraffts were a prominent Boston family of candy makers, and it was one of their salesmen, Frank G. Shattuck, a farm boy from upstate New York, who dreamed up the idea of a new kind of restaurant. Frank, having tried his hand at working on the Erie Canal and on a tugboat in New York harbor, took a job with W. F. Schrafft and Sons of Boston in 1897. The next year Mr. Schrafft decided to expand his candy sales beyond the New England canton and gave young Shattuck the exclusive New York franchise to his merchandise. Shattuck, who owned his store outright, prospered happily and soon summoned his sister Jane and a fledgling businesswoman, Anna M. Herlt of Tyrone, Pennsylvania, to assist him. He opened a few more shops and then went back on the road again, leaving the girls to mind the store.

For a traveling salesman, eating on the road could be a dismal experience. The average restaurant may have advertised "home-style" cooking, but in cold actuality the fare consistently tasted like the grease and flour of yesteryear. Returning to the city grim, disgruntled, and several pounds lighter, Shattuck determined to do something positive in the way of remedial action and, by 1906, had a bustling restaurant operating in Syracuse. Others were rapidly opened in New York and Boston. The purpose of these restaurants was to serve good, wholesome American meals at reasonable cost to good wholesome Americans. Mr. Shattuck's modest ambitions, however, exceeded all human expectation. Several genera-

Lindy's. *Courtesy of the New York Public Library.*

tions of wholesome New York children gorged themselves on hot fudge and butterscotch sundaes while their elders were pacified by egg salad sandwiches or chicken à la King, and rafts of unwholesome businessmen cured their hangovers with flagons of fresh grapefruit juice or fastidious hairs of the dog. The setting was usually one of timid but tranquilizing traditional American nondecor: Grand Rapids cherry veneer and suitable-for-framing prints.

The Schrafft's at 21 West Fifty-first Street (it runs completely through the block, and consequently faces its opposite number on Fifty-second Street) was the sight, some years ago, of a party for the people working across the street. The people across the street have a sort of reciprocal agreement, and when Mrs. Shattuck had a birthday, she celebrated it with them. Mac and his brothers sang Happy Birthday (George Jean Nathan in the humor of his period always said Mac's family sings were a "must"—"You must blow the joint as soon as they start singing," he said), and then someone else cried "throw the cake over your shoulder for good luck," and she did, and everyone laughed and laughed, and then went into the Laughing Room and laughed some more.

Jack and Charlie's "21," already recognized as an institution before repeal, became a legal institution after 1933. Its co-founder John ("Jack") Kreindler was the newphew of the most fastidious saloonkeeper on the Lower East Side. His uncle took care of the customers, and, if beseeched by an irate wife, would not grant the husband in question more drinks than she would allow. He served only the highest quality beer and charged ten cents a glass, rather than the going nickel. Not only did he pass on his implacable standards to Jack, but also sent him to Fordham College.

In 1922, when he was twenty-four, Jack and his second cousin, Charles Berns, opened their first speakeasy in Greenwich Village. It was called the Red Head. By 1930 they had moved up the ladder to 21 West Fifty-second Street. As Jack's brothers matured, each one came into the business—Maxwell (who trained as a singer), Robert, and finally Peter. The family style was set by Jack, known as "the Baron," whose taste ran to sable-lined overcoats, cutaways, boxes at the opera, and Havana cigars. Louis Sobol described him in the *New York Journal* as "Two Trigger Jack" who "in chaps and ornate cowboy shirts, dusting his $10,000 silver saddle with an imported monogrammed handkerchief, will stand at a bar, preferably in Palm Springs or Tucson or Phoenix, tossing off drink after drink with you, leading the choral assembly in cowboy or nostalgic chants, riding his trusty mustang hard and true and relentlessly with no sign of fatigue." But the dedicated vocalist of the family was always Mac, who also distinguished himself in the Air Force. It was of Mac that John Steinbeck, one of the masters of American prose, said, "He served his country well. Some even say he served his country right." In tribute to Bob, who sweated out the war as a marine in the Pacific, Steinbeck said, "He did not lead the charge of San Juan Hill but this does not mean there will be no charge."

Steinbeck was not the only literary figure to frequent the "21." The brothers were also on friendly terms with John O'Hara, Ernest Hemingway, Sinclair Lewis, Somerset Maugham, H. G. Wells, and Robert Sherwood. The list of the contributors to *The Iron Gate*, a memorial to Jack Kreindler for the benefit of the New York Heart Association, includes Damon Runyon, Paul De Kruif, Lucius Beebe, Louis Bromfield, and Robert Benchley, as well as almost every figure in the world of journalism who was old enough to sit up at a typewriter before Jack's death in 1947. But that is speaking of the literary world alone. In fact, virtually every notable of the mid-twentieth century appears to have dined at the "21," from Lords Halifax and Beaverbrook to Mae West and Polly Adler, and it ranked with the Colony, the Stork Club, and El Morocco as one of Café Society's major havens in the gloom of the Depression.

It may be the "21's" clannish atmosphere that has attracted the modishly madding crowd. The clubhouse itself was never extraordinary: a brownstone exterior ornamented by New Orleans iron grillework, a comfortable foyer with overstuffed chairs, a crackling fireplace, and fresh flowers (and a professional greeter who warmly recognises "the right" guests and claims to turn away seventy people a night as undesirable), a barroom with dark red leather banquettes, oak panels and beams, rough plastered walls embedded with decorative wooden plaques, and red and white checked tablecloths. Above lies the main dining room, the smartest, with similarly rough plastered and wainscoted walls, and two auxiliary rooms, the Tapestry Room and the Bottle Room, decorated as indicated.

The owners of "21" were always incurable collectors. In the foyer and private dining rooms are the collections of Remington's paintings of the "Old West" started by Jack, who loved all things western. He also collected fancy shirts, saddles, and boots. The dining rooms house a huge collection of steins, a collection of pewter ware, and Georgian silver plates, urns, cups, samovars, trays, and candelabra. Out in front is the Kreindler collection of Jockey hitching posts, each painted in the riding colors of some patron—the Whitneys, the Vanderbilts, the Sassoons. The bar became what might be called the storeroom of some sophisticated squirrel. One day a marine friend of Bob's gave him a model of an S.B.D., an early bomber. Bob hung it on the ceiling. Such an act could lead to only one consequence: another collection. The bar ceiling came to resemble a mechanized Mammoth Cave, and pendant from it are models of trucks, tractors, helicopters, trains, cars, everything outside of the animal kingdom that moves. Also collected in the bar were university seals, mugs, coach lanterns, old firearms, bronze statues of horses and game, iron keys, original cartoons, Waterford glass, medallions, and wood carvings, all for the sake of collecting.

Above the public dining rooms are private dining rooms, like the Remington Room, and above these, two inner sancti for good friends only, a gym, and a barber shop. Damon Runyon always maintained there was a Laughing Room, well insulated for sound, where the brothers got together to plan the day's price list, but that was not to be found on inspection.

One had the feeling when speaking to a Kreindler or Berns not so much of life but of the deliberate magnification of it. Like Kosher salt and Christmas whiskey, it came to them in larger chunks and brighter packages than the regular kind. It was their talent to transmit this grandiose excitement to hand-chosen patrons that automatically led the twentieth-century New Yorker to associate expensive wine, expensive women, and mercifully infrequent song, with the number "21."

Le Pavillon, the latest comer of the institutions, ranked as the only restaurant in New York that could claim to hold the position in this century that Delmonico's held in the last. In his day Henri Soulé, its founder, trained as many chefs, impressed as many New Yorkers, and produced as many gastronomic masterpieces as the magnificent Lorenzo.

Soulé's life story is a typical rags-to-riches-via-the-hotel-kitchen tale, almost classic in outline. At fourteen he left his birthplace, a small French village near Bayonne, where his maman created a superb *brandad de morue,* a purée of salt codfish with sliced potato, vinegar, oil, garlic, and chopped parsley which he was later to recall as the epitome of culinary perfection. Traveling to Biarritz, he became busboy in the Continental Hotel, and by the age of twenty-three could claim to be the youngest captain of waiters in Paris. At thirty-six he took charge of the French Pavilion restaurant at the 1939 World's Fair, which was so universally acclaimed that two years later he opened his own austerely luxurious restaurant, Le Pavillon, on East Fifty-fifth Street. His staff included many who were with him at the French Pavilion. Isolated like Brillat-Savarin before him, alone in a foreign country with his exile group of aids while Paris trembled under the heel of the Occupation, he was to French cuisine what De Gaulle was to the Third Republic.

But it was Napoleon whom Soulé resembled, in both his height and his attention to detail. A florist was employed to spend six hours daily arranging flowers for the Pavillon's interior, and fifteen thousand dollars worth of roses were bought yearly to compliment the cerise damask upholstery of its chairs. Stories about Soulé lead one to the conclusion that he was a martinet and a snob, but above all a perfectionist who lived for his art. His art, according to Craig Claiborne (writing two decades later in the *New York Times*), was best expressed by mousse of sole "tout Paris," pilaff of mussels, pheasant with truffled sauce, and oeufs à la neige, a breathtaking caramel, custard, and meringue confection. Success did not turn Soulé's head. The Pavillon served one hundred and fifty people at lunch and seventy-five at dinner, but Soulé claimed that, in order to assure the most exquisite perfection of service, he would far rather serve no more than a third that number. As it was, fully half the tables at Pavillon were permanently reserved for steady customers—the Duke and Duchess of Windsor, Henry Ford II, Cole and Linda Porter, Mrs. Millicent Hearst, Mainbocher, and others. Only if they had not appeared by, say, one thirty, were their luncheon tables made available to others, also well known to Soulé.

It was the great World's Fair of 1939, more than anything before, that attracted the eyes not only of the nation, but the world, to New York. Soulé's Pavillon and another flawless restaurant, the Belgian Brussels, remained as small mementos of the event which stated New York's position as the metropolis of the twentieth century, once and for all. Probing searchlights were turned on structures that showed an awed audience of millions what modern architecture could do to reshape the world of the future. The visitor found himself in a never-never land of international friendship, a world in which the large corporations ranked as nations, and both were benevolent—Remington Rand, General Motors, Soviet Russia, Venezuela. Dinner was served by every country in the world, and homage was paid to the colors—orange and blue—of New York. The theme of the Fair was "The World of Tomorrow," and where else, but in New York, with its Rockefeller Center and Empire State Building, could the world of the future possibly be situated? And wasn't the future rosy? Wasn't the Depression on the wane? What if soapbox speakers ranted about fascism and world capitalism in Union Square. What if Nazi storm troopers marched along Yorksville's Eighty-sixth Street, right by the Drei Maederl Haus shouting "Heil Hitler" and breaking Jewish shop windows. The police could be counted on to deal with unsightly demonstrations, involving, after all, a small number of people. Meanwhile millions, out at the fairground, were applauding ceremonies held in the "Court of Peace."

Chapter VI

---◦⟨∞⟩◦---

1940-1973

Little Birds are tasting
    Gratitude and Gold
    Pale with sudden cold

Pale, I say, and wrinkled
    When the bells have tinkled
    And the tale is told.

Driven from their homelands by the bleak winds of a Second World War, countless uprooted Europeans escaped to America. Not since the French Revolution had so many talented, prosperous, and well-educated refugees gathered in New York at the same moment in history. Together they managed to reproduce a civilization in microcosm as artists, aristocrats, teachers, bankers, businessmen, and other representatives of a disrupted society reconvened in a foreign land where friendly natives welcomed them with warmth. Their presence contributed greatly to the status of New York, establishing the city once and for all as the cultural capital of a disjointed world, yet at the same time, the refugees' arrival stirred some ambivalent feelings—how long could Washington keep us out of this war, how long could the English stand alone?

Concerned New Yorkers hastily organized committees for the assistance of war victims. Bundles for Britain snapped into action, and at the English Speaking Union members excitedly poured tea for British military officers on brief leave. In Central Park, uniformed English nannies clustered in staunch little knots, furiously knitting long woolly scarves for His Majesty's Navy while their charges napped in stately lacquered prams.

Scuttling homeward at the approach of yet another major war came a chastened Elsa Maxwell. She later insisted that "Upon returning to the United States in 1939, I put frivolity, but not gaiety, behind me for the duration. Laughter is the best therapy for despair when reason and sanity seem to have disappeared from the face of the earth. In the next six years I gave, and helped to arrange, innumerable parties and balls."[12] Only one of these entertainments, she righteously explained, a joyous celebration sparked by the liberation of Paris, could have been labeled a "purely social affair." Although millions of dollars were said to have been raised at these parties, Miss Maxwell preferred to think that a more crucial contribution to the war effort had been hers, specifically the vehemence with which she stumped the country "pleading for all-out aid to the Allies and American participation in the war."

The last was soon enough to come when, at the close of 1941, a reeling United States declared war on Japan, Germany and Italy. New York blacked out, and city dwellers marveled at the soaring velvet silhouettes against a field of stars. War posters and official notices appeared everywhere, exhorting New Yorkers to buy bonds, save waste paper, and convert old bacon grease into glycerine for explosives by taking "your fat can to your butcher's." Others, more grim, warned compulsive gossips that their weakness could lead to the sinking of convoys and the wholesale death of troops. Enemy agents and saboteurs menaced the land; in their hands the most trivial sounding snippet of information could lead to disaster. In Yorkville, the German-American quarter centered on East Eighty-sixth Street, a cell of Axis sympathizers were apprehended in the upstairs dining room of a cozy Bavarian restaurant, while a few blocks to the north an espionage ring continued to ply its trade out of a respectably dowdy brownstone until the F.B.I. ran it to ground, a cheering denouement applauded by Hollywood in *The House on Ninety-Second Street.*

Morale-boosting movies braced weary audiences with the adrenalin of patriotic fervor, as did the tirelessly paternal voice of radio. New Yorkers from all walks of life rallied at war bond drives and daily rounds of benefits. School children collected defense stamps, old toothpaste tubes, and other scrap metals, penthouse dwellers rose at dawn to weed their miniature victory gardens, hysterical sirens heralded air raid drills as grandmotherly wardens in military helmets scanned the heavens for enemy craft through stout official lenses. By governmental decree, sugar, coffee, meat, butter, and canned goods were to be strictly rationed, ice cream flavors reduced to ten, and all production banned on metal asparagus tongs, beer steins, spittoons, popcorn poppers, and lobster forks. In their windows, housewives were urged to post an Office of Price Administration sticker portraying a gallant young matron in ruffled apron, her right hand held aloft to voice the Home Front Pledge: "I Pay No More Than Top Legal Prices. I Accept No Rationed Goods Without Giving Up Ration Stamps."

At the automat, the individual beef pot pie could still be wrested from behind its locked window for a mere fifteen cents, and at drugstore counters office workers continued to lunch cheaply on tuna salad, grilled cheese, or fried egg and catsup sandwiches washed down with cherry cokes or a "cuppa java." Europeans accustomed to afternoon levity clustered around the elegant tea tables of Rosemarie de Paris and Rumpelmayer's, amidst a welter of beribboned bon-bon boxes, silver pastry trays, beady-eyed fox furs, and tiers of glassy counters guarded by giant stuffed toys. Inevitably, a note of sobriety would be struck by a soignée volunteer seated near the doorway rattling a can of coins for a worthy cause.

Finding accommodations for the mass invasion of refugees and wartime personnel became increasingly difficult. Along with rationing and price controls, restrictions had been imposed upon Manhattan hotels limiting to five the number of days a transient might occupy a room. The result was a frustrating citywide game of musical beds. Ration coupons clutched in hand, New Yorkers learned to stand on line with the bland resignation of so many native Englishmen. Cigarettes and liquor were pursued with particular patience and cunning. The ferreting out of restaurants and shops where rare luxuries could be had preoccupied gourmets and gourmands alike. At parties it was whispered confidentially that authentic butter still melted in the mouth at a modest French bistro called Steak Pommes Frites or that hardly a dent had been made in the Lafayette's wondrously mellow cellars. Savants also knew that beef of mythic beauty could be savored regularly in the somewhat psychotic atmosphere of Marcel and Louise's Au Cheval Pie on East Forty-fifth Street, where a constant air of nervous electricity was generated by the Punch and Judy marital incompatibility of the owners. Marcel, a rabid Gaullist whose restaurant walls glared with photographic testimony to his idol's gawky grandeur, kept a vast box behind the bar into which, in cacophonic counterpoint to Louise's shrieks, he would viciously smash bottles as soon as they were empty. Louise's wrath would only occasionally express itself in direct assault and battery, leaving traces of dark blood to dry slowly on Marcel's apoplectic face. Nevertheless, the beef was the best in the city, an overriding consideration from any point of view, and, practically speaking, it was a wise diner who came to ignore the management's sporadic bursts of mayhem. The source of Marcel's treasures remained a mystery despite the disconcerting visitation, one evening during the dinner hour, of a shrouded figure bearing a trussed live sheep in his arms. As this apparition padded silently into the kitchen and out again and into the night, Louise's flintily blank gaze effectively quelled any unwelcome queries.

Throughout the war, midtown restaurants such as Au Cheval Pie were patronized heavily by "Madison Avenue," the New York advertising fraternity whose business occupation over the years had gradually assumed the patina of a "glamor profession." Madison Avenue had a lot of money to spend and spent it freely, whether entertaining clients or merely themselves. Two leading admen, Frank Dougherty and Chet Laroche, together with Ted Patrick, founder of *Holiday* magazine, launched a private lunch club for the use of their friends, calling it the Hapsburg House and commissioning Ludwig Bemelmans to cover its walls with gaily painted black and white murals of Old Vienna. In contrast to the Hapsburg House's evocation of prewar Austria's kitchens, Third Avenue's Joe and Rose's hamburger and steak ambiance recalled nothing so acutely as a collegiate bar and grill. At Joe and Rose's the honor system prevailed, and it became a common sight to see a vice president of Young and Rubicam or J. Walter Thompson behind the bar mixing his own martini, ringing up the sale on the cash register, and punctiliously settling his bill. So scrupulous were these amateur bartenders that the management found that they turned a higher profit than they did with professionals at the helm. This revelation led to the encouragement of fraternal service to others as well as one's self, with the end result that convivial legions of copy writers and art directors might spend as much time behind the bar dispensing drinks as they would consuming them at tables.

As well as patronizing "21," Toots Shor's, and the Cub Room, advertising men lifted elbows with *New Yorker* staffers and other various convivial literati, foreign correspondents, sports writers, playwrights and novelists at Tim and Joe Costello's bar and restaurant, immortalized in John McNulty's riotous *Third Avenue, New York* stories. Hemingway left a shillelagh snapped in half as a test of strength, hanging over the bar, and Thurber decked the walls with his harridans and craven males, beamish bassets, and a grim football scrimmage line, above the gripping legend "The Day Vassar Beat Harvard."

The Costello brothers hailed from County Offaly in Ireland. Tim, an I.R.A. man who had driven a taxi in Dublin, mourned the days of Prohibition: "The boss of this saloon on Third Avenue," recalled McNulty in one of his Costello tales, "often says he wishes there was such a thing as a speakeasy license because when all is said and done he'd rather have a speakeasy than an open saloon that everybody can come into the way they all are now. Not that he is exactly opposed to people coming in. They spend money, no denying that. But a speakeasy, you could control who comes in and it was more homelike and more often not crowded the way this saloon is now. Johnny, one of the hackmen outside, put the whole thing in a nutshell one night when they were talking about a certain hangout and Johnny said, 'Nobody goes there any more. It's too crowded.'"[12]

At the end of the day, war anxieties and battle fatigue could be combated in one of New York's many packed night clubs where, in grateful oblivion, servicemen and their girls jostled their way through a crush of newly poor European titles and newly rich defense tycoons onto napkin-sized dance floors. The U.S.O. threw open its doors for lonely enlisted men; rich college boys, having picked up their dates under the clock at the Biltmore, vied for the best tables at Larue's, as smart now as the Stork Club; juke boxes in taverns all over town blared "Praise the Lord and Pass the Ammunition" and "Comin' in on a Wing and a Prayer"; and at a smart little East Side boite called La Vie Parisienne, Celeste Holm, the insouciant blond star of Rogers and Hammerstein's *Oklahoma!*, summed up a common wartime complaint when nightly she took the tiny stage to sing: "They're Either Too Young or Too Old!"

Close by at the Persian Room, suffering long satin gloves while accompanying herself at the piano, "The Incomparable Hildegard," a self-styled Parisienne from Milwaukee, or "chantootsie" as the columnists might have it, lamented her brief yet unforgettable Gallic past—"The Last Time I Saw" . . . twinkle, trill, sigh . . . "Paris, her heart was young and gay," a reverie inevitably followed by the singer's musical trademark, a ballad of linguistic

despair in which an American soldier wooes his inamorata in frac-
tured French.

Music of a completely different order was being heard for the
first time by New Yorkers in the sophisticated environs of Max
Gordon's Village Vanguard, the Blue Angel, and both the Uptown
and Downtown Café Societies. Legendary blues and folk singers,
including Leadbelly, Josh White, and Woody Guthrie, hoisted
their guitars, lowered their heads, and belted out unvarnished
truths from a dark and unfamiliar America. New Yorkers loved it
as indiscriminately as they did the syncopated barrel-house piano
pounding of Albert Ammons, Pete Johnson, and Meade Lux Lewis
whose rambling, rumbling jazz patterns inspired painter Piet
Mondrian's final abstract masterpieces, the "Broadway" and "Victory Boogie-Woogies."

In Greenwich Village the prodigally talented young team of
Betty Comden, Adolph Green, and Judy Tuvim, who would later
appear on Broadway marquees as Judy Holiday, were experimenting in night clubs with a fresh form of satirical comedy remarkable
for its uncustomary intelligence, imagination, and wit. A brief
taste of their style can be sampled in this, one of their merciless
condensations of Great Books:

> *Les Miserables*
>
> *Jean Valjean no evil doer*
> *Stole some bread 'cause he was poor*
> *Detective chased him through a sewer*
>
> *The End.*

In the grill of the Hotel Roosevelt, the reassuring drone of Guy
Lombardo's saxophones mollified conservative dancers unmoved
by the rumba, conga, and jitterbug crazes. The hotel itself was
purchased early in 1943 by Conrad Hilton to get in practice, he
explained, for buying the Waldorf.

Habitués of the Roosevelt expressed their qualms about a parvenu cowboy saloon keeper, or worse, commandeering their
chosen territory. Up in Albany, an alarmed state assembly was
driven, rashly, to verse:

> *From what we hear it would appear there's been some changes made,*
> *That great hotel, the Roos-e-velt, is mixed up in a trade.*
> *Some bird named Connie Hilton from his California nest*
> *Is gonna show New Yorkers how they do it in the west.*[11]

Little did he know.

One friendly Manhattan real estate broker had tried to dissuade Hilton from saddling himself with the twenty-three-storey Roosevelt, insisting that "In New York hotels are dead. There are too many of them." In his memoires, Hilton later recalled that "That was the attitude about hotels all over the East. They had watched them go down, and further down, during the Depression. They had seen what the war brought, increased patronage perhaps, but nothing yet to offset the difference between rising costs, wages, shortages and the OPA ceilings. Maintenance was difficult and not only did the financial picture show strain but the physical plants themselves were deteriorating."[11]

A few months later found an undaunted Hilton, in partnership with an investment trust, buying the Plaza, the dowager duchess of New York's luxury hotels for seven million, four hundred thousand dollars, many millions less than it originally had cost to build in 1907. Across the country, newspapers ran overwrought stories beneath Horatio Alger headlines—"From bellboy to the owner of the Plaza!" marveled one typically misinformed reporter. Before the war's end Hilton had launched a six million dollar renovation program, the prospect of which, with some justification, alarmed the Plaza's full complement of reactionary residents. In deference to Hilton's credo—"Make the Space Pay!"—the glorious Tiffany glass dome arching luminously over the Palm Court was condemned to death with lucrative new office space usurping its lofty upper regions. On the brighter side, when Hilton discovered that the brokerage firm of E. F. Hutton was still occupying main-floor headquarters overlooking Fifty-ninth Street and the park for the laughable sum of four hundred and sixteen dollars a month, he ordered them out and transformed the vacated space into an instant institution, the Oak Bar. Within an astonishingly short time, the bar's annual gross reached a quarter of a million dollars, supporting Hilton's grandiose predictions for the future—"How big can you dream? The sky is the limit!"

The postwar boom confirmed Hilton's optimism and, for a generation, the country's economy spurted upward into previously uncharted stratospheres. With the Fair Deal came a steep hike in the legal minimum wage—from forty to sixty-five cents an hour, and the average salary soared to more than three times that of 1918, a vintage boom year. By the early 1950s Americans were spending over seventy billion dollars a year on food alone. Observers felt the key to this unparalleled prosperity to be an explosive increase in both population and productivity. In any case, the emergent "Affluent Society," as the economist and diplomat John Kenneth Galbraith dubbed it, had been born full-blown as a cabbage rose on a war memorial, an unnatural hybrid, in the traditional American mating of luxury and squalor, of private wealth and public poverty. The line of demarcation between the two, wrote Galbraith, was "roughly that which divides privately produced and marketed goods and services from publicly rendered services."

The boom, then, presented drawbacks as well as advantages. As early as 1948, E. B. White, unabashed lover of New York City, was quick to recognize its insidious side effects:

The city has never been so uncomfortable, so crowded, so tense. Money has been plentiful and New York has responded. Restaurants are hard to get into; businessmen stand in line for a Schrafft's luncheon as meekly as idle men used to stand in soup lines. (Prosperity creates its breadlines, the same as depression.) The lunch hour in Manhattan has been shoved ahead half an hour, to 12:00 or 12:30, in the hopes of beating the crowd to a table. Everyone is a little emptier at quitting time than he used to be. Apartments are festooned with No Vacancy signs. There is standing-room-only in Fifth Avenue buses, which once reserved a seat for every paying guest. The old double-deckers are disappearing—people don't ride just for the fun of it any more. . . . By comparison with other less hectic days, the city is uncomfortable and inconvenient; but New Yorkers temporarily do not crave comfort and convenience—if they did they would live elsewhere.[13]

What New York could supply, on the other hand, was all its old excitement and energy, its capacity to thrill. New York—with the seal of the U.N. freshly pasted to its mailbox, bringing final, official recognition as center of the world—New York was the city of life.

Brendan Behan, that boiling Irish broth of a poet, playwright, and patriot, understood this well, and before his death in 1965, echoed Nat P. Willis' dark sentiments concerning private dining in nineteenth-century grand hotels: "We don't come to a city to be alone, and the test of a city is the ease with which you can see and talk to people. A city is a place where you are least likely to get a bite from a wild sheep and I'd say that New York is the friendliest city I know."[14] Behan loved the skyscrapers and neon lights, finding them "so homelike, safe and reassuring . . . Manhattan is a mother clasp," and endorsed the city's singular tonic powers: "I knew an old Irishman who went there when he was seventy-five and ill and like a Lourdes of light, New York cured him and he lived for years afterwards, a healthy and happy old man. He painted my wife's grandmother and his name was Jack Yeats, the father of William Butler Yeats, the great poet of Ireland and the world, and Jack B. Yeats, a great painter himself."[15]

The resonance of the postwar boom jarred New York into unfamiliar new visual shapes and social patterns. Metamorphosis was gradual yet severe. Rather than collapsing overnight, many seemingly invulnerable institutions and landmarks, along with personal values and public traditions, slowly sank in shifting social tides or crumbled to dust through sheer neglect.

Once again New York celebrated the end of war by submitting to major surgery. With the easing of building controls, the wrecker's ball and eager construction crews roared into action. Great raw craters were blasted in Manhattan's bedrock, and out of these eventually arose shimmering towers of glass and steel like giant mineral prisms or, depending on the depths of one's nostalgia, like silent crystal tombstones in the graveyard of the past.

Internationally speaking, the most auspicious new buildings of all were those designed to house the United Nations. Manhattan was considered too crowded for the U.N. and had not been seriously studied; it was not until John D. Rockefeller, Jr. offered eight and a half million dollars for the purchase of a rundown waterfront site along the East River, a shambles of sagging slaughterhouses, light industry, and a railroad barge landing, that the final decision to settle in New York was accepted by a large majority of the General Assembly. Excavators began blasting in 1948 and on August 21, 1950 the first members of the Secretariat moved into their new home.

Indisputably, the favored residential hotel of well-heeled U.N. delegates has been the Waldorf Towers. Comprising the twenty-first to forty-second floors of the Waldorf-Astoria, the Towers has a private lobby, private elevators, its own bellmen, and other staff. On the walls of its most elaborate suites, discreet bronze plaques attest to the illustriousness of former residents and transient guests; the great seal of the United States surmounts the entrance of one such apartment, the official residence of the United States Ambassador to the United Nations.

As many as fifty delegations have been accommodated in the

Towers at the same time, a chaotic situation for the flag room, where more than a hundred flags are kept to be unfurled not only when a dignitary is in residence, but on a country's national holiday. In times of stress, such as the opening session of the U.N., the flags must be dutifully rotated on five poles around the hotel in strict accordance with protocol.

A more delicate crisis arises when insufficient room is found at this rarefied inn (triumphantly acquired by Hilton in 1949), and diplomatic tempers have been known to simmer dangerously. Other minor disasters can occur; it was here at the Towers that a volatile Nikita Khrushchev, accompanied by the hotel's ashen manager, was trapped in a stalled elevator prior to attending a reception at the Hotel Plaza, where he was roundly booed in the corridors by two azure-haired American dowagers. Turning upon the ladies, Comrade Khrushchev stuck out his tongue and gave them a juicy Bronx cheer.

This and other misadventures seemed to daily plague the summit conference of 1960, attended by Tito, Nehru, and Nasser, among others, with the ubiquitous Khrushchev haranguing crowds from a Park Avenue balcony, pounding a sturdy Soviet shoe on his delegation's U.N. desk, and Fidel Castro being ejected from a Murray Hill businessman's hotel for allegedly plucking and cooking chickens in a suite with no stove.

"There are times," observed New Journalist Gay Talese, "when it seems the whole city of New York is capable of going mad, of exploding into riot."

On Tuesday, September 20, 1960, when Khrushchev, Castro and other foreign leaders visited the UN, everybody in New York seemed mad at everybody else. The Ukrainians demonstrated against the presence of Khrushchev, Khrushchev complained of police brutality, many of the police were mad because they had to work through the Jewish holidays; New York's rabbis blamed it all on Police Commissioner Kennedy, who blamed it on Khrushchev. Outside the UN the Greeks cursed the Albanians, nihilists blasted pacifists, British Guiana students scorned England, and a group of rioting anti-Castro Cubans paraded up and down shouting, "Fi-del-ist . . . Com-mun-ista!" Outside the Waldorf, staff members of the *Catholic Worker* picketed against the American Banking Association's convention, and on East Fifty-fifth Street a truck driver named Tom Horch denounced the National Biscuit Company and demanded higher wages. All over town sirens blared, plain-clothesmen stood like gargoyles on rooftops, and cab drivers insulted everybody. And on Forty-fourth

Street, Mrs. Sylvia Kraus of 25 East Seventy-seventh Street carried a placard reading: "Americans Awake—Germ Warfare Has Begun."

"I know that people are putting things in my food," she told crowds in the street. "They've been trying to eliminate me since 1956, but I know how to combat it." Then she disappeared into the crowd without telling how.[15]

The easing of controls by President Truman in 1947 might have been taken as a symbolic portent, a small-craft warning for social navigation, had anyone been able to foresee the rampaging tides of the future. The first symptoms of social revolution seem almost benign in contrast to the tumult of a later day. The 1947 Broadway production of Tennessee Williams' *A Streetcar Named Desire,* for example, with its sexually obsessive themes, managed to violate a whole catechism of Broadway taboos. Twenty-five years later, in the same theatrical district, live performances of a type once restricted to the fleshpots of Port Said and other similarly inaccessible Sodoms, would be conveniently scheduled so that office workers might attend at lunch hour.

The seeds of a more bleak trend were sown when a few prominent jazz figures switched from "reefers" to hard narcotics. "Boppers" of the 1940s had cut traditional jazz from its moorings, dispersing it into "cool" and "way out" mosaics of sound that alienated older musicians, among them Louis Armstrong, who damned bop as "modern malice." In New York, the prince of the bop world, and Birdland, its royal palace, was the horn-playing band leader Dizzy Gillespie, whose black beret, sparse goatee, leopard-skin vest, suede shoes, and indispensable dark glasses, together with his goofy jokes and drug-oriented argot, epitomized the public image of the quintessential bopper.

The widespread use of narcotics by jazz musicians apparently harks back to the 1920s; according to jazz historian Gary Kramer they "needed something to offset the desperate condition in which they found themselves after 'hot' jazz all but died in the late twenties. Marijuana helped them reconcile themselves to an intolerable reality, they insisted." However, with bop came a high incidence of heroin addiction, and it was not long before "the *hipness* of drug use . . . was so smart that some clubs all but advertised publicly that addict-musicians made their place their headquarters," a disastrous turn of events in terms of the future. In their doomed rush to enter "the cool world," worshipful fans could not discern what Kramer could see all too clearly—"The surface level gaiety and insouciance of the boppers and their casual, hip use of drugs represented only one side of the coin; fanaticism, malice, deep-seated anxiety, expressions of inadequacy and a wish for withdrawal from reality were a part of bop, too."

A more salutary mode of escape presented itself to Americans with the advent of the commercial jet airplane, a technological breakthrough democratizing travel once and for all. Not only could Café Society rise phoenix-like from the ashes of war as the peripatetic Jet Set, but soon it would appear as if all walks had become all flights of life.

Fueled by the cultural and economic booms, the national passion for flight amounted to mania, with wily "fly now and pay later" schemes making it possible for both white- and blue-collar workers to abandon their annual outings at the nearest peak or shore for a dizzying two-week spin through as many countries as energy and digestive equilibrium might permit.

With money in the bank and time on their hands, Americans turned to an enticingly unfamiliar agenda of pursuits and interests. At the head of the list stood the arts, travel, good food and good drink.

During the 1950s certain of the city's leading citizens began to feel that the performing arts were not well housed in New York. The "yellow brewery" of the Metropolitan Opera House looked more like a grimy warehouse than a citadel of the arts, although its auditorium was still one of the finest in the world, and its red-damask Sherry's Restaurant, the last Sherry's institution of that name, with its life-size portraits (Ezio Pinza staring sardonically down as an earringed Don Juan), exuded more nostalgic atmosphere than any other public place in New York. It now stood on the upper reaches of the garment district and in embarrassing proximity to the leering pornography shops that had invaded the Broadway area. Carnegie Hall, with its peerless acoustics, grew shabby and stood in constant danger of demolition. The Byzantine-Masonic atmosphere of the City Center, crushed into the side street of Fifty-fifth, served as a pleasant enough setting for ballet but did not supply the grandeur necessary to the city that had outstripped both Paris and London to become the world's center of performing arts (in terms of both attendance, and the number and variety of performances available).

Again it was a Rockefeller who initiated a vast construction project for the city, this time John D. Rockefeller III, who envisioned a cultural complex that would help fulfill "some needs of an anxious age." In 1956 the Lincoln Center for the Performing Arts was incorporated, not without controversy. Champions of the old Met pointed out that it was far from worn out despite its scabrous exterior, and lovers of Carnegie Hall were prepared to throw themselves physically in the path of the oncoming bulldozer.

Despite the outcries, however, the Metropolitan Opera Association, the New York Philharmonic Symphony Society, and the Julliard School joined the project in quick succession, and in a remarkably short time, private donations to the sum of over a hundred and forty million dollars were raised (and supplemented by over thirty seven million in government funds). A site in the West Sixties, just north of Columbus Circle was purchased, and by May 14, 1959, all was ready for President Eisenhower to come to New York and break ground.

If ground was broken during the Republican fifties, the new Philharmonic Hall opened on September 23, 1962, to the full blaze of the glory of the New Frontier, with Jacqueline Kennedy in attendance. CBS pronounced the opening "an unprecedented moment of living history," and Leonard Bernstein conducted a performance of Beethoven, Copland, Vaughan Williams, and Mahler with four hundred voices. The monolithic volumes and clean lines of Max Abramovitz's new hall came as an unpleasant jolt to some, while to others it was just "a gracefully simple building of concrete and glass." Unfortunately, the acoustics were off, and the gold and blue auditorium underwent a series of painful operations. Overly smooth walls were pulled out and replaced with curved, reflecting surfaces. Carpeting was taken up and new chairs installed to reduce the amount of sound-absorbing fabric. The shape of the ceiling was altered. New air conditioning and lighting systems were required. Finally, the entire room was redecorated, and its predominantly blue color scheme changed to one of red and shades of natural wood "in order to build rapport between artist and audience."

Meanwhile, one after another, the New York State Theater, built especially for George Balanchine and the New York City Ballet by Philip Johnson, The Vivian Beaumont Theater, and the new Metropolitan Opera House opened around the central plaza and fountain, to combined adulation and catcalls.

Lincoln Center, like Rockefeller Center, became a city within a city, with the addition of Alice Tully recital hall, the Julliard School of Music, and the Library and Museum of Performing Arts, and in the various theaters areas were blocked off by barriers and palms to serve as cafés, although the glamor of the hellish red glow of Sherry's had departed. Like a magnet, it attracted business, most especially the entertainment business, to the area. The neighborhood had decayed badly since the days Reisenweber's drew crowds to Columbus Circle, but now tenements were pulled down and office buildings went up, housing television studios, the media, and commercial music enterprises, and, of course, the fresh awnings of restaurants extended over the side streets, although Vorst's, one of the neighborhood's ancient German restaurants, was bought out by the Mormon Church.

The nearby Ginger Man staked out the first, loudest, and most persistent claim to the attention of Lincoln Center regulars. At the Ginger Man, Metropolitan stars and administrators rubbed shoulders with rock singers and film festival movie buffs, few of them recognizing the other. On Mondays, Society's traditional night at the Opera, the late crowd would usually be divided between parties in black tie and others in dungarees.

The Ginger Man's owners, Mike and Patrick O'Neal, had no claims whatever to the mastery of cuisine. They had envisioned something like a standard P. J. Clarke's hamburger-and-chili menu, but while waiting for a liquor license, Mike decided to take a few lessons from the English *grande dame* of the kitchen, Dione Lucas. When he explained his intentions to his new teacher, she unexpectedly announced that she would like to work with him. And so for three years Dione Lucas herself sat whipping up omelettes in the Ginger Man's increasingly reputable kitchens, for the nominal fee, at first, of a hundred and eighty-five dollars a week. But working with Dione was not always easy. Not only was she a professional cook of the most exalted rank, but an advanced hypochondriac, given to elaborate fainting spells for her most startling dramatic effects, she would powder her face with flour. Eventually, the tenuous and never formalized partnership broke up, but in the meantime Dione Lucas' light luncheon and after-theater supper menus had set a high and frequently imitated standard for the Lincoln Center restaurant community.

LUNCHEON

Potages—
Soupe du jour *1.00*
Vichyssoise froid *1.00*
Soupe à l'oignon gratinée *1.50*
Gazpacho *1.45*

Hors d'oeuvres—
Pâté maison *1.20*
Quiche Lorraine *1.25*
Champignon á la Grecques *1.50*
Salade de Tête d'veau *1.00*

Salades—
Tossed French salad *1.00*
Sliced tomatoes vinaigrette *1.00*

Plat du Jour—
Coquilles St. Jacques *3.15*
Ragout de Boeuf Bourguignon *3.65*
Large Portion of Prosciutto and Melon *3.00*
Caesar Salad *2.85*

*French Omelettes—*Caviar and sour cream *4.50*
Eggplant and tomato *2.65*
Chicken livers *3.00*
Fine Herbs *2.30*
Mushroom *2.85*
Cheese *2.75*
Bonne Femme *2.95*
Grandmére *2.80*
Watercress and sour cream *2.65*
Ham *2.75*
Omelette Special: Benedict *3.85*

From the Grill—
Sirloin Steak *4.65*
Chopped Steak Provençale *4.15*
Cheeseburger *1.75*
Hamburger *1.50*

Saladier—
Poulet Froid, Garni *3.85*
Steak Tartare *4.95*
Chef's Salad *2.75*
Spinach, raw mushrooms, bacon salad *2.85*

Deserts—	Roulade aux noix–Austrian nut roll *1.20*
	Roulade Léontine–Chocolate roll *1.20*
	Chocolate Mousse *1.00*
	Créme au Caramel *1.10*
	Melon Glacé *1.20*
Beverages—	Coffee, Tea or Sanka *.35*
	Pot of Espresso (for two) *1.20*

The O'Neals also opened a larger, less expensive, bent-wood-chair-and-hamburger cafe directly across the street from the New York State Theater. It became immediately popular with the chorus girls and boys from Lincoln Center, and was called by the jaunty name of O'Neal's Saloon. All of this might have seemed innocent enough, but not to the New York State Liquor Authority. To most New Yorkers the old pre-Prohibition "saloon" presents about as immediate an image as Washington Irving's "Old Bull's Head." But as a splendid, almost awe-inspiring tribute to history, the very word "saloon" has been pronounced legally anathema. To quote the New York State Liquor Authority Digest: "It shall be against the public policy of the State to permit the selling or serving of alcoholic beverages for consumption in such premises as were commonly known and referred to as saloons, prior to the adoption of the eighteenth article of amendment to the Constitution of the United States of America." On repeal day, thirty and more years before, Roosevelt had asked that there be "no more saloons," and so, in the presence of three network television cameras, O'Neal removed the "S" from the sign over his front windows, and replaced it with a "B," laying to rest at long last the ghost of Carry Nation.

In Manhattan, a populous new class of cosmopolites, composed of residents and visitors alike, began to exercise their recently acquired gastronomic expertise in the island's inexhaustible collection of restaurants. How, they might ask, did New York's great restaurants really compare to those abroad?

An honest assessment could lead to one conclusion only. Despite the bold predictions of James Fenimore Cooper's "Traveling Bachelor," France had yet to relinquish its position of epicurean preeminence to America. And the chances of its doing so seemed fairly dim. Cole Porter and Elsa Maxwell might claim Le Pavillon to be the finest French restaurant in the world, but what about Cocteau and the Viscomtesse de Nouailles, would they concur? Assuredly not. Once again, what New York had to offer was its endless excitement and variety. A kaleidescopic city where age might wither, but custom never staled, New York was the only place on earth renowned equally for its osso bucco, ratatouille, Swedish pancakes, Russian babka, paella, wiener schnitzel, and matzoh balls. New York listed more than forty-five thousand eating places, night clubs, and bars in its yellow pages, with hundreds more opening each year (although an estimated 80 per cent of the newcomers would close after a brief and unsuccessful period). Certain ethnic dishes, such as pizza, bagels, frankfurters with sauerkraut, egg rolls, Italian ices, enchiladas, and crêpes had become so integral a part of daily life that in most instances they could be bought at open-air counters or pushcarts on the street.

As a tradition, the Lower East Side's gift to New York society, the Jewish delicatessen, had entrenched itself as firmly as the chop house. The basic menu of a restaurant-"deli" generally would embrace borscht and chicken soup with kreplach, liver dumpling or matzoh balls; pot cheese, vegetables or fruit with sour cream; lox and other smoked fish, chopped liver, herring, dill pickles, turkey wings and drumsticks, boiled beef with mild horseradish sauce, chicken in the pot, potted steak with potato pancakes, kasha, knishes, Kosher frankfurters "with individual can of beans," apple strudel and "Danish" (pastry). Under "Beverages and Cold Drinks" one would find prune, tomato, and citrus juices, glasses of tea, Dr. Brown's Cel-Ray, cream or cherry sodas, Saratoga Geyser water, Lo-Cal, buttermilk, Postum, coffee, local and imported beer.

With the rise of the city's Latin-American population came *comidas criollas,* literally "creole meals," suggesting rice, either plain or saffron tinted, with chorizo sausage, black or pink beans; fried plantains; *ropas viejas* ("old rags"), a chewy ragout of beef and peppers; chicken, pork, or shrimp; avocado, lettuce and onion salads and sweet, delicate flans. To drink would be *tamarindo,* colas, and an occasional "egg cream," a relic of the Jewish Lower East Side blending syrup, seltzer, and milk—a kind of ice cream-less ice cream soda.

A new breed of Italian restaurant had sauntered suavely onto the scene, bringing an elegant aura of *dolce far niente* to a frenetic Manhattan—the worldly Quo Vadis, the expansive Romeo Salta, the scintillatingly sexy brothers Orsini, and Gino's—championed by Fred Allen and an enduring clubhouse for the international set.

Japanese steak houses, Japanese tea houses, Japanese gift shops abounded. Naya Tolischus' exhuberant Athena East introduced New York to its first taste of Greek shipping magnate highlife. Elsewhere the tintinnabulation of finger cymbals, and the thrashing of sequinned abdomens at the Egyptian Gardens took Levantine minds off their stuffed vine leaves, while at smorgasbord restaurants further uptown, Nordics soothed themselves with smoked eel and akvavit. At the Russian Tea Room on West Fifty-seventh Street, blini, Beluga, and glacial Slavik vodkas fueled the artistic visions of such noted neo-Romantics as pianist Arthur Rubinstein, critic Leo Lerman, and dancer Rudolph Nureyev.

The 1964–65 New York World's Fair came as something of an anticlimax, as it needlessly underscored the city's international character. According to its president, the purpose of the Fair, which had been refused the blessing of the Bureau of International Expositions, was to serve as " 'Olympic Games,' where the peoples of the world send their best products, scientific development, art and culture for examination and competition in much the same way as athletes are sent to compete in the many and varied events of an Olympic schedule." Translated, this meant that New York was to be treated to a glorified global trade show, and an imbalanced one at that in view of the notable lack of participation on the part of national giants such as Russia and France. Nevertheless, according to Kate Simon in her guidebook to the Fair, the project's entrepreneurs had promised prospective exhibitors something like "a composite of Eden, Harvard, the United Nations, the Tivoli Gardens, a most expensive, permissive nursery school, the Louvre—all housed in a lustrous string of modern Taj Mahals managed by the managers of Disneyland."[16]

Food lovers were not disappointed and, after the opening ceremonies, settled like scattered birds of passage in the leafy shadows of an African tree restaurant, on Moroccan pillows, Japanese tatami mats, teetering Mexican barstools, and functional Danish chairs. One American dining retreat fearlessly proclaimed itself the World's Fair Festival of Gas, but serious critics had to agree that the truly exceptional contribution to gastronomy had been made by the Spanish Pavillon. The Pavilion itself was the jewel of the Fair, and its formal restaurant a triumph. Most New Yorkers had previously equated Spanish food with bland *arroz con pollo* of Puerto Rican extraction, Greenwich Village *paella,* Fourteenth Street garbanzo salad, or, at their most benighted, Mexico's chili pepper repertoire. The impeccable *alta cocina* and noble wine list of the Spanish Pavilion, under the supervision of Alberto Heras, came as a revelation. Boned partridge in a white wine sauce made mysterious by a mite of chocolate and sea bass in a rosy essence of shell fish were only two of Heras' sumptuous surprises. Soon, long lines formed at his door.

The inspirational success stories of Le Pavillon and the Brussels, both of which had been launched at the 1939 World's Fair, encouraged Heras to stay on in New York after 1965. Elegant new quarters were found in the Park Avenue Ritz Towers, and there the Spanish Pavilion authoritatively ensconced itself as a member of the city's loftiest restaurant aristocracy. Immediately the Spanish diplomatic corps and local New York colony took up unofficial residence, with Salvador Dali feeling sufficiently at home to commandeer a chafing dish and whip up a dish of his own Surrealist invention, flamed striped bass with pears and grapes in a puree of green peas. This to be washed down with the house's famous sangria, the recipe for which has been divulged by the helpful Señora Heras:

Sangria

1 bottle of red Spanish wine	2 tbsp. sugar
1 12 oz. bottle of club soda	2 ice trays of ice cubes
1 shot glass of Spanish brandy	1 lemon, cut in slices
1 shot glass Cointreau	1/2 orange, cut in slices

1. Empty bottle of red wine in large pitcher. Add sugar and mix well so that sugar disolves.
2. Add slices of lemons and oranges, stir.
3. Add Cointreau, brandy, ice cubes and soda, stir.
4. Sangria should be made 15 minutes before serving so that it is cold when served. Serve Sangria only, leave fruit in the pitcher.

Serves four.

The birth of the Spanish Pavilion roughly coincided with the last halcyon days of haute cuisine in New York. Henri Soulé had already spawned a sprawling dynasty of former employees and ex-disciples to carry on in their own arrogant preserves—La Caravelle, La Grenouille, Lafayette, to name a few, as well as such

Salvador Dali preparing at the Spanish Pavilion. *Courtesy of Oscar Abolafia*.

cheerfully relaxed establishments as L'Escargot and La Cocotte. Soulé often spoke of his own "indispensability" to anyone within earshot, a professional evaluation that time would prove correct. Perhaps one of his last dinner menus might serve as an epitaph and an indication of the inspiration he afforded his dynasty; the following, in its classic grace of composition, reflects the mortal man.

DINER

Caviar Malossol
Jambon de Bayonne *2.50*
Little Necks *1.50*

Saumon Fumé *3.25*
Foie Gras Truffé *6.00*
Cherrystones *1.50*

Anguille Fumée *3.00*
Grapefruit *1.00*
Melon *1.50*

Cocktails: Lobster *5.00* Shrimps *3.00* Crab Meat *4.00*

POTAGES

Petite Marmite Pavillon *2.75*
Bisque de Homard *3.50*
Ox-Tail Clair *3.00*

Consommé Double Julienne *2.00*
Soupe à l'Oignon Gratinée *2.50*
Billi-Bi *3.00*   Madrilène en Gelée *2.00*

Saint Germain Longchamp *2.00*
Tortue Verte au Madère *3.00*
Germiny aux Paillettes *3.00*

POISSONS

Délices de Sole Polignac *4.50*
Moules au Chablis *4.25*
Goujonnette de Sole, Sauce Moutarde *4.50*

Timbale de Crab Meat Newburg *5.50*
Suprême de Striped Bass Dugléré *4.50*

Homard Xavier *7.50*
Grenouille Provençale *4.75*
Truite de Rivière, Beurre Noisette *4.75*

Specialties: Homard *7.50* Sole Anglaise, Moules *4.25*

ENTRÉES

CUISSEAU DE VEAU PAVILLON 6.00
Suprême de Pintadon Carmen *5.50*
Caneton aux Cerises (Pour 2) *15.00*
Côte de Volaille Pojarsky *5.25*

LA POULARDE ETUVÉE AU CLICQUOT (Pour 2) 16.00
Coeur de Filet Clamart *9.00*
Medaillon de Veau Smitane *5.00*
Rognon de Veau Ardennaise *5.00*

Pigeonneau aux Olives *6.00*
Ris de Veau Meunière *5.25*
Foie de Veau à l'Anglaise *4.75*

Specialties: Chateaubriand (Pour 2) *18.00*
Volaille (selon grosseur), Ris de Veau *6.50*

RÔTIS

Poularde   Reine   Grain   Poussin   Canard   Pigeon
Selle d'Agneau   Carré d'Agneau

PLATS FROIDS

Poularde à la Gelée à l'Estragon *5.50*   Langue Givrée *3.25*
Terrine de Canard *4.50*   Boeuf Mode à la Gelée *5.00*

Jambon d'York *3.50*
Terrine de Volaille *4.50*

LEGUMES

Coeur de Céleri à la Moëlle *2.50*
Courgette Fines Herbes *2.00*
Laitue Braisé au Jus *2.00*

Epinards au Velouté *1.50*
Haricots Verts au Beurre *2.00*
Champignons Grillés sur Toast *2.75*

Petits Pois aux Laitues *2.00*
Choux-Fleurs au Gratin *2.50*
Artichaut à l'Huile *2.50*

ENTREMETS

Patisserie Pavillon *2.50*   Soufflés Tous Parfums *3.50*
Désir de Roi *2.50*   Cerises Jubilée *3.50*
Pêche Melba *2.50*   Macédoine de Fruits aux Liqueurs *2.50*

Crêpes Pavillon *3.50*
Poire Hélène *3.00*
Coupe aux Marrons *2.50*

Glaces: Vanille *1.25*   Chocolat (Menier) *1.25*   Moka *1.25*   Fraise *1.25*
Citron *1.25*   Framboise *1.25*

Café *.70*   Demi-Tasse *.60*   Bread and butter *1.00*

But jaded New Yorkers were beginning to yearn for something other than traditional haute cuisine in a traditional setting. "Packaging" was the watchword of the fifties and sixties. New Yorkers wanted to be transported, and entertaining became an environmental art. People suffered from a form of Romanticism that looked back to the turn of the century, to Murray's Roman Garden and the Café de l'Opera, and past that clear back to Pompeii. Gone were the arid years of chrome and clean lines, the photographic montage mural so popular in the restaurants of the forties, the bland scenic paintings that reached their apogee with the Champs Elysées views at the Pavillon. Moreover, for the "atmosphere of the unusual," New York had come a long way from the Pirate's Den.

What in fact developed was a distinct style of American Eclectic, a kind of cross between the Palm Court of the Plaza and Grandma's attic. It was characterized by brick walls, sky lights, palm fronds and giant house plants, columns, leaded windows, tiles, bent wood chairs, marble-topped tables, Tiffany glass, and, beyond that, anything from stuffed gay-nineties costumes to the signs of the cabala.

A prime example, prime in the sense that it was not only among the first, but a founder of the movement, was Nicholson's Café, which opened originally in 1949 in what had been Jo Davidson's studio on Fifty-eighth street; it then moved to larger quarters on

Fifty-seventh, before returning once again to its original site. It was hard to tell whether the interior of Nicholson's was intended to suggest Singapore Grand Hotel or Balkans Intrigue Palace. When the question was put to him, Mr. Nicholson patiently explained that it represented the *fin de siècle* Caribbean of Cuba in 1944. This might explain the yellow Dionysiac tile on the walls, the globe lights with their slowly revolving horizontal fans, the Victorian marble and oak bar, the gilt mirrors (from the ballroom of a Saratoga hotel, as it happens), the marble nudes of the decadence, and the parrot (Lolita), although it did not explain the bronze slave boy, the ceramic Imperial busts, and the sphinxes. But when the candles were lit and the huge central table loaded with fruits, wines and china, silver bowls piled high with lemons, and baskets bearing monumental constructions of pure white eggs, it really couldn't matter less.

At Nicholson's there were no menus, and the choice was small but excellent. These innovations originated with Nicholson and were wisely adopted by several New York restaurateurs who discovered that in the day of rising labor costs it was no longer possible to ensure the quality of every dish on a huge menu.

At Serendipity 3 the eccentricity extended to the menu. It called itself a general store, and if that classification denotes an emporium where one can order a filet mignon stuffed with steak tartare, Apricot Smush, a rhubarb omelette, Miss Milton's Lovely Fudge Pie, Ftatateeta's Toast (cream cheese and jam on French toast), one of Aunt Buba's sand tarts, or a Foot-long Hot Dog, try on Garbo hats or a maribou-trimmed garden party dress, splash oneself with Summer Camp cologne, fondle a Shirley Temple milk mug, sniff cucumber soap, price a statue for the patio or a Tiffany shade for the den, pick out a cross-eyed calico rooster for the baby, and drink martinis out of tea cups, then a general store it most certainly was.

The 3, Calvin Holt and Patch Carradine, Arkansas travelers both, and Stephen Bruce, a native up-state New Yorker, banded together in 1954 and opened the first Serendipity in a basement on Fifty-eighth Street. A few unmatched tables and chairs, a jumble of gift items such as cocaine cannisters, artificial flowers under glass bells, colorful strings of Haitian pod beads, Victorian hand soap dishes, Andy Warhol silk-screened cats, and a practically non-existent kitchen out of which were trotted coffee fantasies, frozen desserts, and little else were all they had to offer. But that was

Café Nicholson, foliated temple of American Eclectic.

enough. It is hard to realize today that Serendipity's brand of calculated whimsey ever looked startlingly new. Yet it did. Bizarre but never quaint, blatantly chi-chi but self-derisive, frightfully "in," "campy," and "kicky," Serendipity epitomized a trend in the taste of New Yorkers who had seen too many late, late shows and were beginning to feel a tugging nostalgia for the twenties and thirties. So much so that within four years, the Messrs. Holt, Bruce, and Carradine packed up their espresso machine and potted palms and moved into much grander quarters at 225 East Sixtieth Street. This success was prophetic of a change in public mood, a willingness to taste the "kooky," the "off-beat," the "far-out," and suddenly everyone found himself entangled in yards of beaded curtains and stirring his cappuccino with a cinnamon stick.

With Restaurant Associates, two trends of the fifties and sixties collided—the yearning for a new and more *outré* "packaged" food and atmosphere, and a tendency for restaurant management to take on the dimensions of big business. Chain restaurants, of course, date back to the end of the last century and the early years of this, when Child's, the automats, and Schrafft's discovered a viable pattern, and literally cloned. Schrafft's, with its twenty-eight branches in Manhattan alone, became so large an organization that it required its own newspaper to generate esprit-de-corps among its thousands of employees with headlines such as "How does it feel to give blood? Be a proud Schrafft's donor," and up to the minute flashes: "Congratulations to our Seniors: Marie Sobol, head candy girl, 20 years . . . Rose Mallory, coffee girl, has taken up hairdressing . . . We'll be glad to see Stan Garvey, dishes, back and well again. . . ."

Longchamps attempted a rather more pretentious cuisine in a setting of red, yellow, and gold, with edifying scenes of Indian life and the west. And Chock Full o' Nuts, to be found on seemingly every corner of the city, offered the public a quick and tasty lunch menu of pie, brownies, chocolate or orange drinks, their famous coffee, and sandwiches "never touched by human hands."

But the restaurant with a special atmosphere, distinctive or foreign food, the unusual or eccentric, was always an individual undertaking. One could hardly imagine a chain of Pavilions or Nicholson's. One man, however, rushed in where angels feared to tread.

While reading a collection of *haiku,* Joe Baum suddenly envisioned the Four Seasons. Upon reading Apicius, he invented the

Forum of the Twelve Caesars. It is not known what he was reading when he conjured up the Fonda del Sol. The original Restaurant Associates was the firm name of the chain of coffee shops known as Riker's, acquired by Mr. Abraham Wecksler for his coffee company in 1942. Restaurant Associates, which had run a snack bar in the old Newark Airport, was asked by the Port of New York Authority to open a luxury restaurant there when the airport was rebuilt in 1952. At that point R.A. found Mr. Baum, then in his early thirties, and working as director of food and beverages for the Shine chain of hotels in Florida. He was brought north to run what might well have been a failure—an expensive restaurant in a busy commuter airport at a time when the public image of an airport restaurant was not the most distinguished. But Mr. Baum was nothing if not original. He turned the Newark into something like the proverbial three-ringed circus. Oysters were served by the sevens, lobsters came with three claws, and everything was flambé: shashlik, steak chunks, scallops (wrapped in bacon), and brandied coffee. There was even a dessert illegal in New York because of fire regulations. According to Gael Greene, Baum's explanation was: "The customers like to see things on fire . . . it doesn't really hurt the food much."

The Newarker was a success and the others followed, each one an extraordinary enough flight of the imagination to content one restaurateur for a lifetime, and all straight out of the head of the master—the management of the Hawaiian Room in the Hotel Lexington, then the Forum of the Twelve Caesars, then the Four Seasons and the Brasserie, and then, one by one, the others—the Fonda del Sol, the Tower Suite, the Trattoria, Zum Zum, and Charley Brown's (these last three in the Pam-Am Building), Paul Revere's Tavern and Chop House on Lexington Avenue, Charley O's "Bar and Grill and Bar" in Rockefeller Center, and the Fountain Café and Tavern-on-the-Green in Central Park. Except for the Four Seasons, the Tower Suite, the Fountain Care, and of course the Tavern-on-the-Green in Central Park, each has a separate regional or historic flavor, running the gamut from Polynesian, French, Italian, Latin American, German, English and Irish, to Imperial Roman and Colonial American, and every one a packaged masterpiece. Each was approached by Mr. Baum and a team of experts as a special project, a sort of cross between an "on location" M.G.M. costume epic production and a State Department study of underdeveloped areas. (It might be added here that in 1959, R.A. assumed the guise of Mama Leone.)

At the Forum of the Twelve Caesars, the final effect was achieved by larger than life silverware, slightly lowered tables, waiters uniformed in the imperial colors of Phoenician purple and red, and seventeenth-century Italian paintings of the Caesars from Julius through Domitian. The menu, its Latin all carefully corrected by a Hunter College professor, included "Lentils and Sausage, Sweet and Sour; Wild boar Pâté, Sauce of Damascus Plum; Snails in Dumplings, Green Butter Sauce; and Great Mushrooms stuffed with snails, Gallic cheese and walnuts, glazed." Among the main courses were "Capon Fronto, with Leeks and Coriander; The Wild Fowl of Samos cooked in sherried tomatoes, under a mantle of crusty corn; Cutlet of Wild Boar, deviled in mustard seed, with apple nuggets," and "Truffle Stuffed Quail, Cleopatra —wrapped in Macedonian vine leaves, baked in hot ashes." For those who wanted a real show, and were willing to pay for it, there was "Pheasant of the Golden House on a Silver Shield in Gilded Plumage, roasted with an exquisite sauce." But at some point our Roman's tongue made its way into his cheek. Among the *hors d'oeuvres* were "Mushrooms of the Sincere Claudius—an Emperor's delight" (for the sake of those who have forgotten, Claudius was poisoned, intentionally, with a mushroom), and among the entrees were "Fiddler Crab Lump à la Nero" (flaming, of course), and "Sirloin in Red Wine, Marrow and Onions—a Gallic recipe Julius collected while there on business." The salad list included "The Noblest Caesar of Them All," and for dessert there was "Nubian Chocolate Roll" and "Tart Messalina."

According to the *New Yorker*, a food expert who complained of his boiled beef at the Forum was told by its director, "You just don't understand. We have studied Apicius." (In actual fact, there is only one recipe for unminced beef in that treatise. Quoted verbatim it reads: *"Vitulinam sive bubulam cum porris cydoneis vel cepis vel colacasiis: liquamen, piper, laser et olei modicum."* An exact translation would be: "Veal or Beef with leeks, quinces, or onions, or taros. Prepare with *liquamin* [a sauce made from small fish and fish entrails salted and left in the sun for two to three months], pepper, asafoetica [a fetid gum resin from the Black Sea imparting a peculiarly putrifactive flavor], and a little oil.")

At Zum Zum, Restaurant Associates came up with kegs of beer and twenty-nine kinds of wurst and so successful a blueprint that

The Forum of the Twelve Caesars: "Pheasant of the Golden House on a Silver Shield in Gilded Plumage, Roasted with an Exquisite Sauce (serves two).

it launched a booming chain. They served torta rustica at the Trattoria, roast ribs of beef with horseradish cream and Yorkshire pudding at Charley Brown's, Limerick ham steak and smoked kippers at Charley O's, and chicken breast and asparagus in rose aspic and coeur à la crème with berries at their café in Central Park. Joe Baum wrote all the menus and at this point ranks as a classic author.

But the Four Seasons was the jewel in Restaurant Associates' crown. Located in the ground floor of the Seagram Building, it occupied a light and spacious setting; at staggering expense, everything changed with the Seasons—the menus, china, glassware, tablecloths, waiters' uniforms, and, above all, the horticultural decorations, which include full-grown trees and beds of flowers—at a cost of seven thousand dollars for overtime on season-changing day. A large pool in the middle of one dining room could be frozen over when "weather permitted." From a financial point of view, it ran as four restaurants in one setting.

The Four Season's resemblance to Lincoln Center was overpowering, and was due largely to its architectural designer, who was responsible for the State Theater. It may also be due to a sense of hiding behind a curtain of fine, quivering brass chains and cowering under the overhang of a shimmering, gold-dipped Leppold construction, as at Philharmonic Hall. However, to relieve the clean lines and atmosphere of "less is more," three Miro tapestries, worth ten thousand dollars apiece, were hung at the entrance, along with a Picasso *corrida* painted on a theater curtain and valued at seventy thousand dollars.

The imaginative menu, two feet long and very expensive, one could only be treated with a kind of flabbergasted respect.

D I N N E R   A T   T H E   F O U R   S E A S O N S
(WINTER)

Cold Appetizers

Small clams with Green Onions and Truffles   *1.85*
Large Chincoteagues, a Platter   *1.95*
A service of Scottish Smoked Salmon   *3.75*
Winter Hors d'Oeuvres, a Sampling   *2.50*
Coriander Prosciutto with Pineapple or Melon   *2.75*
Tranche of Sturgeon and Caviar   *6.50*

Stuffed Leeks, Orientale   *1.65*
Lobster Chunks on Dill   *3.75*
Iced Brochette of Shrimp   *2.50*
Today's Melon   *1.25*
Caviar, per serving   *9.00*
Mousse of Chicken Livers   *2.25*

Oysters, Horseradish Ginger   *1.95*
Winter Farmhouse Terrine   *2.25*
Virginia Blue Crab Lump   *3.25*
A Tureen of December Fruit   *.95*
Orchid Grapefruit, Honey Dressing   *.95*
Little Neck or Cherrystone Clams   *1.65*

Hot Appetizers

Our Coquilles St. Jacques   *2.25*
Deviled Oysters on the Half Shell   *2.25*
Spiced Crabmeat Crêpes   *2.45*
Baked Clams with Peppers   *1.95*

Tiny Shrimps in Shoyu, French Fried   *2.50*
Moussaka Orientale   *1.65*
Steamed Mussels in Crock, Honfleur   *2.25*
Crisped Shrimp filled with Mustard Fruit   *2.35*

Beef Marrow in Bouillon or Cream   *1.85*
The Four Seasons Mousse of Trout   *2.50*
Snails in Pots, Dijonnaise   *2.25*

Soups and Broths

COLD

Beet and Lobster Madrilène   *.95*
Watercress Vichyssoise   *1.25*

HOT

Cream of Leeks and Potatoes   *1.10*
Double Consommé with Madeira   *1.25*
Onion Soup with Port, Gratinée   *1.35*
A December vegetable Potage   *1.10*
Vermont Cheese soup   *1.35*

Sea and Fresh Water Fish

Barquette of Flounder with Glazed Fruits   *4.95*
Frogs' Legs Provençale or with Vermouth and Truffles   *5.50*
Lobster Aromatic Prepared Tableside   *6.50*

Winter Sole, Four Seasons   *4.95*
Braised Striped Bass, for Two   *10.50*
Crabmeat Casanova Flambé   *6.00*

Red Snapper Steak—Grilled   *4.65*
Broiled Maine Lobster   *7.50*
The Classic Truite au Bleu   *5.75*

This Evening's Entrées

Crisped Duckling with Tangerines, Flambé   *6.50*
Seasoned Pheasant en Salmis   *6.75*
Rare Filet Stroganoff   *6.75*
Mallard Duck à la Presse   *8.50*
Venison Cutlets Sautéed with Juniper, Sauce Poivrade   *5.85*
Two Quail en Brochette, Chestnut Gnocchi   *7.75*
Roast Rack of Lamb Persille   *14.00   for two;*
   or with Cepes and Potato slices   *16.00*

Carré of Meadow Veal, Mushrooms in Cream   *5.95*
Noisettes of Young Lamb, Minted Flageolets   *5.85*

Breast of Chicken with Lobster, Nantua   *5.75*
Filets of Dover Sole, Sautéed with Tiny Shrimp   *5.95*

Steaks, Chops and Birds
BROILED OVER CHARCOAL

Three French Lamp Chops   *5.95*
Jersey Poularde   *4.50*
Sirloin Vintners Style   *7.75*
Calf's Liver, Thick, Sage Butter   *5.25*
Amish Ham Steak, Apricot Glaze   *4.85*
Sirloin Steak or Filet Mignon Served for One   *7.50; for two   15.00*

Butterfly Steak Paillard, Four Seasons   *6.75*
Côte de Boeuf, Bordelaise, *for Two   18.00*
Entrecôte à la Möelle   *7.75*
Filet of Beef Poivre, Flambé   *8.00*
Skillet steak with Smothered Onions   *7.75*

SPIT ROASTED WITH HERBS

Marinated Lamb with Cracked Wheat *5.85*
Saddle of Venison, Wild Game Sauce, *for Two* *14.00*
Roast Sirloin of Beef, Sauce Perigord *6.75*

Twin Tournedos with Woodland Mushrooms *7.00*
Larded Pigeon with Candied Figs *6.25*
Skillet Steak with Smothered Onions *7.75*

Winter Salads
AS A MAIN COURSE

Bouillabaisse Salad *5.25*
Julep of Crabmeat in Sweet Pepperoni *5.50*
Julienne of Turkey Breast and Pineapple *5.25*
Dariole of Lobster with Spiced Mushrooms *6.50*
Beef in Burgundy Aspic *5.50*

AS A DINNER ACCOMPANIMENT

Belgian Endive and Grapefruit *1.50*
Winter Greens *1.25*
Raw Mushrooms, Malabar Dressing *1.75*
Nasturtium Leaves *1.50*

Cucumbers and Dill *1.25*
Beefsteak Tomato, Carved Tableside *1.25*
Zucchini and Hearts of Palm, Lemon Dressing *1.50*
OUR FIELD GREENS ARE SELECTED EACH MORNING AND WILL
VARY DAILY
Salad Dressing with Roquefort or Feta Cheese *.50 additional*

Vegetables and Potatoes
(Seasonal gatherings may be viewed in their baskets)

Brussels Sprouts and Bacon Cracklings *1.25*
Souffle of Artichoke, *for Two* *3.85*
Broccoli Flowers, Hollandaise *1.95*
Bouquet Platter, *per Person* *1.50*
Cauliflower Gratinée *1.35*

Squash with Lemon Butter *1.25*
Braised Celery *1.25*
Wild Rice *2.50*
POTATOES: French Fried *.95*
Mashed in Cream *.95*

Country Style .95
Glazed Sweets .95
Baked in Jacket .95

The desserts may be imagined. They included Candied Harlequin crêpes, Rose-petal parfait, Pomegranate sherbet, and Chocolate Velvet.

Gael Greene, *New York Magazine*'s "insatiable critic," vividly described the heroic Baum, a gourmet and impresario to make Niblo pale, as "the man whose ego, taste, drive, showmanship, and capacity to terrorize and ingratiate had set a new style in American restaurants.... Though what Baum created was grandly intimidating, his theme was simplicity: eliminate phony Frenchities; dispense with sommeliers and the maitre d' hotel; update cuisine classics. Style was crucial. The style of pouring wine at the Forum is different from the style of pouring wine at the Four Seasons...."[17]

By the mid-1960s Restaurant Associates was grossing close to a hundred million dollars a year. But toward the end of the decade the empire suffered a Decline and Fall of Gibbonesque proportions. It had expanded, diversified. It had taken on the Barricini Candy Company, Treadway Inns, and the management of three hundred Big Alice Country Kitchens, among other enterprises. Restaurant Associates issued its own credit cards. Like the Roman Empire it became unwieldly. Joe Baum, its Nero and Marcus Aurelius, was deposed, and a series of strikes and failures of ventures outside the Eternal City itself led to calamitous failures.

Meanwhile, a new giant had lumbered onto the scene. Larry Ellman, owner of the Cattleman, a riotously successful steakhouse, purchased the Longchamps chain, and partially metamorphosed it into his new bonanza, the Steak and Brew. The Steak and Brew, serving a cut of beef and all the salad and beer one could eat at a set price, presented what was thought to be a new concept. It became a popular one, and "all you can eat" menus spread across the country—as much a novelty in its day as the à la carte menu, according to which, when it first appeared, one was charged for the quantity one ate. The Steak and Brew itself served one quarter of a million meals per month, nationwide, in 1971.

Meanwhile the highest pinnacles of subtle culinary excellence were being scaled in the city's new strain of Chinese kitchen. Until the early 1960s, Cantonese cooking, undistressingly bland and economically segregated into "Groups A and B," ruled the American roost. With a few exceptions, such as "Lobster Cantonese," prices were usually very low and restaurants offered bargain "family dinners," the remains of which could be carted home in gooey paper containers. But with the ascendance of Chinese restaurants in the midtown business district, where a lunchtime expense account society demanded cocktails "on the rocks," another recent development, along with their barbecued spareribs, it was inevitable that prices would soar.

The better Cantonese restaurants of Chinatown and the upper West Side had permitted financially pinched intellectuals and artists to flex their epicurean muscles in public, an exercise denied them in the glades of haute cuisine, and consequently there arose a fanatic cult devoted to the appreciation of Chinese food, as snobbishly intransigent in its demands as a Burgundian Chevalier du Tastevin. The time was ripe for the introduction of the lesser known cuisines of Shanghai, Peking, Szechuan and Fukien, each considered by old Chinese hands to be of a "higher class" than Cantonese. Cultists flung themselves onto the banquet table without a backward glance, and jealously traded the latest obscure addresses where Szechuanese Dry-fried Crispy Shredded Beef with Red Peppers, Drunken Chicken Drowned-in-Wine, or Manchurian Date Nut Cake with Lotus Seeds and Salted Duck Eggs daily graced the menu.

An exile's nostalgia for a lost way of life was part of a general thirst for "romance." This incurable form of homesickness propelled many a restaurateur, hotelier or night club impresario to success. A classic example is the career of Serge Obolensky, Oxford-educated Russian prince, Officer of the Imperial Chevalier Guards, fighter of Bolsheviks, husband of Romanovs and Astors, dancer extraordinary and daring World War II Colonel of the U.S. Air Force, for which, in his late fifties, he parachuted into Italy behind German lines. To New York, Obolensky brought impeccable Old World manners and a talent for re-creating the glamor of pre-Revolutionary Russia. When his Astor brother-in-law repurchased the St. Regis in 1935, *bon viveur* Obolensky was made its president. There he opened the brilliant Maisonette Russe (the Maisonette in a later duller incarnation), where a gold-spotted raspberry and mauve decor, suggested by the Eunuch's swirling pantaloons in a Bakst costume sketch for Diaghilev's *Scheherazade*, glowed in a blaze of flaming shashlik and dancing spirit lamps. The maddening strains of gypsy violins and an artillery salute of champagne corks did nothing to dampen the overall manic effect.

Following the war, the Colonel implanted Czarist-style *boites* in hotels all over midtown Manhattan—the Carnival Room at the Sherry Netherland, the Embassy Club at the Ambassador, the Mon Plaisir at the Drake, and the Rendezvous Room at the Plaza, where, as the times would ironically have it, U.N. Delegate Andrei Gromyko had taken up residence, attended by a force of twelve security men and three cooks who gladdened the halls of the fifteenth floor with the stench of boiling cabbage.

New York's tough façade has always hidden a sentimental heart. Over a hundred years ago, local "nostalgia clubs" were organized to revive stage-coaching and country inns and beefsteak societies tried to recapture the happily uncomplicated gluttony of the good old Kraut Club and turtle feast days. Beefsteak societies had roots in the past extending back to the 1780s, when joints or chops were common fare, but rarely a slab of sirloin. The outstanding exception to the rule was the beefsteak orgy given periodically at a tavern called Shannon's Corner by Catherine Street Market butchers in honor of their favorite customers, ship captains who bought meat for long voyages. No knives, forks, or napkins were permitted, and each guest devoured his allotted three pounds of hickory-broiled and buttered beef between planks of bread. The tradition lasted for decades, with yarns of the sea being spun far into the night in an atmosphere of unlimited ale and greasy-jowled gusto. In the 1880s and 90s, a rash of nostalgic beefsteak clubs broke out in New York, but in the transition the original salty simplicity of a "beefsteak," as the feasts had come to be called, had been lost. Now rich merchants and assorted tycoons clannishly assembled to escape the stultifying formality of their wives' dinner parties. No expense was spared, and stage designers were hired to transform large halls into intimate "old-time taverns." In the twentieth century, beefsteak clubs assumed a somewhat political nature, and wives were invited, spelling the end.

Since World War II, New Yorkers in ever-increasing numbers have turned to the city's welcoming hamburger shelters, steak houses, and Anglophilic pubs for physical and spiritual succor. It would appear that beef consumed in "old-fashioned" surroundings can stave off future shock, whether ground and shaped into compact cakes or trimmed into tidy filets.

"The origin of the hamburger," as delineated in 1972 by James Beard, American cuisine's own Brillat-Savarin, "is undoubtedly European."

Early in this century the idea of a hamburger on a bun caught on after two world's fairs had introduced it . . . From the 1920's on, the hamburger began its reign as America's most popular form of meat. It was considered *au fait* for children's and many grown-ups' parties. It became a "steak" in size, and even such fleshpots as "21," the Colony, and Chasen's began to offer plump patties of prime meat on their menus. . . . The version one usually finds at stands and lunch counters is a mercilessly flattened patty of indefinable flavor, cooked to a state of petrifaction and placed on a cold bun, then to be doused with catsup. If its quality varies, so does its price, which can range anywhere from 19 cents for a quick-order sandwich to $4.50 for an elegantly planked hamburger steak. At its best—juicy, filled with flavor—it is an excellent dish, not to be regarded with condescension.[18]

In New York, America's favorite meat did not become chic until the 1930s saw the flowering of Hamburger Heaven; their formula was simple—tiny premises, open grills, bare wooden counters with swivel stools or children's highchairs with trays, reproportioned for adults, onto which genuinely heavenly hamburgers were plunked by a beaming chef in towering toque, along with bowls of pungent relishes and sweet Bermuda onion, airy layer cakes, fudge-coated gold and devil's food slathered with thick white frosting, and richly perfumed mugs of coffee. By earlier hamburger joint standards, Hamburger Heaven's prices were high, a fact which may have encouraged an early morning following in white tie and tails and ermine wraps to wend its stylish way to their doors. Twenty years later, a calamitous decision to gild the lily by installing a cocktail lounge and expanding the menu brought the miniature empire to a dismal end.

Ever since Hamburger Heaven's dramatic social breakthrough, the silk stocking district has been periodically assaulted by poetic variations on an essentially prosaic theme; at one time or another residents have awakened to La Prima Burger, the Plush Burger, where the specialty of the house was produced beneath crystal chandeliers, and Phoebe's Whamburger, suitably named for its proprietress, a steam-roller of a lady with a weakness for wide cartwheel hats.

New Yorkers' nostalgic longing for the good old saloon days fostered a populous tribe of hamburger emporiums, the pater-familias of which is the indestructible P. J. Clarke's. "Clarke's," as it began calling itself tersely in a rising tide of P. J. O'Hara, P. J. Moriarty, P. J. O'Rourke, and, most recently, P. J. Bernstein, started its long and amazing career as a lower-class Third Avenue bar, tucked away in the grimy shadows of the el and patronized by some of the neighborhood's less jovial drunks. As in the plot line of one of its own movies, Hollywood came calling and offered a contract, albeit not a very flattering one—a seedy location was needed for the shooting of *Lost Weekend*, a film exposing the grimmer aspects of an alcoholic's life. Nevertheless, P. J. Clarke's was made a star, and an extremely tenacious one. Success paid for a discreet face-lift; the old-fashioned mirrored bar became more self-consciously photogenic, with portraits of Abraham Lincoln and Carry Nation lending a dignified air. Red and white checked cloths were smoothed over tables, a blackboard chalkily proclaimed a menu of hamburgers, chili, eggs Benedict, apple pie, and up-to-date records spun in the jukebox. At the bar, imported draft beer and ale foamed over the cool rims of heavy-bottomed schooners as waiters in long white shirt-sleeves loaded their trays with Irish coffee, gin and tonics, stingers on the rocks. Dark tin ceiling, dark stained walls, white tile floor—a prototype had been set for the imitative hordes.

An epidemic of nostalgic English- and Irish-style pubs infected New York throughout the 1960s; particularly susceptible to it were the self-styled "swingers" of the day, conscientiously trendy souls who raced to embrace anything new from the British Isles, be it Beatles, Jaguars, miniskirts, boots, Carnaby Street or West End theater, Rolling Stones or royalty.

Michael's Pub, opened in 1954 by Michael Pearman, an urbane theatrical agent turned restaurateur, spearheaded the movement. Debutantes and celebrities, the columnists duly noted, were soon packing the place day and night. The fickle flight of playboys and girls from their zebra-striped and "Louis Cuisine" period haunts to the jolly gloom of an English taproom was observed by thoughtful competitors. Imitation may be the sincerest form of flattery, but in the restaurant business, at least, the original magic is rarely, if ever, successfully reproduced. Many tried to duplicate Pearman's master stroke, but in the final analysis, there was only one Michael's Pub.

A craving for the food of one's childhood knows no class distinctions. And so it has fallen to those blessed with Old Money to seek out the dishes prepared for them and their forebears, before the demise of large domestic staffs, by vanished generations of uninspired cooks. Security Food for the old guard has traditionally been trundled out in New York's most soothingly sedate hotels—at the Mayfair and Carlton Houses' dining rooms, the Carlyle's Regency Room, the Barberry Room at the Berkshire, as well as in the restaurants of the Longchamps chain during its premature golden age. Lamb is the official security meat for the New York conservative—either pantalooned chops garnished with cool toast points and watercress sprigs or a well-done roast accompanied by gravy *and* bottled mint sauce, small, hard, oven-browned potatoes, and peas made brilliant with bicarbonate of soda. On Sunday evenings, creamily costly scrambled eggs encircled by a snippet of parsley, a collapsing broiled tomato, and two jewel-like breakfast sausages have long been counted upon to induce calm, as has half a cold lobster *mayonnaise* after a hectic morning's shopping for hair nets and golf tees.

At the turn of the 1970s, New York's most venerable sanctum of tearoom "home cooking"—the princess royal of the Zoe Chase, Patricia Murphy, and Jane Davies school of security centers was the bustling Kirby-Allen on Madison Avenue near Sixty-seventh Street. Founded in 1923 by the Misses Nell Kirby and Bertha Allen, two daringly liberated young ladies from southern Ohio, the restaurant, graciously tended over the years by an ageless matriarchy of organdy-aproned waitresses, produced such nostalgic American delights as chicken fricassee, freshly baked cornbread and "Sally Lunn," and a Winter Luncheon Plate of sausage cakes, grits, and glazed apples. The menu patiently clarified any violent departures from the immediately comprehensible: "Lasagna (a very interesting Italian dish)," "Quiche Lorraine (a very interesting cheese dish)." The dessert list could be sung to Stephen Foster —warm Indian pudding à la mode, date nut and caramel cake squares, apple cobbler, coffee bavarian cream, pineapple upside down cake, bittersweet chocolate and lemon chess pies. Clementine Paddleford, food editor of the *Herald Tribune,* was utterly carried away: "You will be dining with phantoms," she cried, "Your taste buds will stretch right back into the past and the very ends will be shaken."

Postwar Greenwich Village preferred its security in liquid form at bars and coffeehouses. Heavy drinking helped obscure the approaching end of Bohemia. "Bohemia could not survive the passing of its polar opposite and precondition, middle-class morality," explained *Village Voice* contributor Michael Harrington from the distant knoll of 1972. "Free love and all-night drinking and art for art's sake were consequences of a single stern imperative: thou shalt not be bourgeois. But once the bourgeoisie became decadent —once businessmen started hanging non-objective art in the boardroom—Bohemia was deprived of the stifling atmosphere without which it could not breathe."[20]

In the meantime, there were the afternoons to be spent in Washington Square, where ancient chess players braved the doleful keening of folk songs by soiled bands of amateur guitarists, and the long evenings to rant away at the San Remo, Chumley's, or the White Horse Tavern.

The defunct San Remo, Harrington recalled,

was an Italian restaurant at the uneasy intersection of Greenwich Village and Little Italy, with bad, yellowed paintings over the bar and the Entr'-Acte from Wolf-Ferrari's *Jewels of the Madonna* on the juke box. In 1949 it was the united front of the Village. There were a few old Bohemians, like Maxwell Bohenheim, the poet and novelist who dated back to the pre-World War I ferment in Chicago and was now a shouting, mumbling, drunken, hollowed-eyed memory of himself. . . . There were seamen on the beach, the most important single contingent from working class Bohemia. Some of them had fought in Spain. . . . They all combined two seemingly antagonistic life styles: the militant and the vagabond. At that time, most of the radical seamen were being driven out of their jobs by the loyalty program and had plenty of time to drink and reminisce about Spain in the Remo. Among the other regulars there were the heterosexuals on the make; homosexuals who preferred erotic integration to the exclusively gay bars then on Eighth Street; Communists, Socialists and Trotskyites; potheads; writers of the older generation, like James Agee, and innovators of the future like Allen Ginsberg, and Julian Beck and Judith Malina, who were to found the Living Theatre.[19]

The old White Horse Tavern managed to ride out the blustering social cycles of postwar Village life; longshoremen, radicals, and Norman Mailer in the early fifties, followed by a self-destructing Dylan Thomas and his fawning coterie, who in turn were replaced by tribes of blue-denimed folk singers, one of whom, the poet Robert Zimmerman, would take Thomas' Welsh first name for his last.

Until her untimely death in 1960, avant garde film maker and voodoo authority Maya Deren, with her husband composer Teiji Ito, continued to stage the kind of party that had made the Village famous in the twenties. Waxen pale and blue eyed beneath a flaming aureole of rich red frizz, Miss Deren set the fashion for a later day; braless, barefoot, and beAfro'd she adopted see-through blouses, long flowing skirts, peasant embroidery, love beads, ankle bells, arched eye-brows, and women's liberation as an extension of artistic self-expression. Stalked by skulking droves of over-fertile cats, her friends dipped deeply into Miss Deren's Haitian punch, a potion enriched with scrapings from a "voodoo love root." To the hypnotic booming of native drums, their hostess would open the evening's revels by falling into a trance and lurching violently amongst the guests, many of whom immediately improvised their own choreography. Dylan Thomas, in one memorable performance, crashed through a beaverboard wall; his wife Caitlan, unimpressed by Miss Deren's psychic sensitivity, announced that she, too, was possessed by spirits and began to smash a towering altar of cult objects and artifacts. "She's destroying my world!" howled the hostess. "Dylan, don't let her destroy my world!" Mrs. Thomas shortly found herself in a husbandly half-nelson. Slipping into a wrist lock, the poet proceeded to swing her around him in ever-widening circles; at the point of maximum speed, he abruptly let go, launching a trajectile blonde across the studio and onto a distant couch. Then to the cheers of the crowd, the party ground back into gear.

The public had to settle for less explosive forms of nocturnal entertainment in Greenwich Village's mismatched collection of night clubs and coffeehouses. Ranging in style from the clamorous Bon Soir, with its suave M. C. Jimmy Daniels and marble-eyed Mae Barnes shouting the pleasures of "my little Paris, Kentucky home," to the peaceful Café Rienzi, with its cappuccino, creaking chairs, chess sets, pomegranate juice, and recorded flamenco guitar.

In 1957, the Village Vanguard, along with funky "jazz clubs" like the Half Note and Five Spot, had been sucked into the vortex of a weird new art form, poetry readings set to jazz, that had ridden out of the West on the motorcycles of the San Francisco "Beats." Mike Canterino, the Half Note's owner, threw caution to the winds and in his window hung the notice: "Poets Wanted." Insisting upon strict censorship ("Anything a bit off-color I cut out.

After all, I run a family place."), he presented the new concept to a typical jazz crowd of beatniks in leather jackets, painters, Bermuda-shorted college girls, leotarded Village sirens, and the Madison Avenue "cool." Like many California wines, the movement had not traveled well, and in the middle of the performance, the jazz group's leader leaped from the bandstand and headed into the night screaming, "I can't stand it, I can't stand it."

The Chelsea of the 1970s is the last outpost, the beleaguered Alamo of Bohemia. A gracefully imposing red brick building, coquettishly veiled with lacey wrought-iron balconies, the ten-story Chelsea originally opened on West Twenty-third Street in 1884 as one of the city's first apartment houses; in 1905 it became a hotel and over the years was patronized by Arthur B. Davies, O. Henry, John Sloan, James T. Farrell, Robert Flaherty, and Thomas Wolfe. Stanley Bard, who assumed his father's role as general manager, has always "given breaks to the artist and understood their dilemma. The Internal Revenue people can't understand this, that I've been given gifts just as gifts and that sooner or later the artists have always come back and paid their bills." The "gifts" referred to works by resident-friends like Larry Rivers, Jim Dine, and Christo hanging in the Chelsea's raffish museum of a lobby. Bard has given over forty of his friends' mementos to museums around the country.

Experimental film makers and stars including Shirley Clarke, Dennis Hopper, Miclos Forman, Jane Fonda, and Viva have often made the Chelsea their home, a sometimes mixed blessing in view of certain incidents. Mercifully, not even the notoriety brought down upon the hotel by Andy Warhol's rancid pornographic movie, *The Chelsea Girls*, could dim visiting Europeans' love for the place; in fact, when Italian *Vogue* devoted a 1972 issue to New York, the Chelsea was selected as the city's most colorfully fashionable inn.

When the "decadent bourgeoisie" began hauling its considerable worldly goods into picturesque Greenwich Village apartments and studios, they hoped to escape the cautious monotony of life in uptown residential areas. What they actually succeeded in doing was to drive rents up and artists out by taking the plunge into Bohemian life.

Economic victims, in all fields of art, turned to the Lower East Side and its drafty lofts for refuge; there the amorphous "underground" movement of the sixties gathered its momentum. St.

Mark's place became the Fifth Avenue of the area, with psyche-delic boutiques peddling everything from antique clothes to hash-ish pipes to "Mary Poppins is a Junkie" buttons. "Flower children" panhandled in the streets, trying to raise enough for an LSD sugar cube or a revitalizing sack of "organic" cereal at one of the city's burgeoning "health food shops." The success of the latter in turn encouraged budding restaurateurs to open health food counters, juice bars, and Zen-oriented hash houses all over the town.

Before the 1960s, it was generally thought that to embrace the health food faith in earnest, one had to be a little bit strange. Health food to New Yorkers meant something vaguely "West Coast," like astrology, beatniks, motorcycle gangs, and Oriental philosophy.

This, as later investigations would indicate, was not necessarily the case. In Manhattan, one of the early unsung heroines of the consumer awareness crusade was Gloria Swanson, the movie star. Since the 1940s, she had been collecting congressional reports and legal briefs documenting the dangers of technologically produced food and drink, chemical fertilizers and insecticides, and hor-mone-jolted livestock, as well as the collective menace of air and water pollution, refined sugar, bleached flour, and sudsy deter-gents. Through the smoked mirror cave of Miss Swanson's Fifth Avenue reception room, into a carnation-scented salon where the indestructible actress enthroned herself upon an antique red vel-vet child's chair, traipsed a gaggle of reporters, food editors, doc-tors, clergymen—anyone she felt might convey her prophetic message of doom to an unwittingly toxified public.

An obsessive interest in diets designed for loss of weight, how-ever, had long held Americans in its thrall. Since World War II, New Yorkers had applied themselves to losing weight with the same determination with which they had shed their inhibitions. This last process had been speeded, at no small expense, by profes-sional guidance, namely that of their ubiquitous analysts. Thwart-ing a tendency to tubbiness, however, was never a goal to be triumphantly achieved and then forgotten; to the contrary, the ferocious instincts of a watchdog are needed to repel unwanted fat. Ever on guard, fashionable New Yorkers were willing to try anything in their pursuit of the body beautiful. A new medical menace, bluntly referred to as "the fat doctor," or "Doctor Feel-good," hung out its shingle. The less responsible of these specialists dispensed appetite depressants which produced a delightful euphoria, a sense of supreme confidence in the anxiously over-weight. These innocuous-looking little pills frequently contained habit-forming amphetamines, which resulted in the addiction of many patients, an unhappy situation recalling a similar one in the past century when opium-based patent medicines were pre-scribed indiscriminately as pain killers and cough syrups. (Not until 1909 was the sale of opiates restricted by the government; until that time the trade had been extremely profitable—even some puritanical Vermont and New Hampshire farmers engaged in poppy cultivation.)

Among the pathetic victims of diet pills was the last of New York's "Glamour Deb No. 1's." Desperate to keep her photogenic silhouette, Joanne Connelly Sweeny Ortiz-Patino, the newspa-pers' "Golden Girl," a luminous blond with a perfect figure, medi-cated herself down to a spectral ninety-five pounds. Having fallen into the "Hollywood cycle" (diet pills, benzedrine to waken, bar-biturates to sleep), she suffered three heart attacks and died in 1953 at the age of twenty-seven. Brenda Frazier fared somewhat better; upon being hospitalized with a nervous breakdown in 1954, it was discovered that she had kept herself on a starvation diet since her debutante heyday in order to satisfy the public's concept of the eternal deb, and like a helpless infant had to be once again "taught to eat."

A favorite philosophical adage of New York fashion editors in the early 1970s was a pensée of much-debated authorship: "A woman can never be too rich or too thin." While the first attribute might prove frustratingly elusive, the second, enviable emacia-tion, lay within every reader's grasp. For those with a will, there were many ways. One could hang like an agitated bat from Kou-novsky's exclusive gymnasium rings or submit to periodic pum-meling at Elizabeth Arden's. At least one fashion leader would devour everything in sight and then disappear like an Ancient Roman matron on mysteriously long missions to the powder room. The easiest technique, once one's stomach had readjusted to its new shrunken size, was the most obvious—eat only for sur-vival.

Accordingly, at luncheon hour the best-dressed lists would con-vene at their favorite watering holes—La Grenouille, La Côte Basque, La Caravelle—and, aside from the water, would order little else. This, at best, was a demoralizing situation for the chefs, but what else could they do in an age when a restaurant's reputation relied on the daily patronage of fashion's overpublicized darlings?

The charismatic presence, for example, of Jacqueline Kennedy Onassis, the Mrs. Astor of the Jet Set, could always assure a restaurant's reputation for at least one season, if not more; witness the skyrocketing popularity of Lafayette after *Women's Wear Daily* photographed her emerging from its portals in a mini-skirt, an historic moment recalling Mrs. Astor's condonement of a shocking new fashion by dining hatless and decolleté on a Sunday evening at Sherry's.

Yet the mini-skirt issue paled beside the Great Pants Debacle. Long after the Four Seasons had adjusted its regulations of dress to the times and permitted women in pants suits and men in blazers with turtleneck sweaters to lunch on the premises, the reactionary managements of La Côte Basque and the Lafayette were waging last-ditch kamikaze-style battle against enemy fashions. James Brady, then publisher of *Women's Wear Daily*, watched with detached irony the recurring guerrilla skirmishes at La Côte Basque's door where, body-blocking the paths of trousered ladies, stood the obstinate figure of Mme. Henriette, the cashier presumed for years to have been Mme. Soulé *même*. (Upon Soulé's death, the legal Mme. Soulé released herself from French anonymity and winged to New York to demand her widow's mite. Mme. Henriette, the soul of tact, patience, and loyalty, bided her time and when it was placed on the market, bought La Côte Basque in order to perpetuate the name of Henri Soulé.) Pants on women, she decreed, were *not* Henri Soulé.

One early afternoon, with Brady noncommittally absorbing the scene, two elderly matrons bowed down by mink pup tents presented themselves to Mme. Henriette. They had reserved a table, they explained, and wished to be seated immediately as one was incapacitated by a slowly mending broken leg. That would *not* be possible, said Henriette balefully, as Madame was not suitably dressed. But she was recovering from a terrible accident! Surely? No. Madame must remove the offending garment or not be served. But how could she eat without pants? In her coat, replied Henriette, firmly closing the matter. The poor woman did as she was bid, and hobbled to the hot, cramped powder room, where, in unbearable discomfort, she thrashed and wriggled her plaster cast free of its tailored cocoon. Perspiring freely in a prison of furs, she rejoined her mute friend and got on with her lunch. Brady's evident amusement had not escaped Henriette and in French she savagely hissed. "Look at him, the assassin! There he is taking it all down to put in his paper and ruin me!" As she continued to rage, an aged blind man inched his way into the room, assisted by a lovely young girl. In pants. The moment of truth had arrived, and each participant knew it. Could she actually throw out a blind man? How would it look? Worse still, how would it look in print? Defeat hung limply in the air as the maître d' was signaled to "show Monsieur and Mademoiselle to their table."

Fashion sociologist Marilyn Bender stated in her clinical study of *The Beautiful People* that "In the Sixties, fashion stopped being clothes and became a value, a tool, a way of life, a kind of symbolism. It became human packaging."[20] Furthermore, "The fashion industry's creed, that appearance controls one's destiny," was now the tenet of a society in upheaval. It no longer mattered who one was or what one did, but what one looked like. A new cast of characters demanded new stage sets, and a young Frenchman stepped forward with a sheaf of fresh designs.

From the days of Corré, Delacroix, and Delmonico to the present, time and again a perceptive European has arrived and found New York atremble with enthusiasm but lacking direction, and has seized the city and shaken it with a series of innovations. In the case of Delacroix and Delmonico, it was confectionery and continental cooking. For Olivier Coquelin it was the discotheque.

The discotheque has a longer and more distinguished history than one would imagine, going back to the *caves* frequented by the existentialists in Paris just after the war. The *caves* were something like the *cafés artistiques* of an earlier period, and in them the poems of Jacques Prevert were set to music, and sung by performers like Juliette Greco. When, toward the end of the forties, the long-playing record came into use, the *caves* collected libraries of the comparatively expensive discs, to be played for their guests, hence *discothèque* as in *bibliothèque*. The discotheque notion was soon adopted by the Whiskey à Go Go (Whiskey Galore). Somebody in the French capital devised the clever plan of opening a tiny night club, lining the walls in tartan, serving Scotch whiskey, which was just then becoming the rage in Paris, and encouraging dancing in between the rather crowded tables to the music of a discotheque. The undertaking provided an inexpensive and intriguing evening (the suggestion of anything English always had much more cachet in Paris than anything French, including cuisine, ever had in London), and the result was a flash success, particularly with those French who found themselves with more dash than money after the war. Within a few years there was a Whiskey à Go Go or a facsimile thereof in every capital in Western Europe. Possibly because Americans had snobbish associations with the jukebox, it took some time for the movement to reach these shores.

Coquelin arrived in New York with panache. He had served in the United States Army in Korea and was covered with medals for "heroism under fire." When he returned to the States he managed restaurants and resorts—in Beverly Hills, Miami Beach, Fire Island, and Sugarbush. But above all, Coquelin loved New York ("Palm Beach is not a city, it is a season."). And he was, like Obolensky, a discerning bon vivant: good-looking, an enthusiastic amateur cook, he claimed to prefer the Korean front to a desk job in Alaska because the social life would be less dreary. In short, he was the kind of man who could say: "The best, the ideal place would be a night club where you would find the most beautiful girls . . . and be able to sit next to a man and talk about everything and anything your soul desires—except sex and money"—with conviction.

Coquelin's career more than any other shaped the social life of New York in the 1960s. In the first year of the decade, he decided that what New York needed was a private dining club. Not since the days of speakeasies had anything on the order of a night club served a severly limited private membership only, and the intimate, "clubby" atmosphere had been recalled with nostalgia for thirty years. Coquelin also felt that his club members should have the opportunity to dance, as well. This notion came at a moment of musical lull. The big "name" bands were by 1960 a thing of the past. Few were still functioning, and still fewer proprietors could afford the astronomic labor costs involved. The obvious answer seemed to be the intriguing innovation Coquelin had seen work so well in Paris—a discotheque. The result was Le Club.

The first effect of Le Club was an assault on the senses: the eyes, because they could see nothing, and the ears, because they could hear a great deal—wistful continental ballads, just then made popular by the growing passion for French and Italian movies, amplified by a superb sound system. When eyes grew accustomed to the dark, they perceived the shadowy settings of a baronial hall or hunting lodge, located somewhere south of the Firth of Forth and north of the Tiber. The enormous baroque paintings, sixteenth-century tapestries (or so they seemed), mounted stag horns, and dripping candelabra represented a considerable change from the Cosmopolitan pinups in the Cub Room of the Stork Club—a taste for which sophisticated New Yorkers, after fifteen years of travel abroad, were unlikely to resume.

For Coquelin, the Jet Set were "the people who want to be with the people who make news. . . . they are magnetic people . . . they are people who are complete in themselves and who have the knack to know what's what." The French expression for what came to be called the "Jet Set" is the *locomotive*, meaning literally that its members follow each other from place to place like railroad cars. And if ever anyone served as the locomotive's chief engine driver it was Coquelin, who was loyally followed by his patrons to L'Oursin, the summer night club he founded in Southampton, to Ondine, his undersea discotheque heavily draped in fishermen's nets, and to the Hippopotamus, a club the decor of which wittily recalled India under the Raj.

Discotheques sprang up all over the city and imagination went rampant. The same old faces were to be found in a series of highly unlikely settings. First one new club would be mobbed, and then another. Fashion could change in the space of a week. At L'Interdit, a Paris gendarme stood at the door, and a non-existent orchestra represented by a three-dimensional Marisol mural, with plaster hands and real instruments, weirdly adumbrated a genuine ensemble. The Garrison, appended to El Morocco, was a British Military Installation, Il Mio was situated in a grotto, and Shepherd's promised a night in Cairo, complete with golden Sphinxes drowsing lazily under a warm sky and horizontal fans to give the bar a breath of air and keep the mosquitos away. It was the day of Jet Set Disneyland. Within a few years the discotheque had virtually replaced the night club in any other form throughout the country.

Meanwhile, in a small and unpretentious gin mill near Broadway called the Peppermint Lounge, some dancers were discovered doing the Twist. It was as grave a moment in history as the night Czar Alexander first danced the waltz at Almacks in London. Within a very short time the beat grew "bigger" in the newly opened discotheques, partners lost their grip on one another, and rock and roll moved in.

If the 1960s are remembered for any phenomenon, it will be the

The Cheetah.

"youth movement" and the "new beat." When Olivier Coquelin opened the Cheetah in the huge, airplane-hangar-like site of the old Arcadia Ballroom, he took it upon himself, with unconscious audacity, to host the entire movement. And with the soul of a true host, a sort of Lorenzo Delmonico to the radical youth, Coquelin gave his guests exactly what they wanted. As they didn't want to sit at night club tables, he gave them ledges to perch on. As they didn't want expensive food and were often not old enough to drink legally, he gave them soft drinks and hot dogs. But they did demand the best in amplified music, and Coquelin invested in sound—booming through the club, on to the sidewalk outside, and down the street. Above all, what they wanted was a "psychedelic trip."

The Cheetah was psychedelic in the fashion of an old-time Hollywood world premiere with spotlights flashing on total pandemonium. To what could it be compared? A Michelangelo Inferno on Judgment Day without a soul over eighteen? The Cheetah was so big it could create its own cycle of weather, and the atmosphere so damp that a raincloud might collect and blot out the multi-colored blinking sunburst on the ceiling before precipitating onto the floor. But a bolt of lightning wouldn't be noticed, and a roll of thunder couldn't be heard. All nature had been vanquished.

The Cheetah also had its imitators. The Electric Circus, situated on St. Mark's Place in the heart of hippiedom, was smaller and, if anything, louder. The Electric Circus turned out to be in its own rasping and tortured way just that—a circus. The fee for entry was reduced fifty cents for those who came barefoot, and the accepted garb was no longer the mini-skirt and bush jacket, but body paint. The room looked like a vast enlargement of some section of the alimentary tract, and while sections of the human anatomy and other telling psychedelic illustrations were projected onto its writhing walls, music blared and strobe lights flashed so as to make any physical movement appear like the jerky gesticulation of a figure in a 1910 movie. But that was not all that transpired under the big top on St. Marks Place. There was an entire program of seemingly spontaneous acts—that is, everyone "did" his or her "thing." To the jubilant acid-rock of the Jefferson Airplane, a girl slowly stripped. She is not a girl. Blackout. An Indian temple dancer from East Orange takes the stage and desperately tries to invoke Krishna by sticking her tongue out and distractedly waving a moulting peacock feather. The spotlight suddenly wheels to the other end of the room and affixes itself to a young man who, atop a platform, is stoically consuming a banana. A surge of excitement ripples through the room, and one is put in mind of an equally stirring performance in the tavern days when a "countryman, for a trifling wager ate fifty boiled eggs, shell and all." A paranoiac maître d', dressed in white tie and tails, performs a hair-raising musical interpretation of King Kong enraged. Finally, a superb trapeze artist twirls over the heads of the audience to the strains of "Up, Up, and Away" and everyone who has not already embarked on a more hallucinatory trip, pads off to the refreshment counter for brownies, milk, and lo-cal coffee sodas.

Pop culture enshrined Coca-Cola bottles and sprawling, room-sized plastic hamburgers: still lifes for the space age. Their creators were now as much in demand for fashionable parties as clothes designers and hairdressers. The "upwardly mobile" were having a field day. No longer must they scale the heap by lashing themselves to committees charitably dedicated to dinner-dancing for a disease. Nor must they attend every political fund-raising dinner at a thousand dollars a plate. Now they could collect art, pop art, and if they proceeded properly, they might be invited to a "happening," an updated version of André Breton's Surrealist "events," or, better still, a party at the Factory, an aluminum-sprayed heaven and hell where the host, artist-film maker Andy Warhol, lunar pale and silver wigged, held constant court.

"In the Factory," wrote art critic Barbara Rose, "there is a kind of monstrous obscene abundance, as if the cream were being spilled off the top of the American boom and channeled into its silver funnel. Elegantly bored debutantes, spaced-out teenage refugees from suburbia, brilliant psychotic dropouts—themselves the excess baggage of a society of waste—pop pills, hold each other's sweaty palms and massage each other's silky skins while Andy films, tapes and preserves for posterity every idle spontaneous word and gesture. The camera is always running in the Factory."[21]

At Truman Capote's masked ball for his old friend Mrs. Kay Graham, publisher of *Newsweek* and the *Washington Post*, camera crews from the major television networks were restricted to a corridor outside the checkroom at the Plaza. Banned from the ballroom, the photographers of the press had to jockey for position at the hotel's main entrance, relentlessly popping away as the guests emerged from a serpentine slither of limousines. Cleveland

Amory, asked for his impressions of the party, helpfully replied: "This is almost like a joke. Fond as I am of Truman, I think we can say that society is not only *kaput*—it is Capote. The Beautiful People may be Society through the eyes of their *maître d's*, their hairdressers and *Women's Wear Daily*, but they are not society in the old sense—or even in the sense of making much."

No one could carp about Director Thomas Hoving's 1970 centennial birthday celebration for the Metropolitan Museum of Art, quite possibly the most beautiful party of the century. The Arms and Armor Court became a waltz ballroom in the 1870 Viennese-cum-Gothic style, serving champagne and oysters, game pie and venison. The sixteenth-century Patio from the castle of Los Velos metamorphosed into a 1910 Belle Epoque Spanish folly, with a tango orchestra, pink shawls draped over balconies, Sargent's portrait of the Wyndham sisters, and a buffet of pastry, sherbet, ice cream, and pink champagne. The 1930 purple, blue, green and gold "moon struck" jazz club was situated, naturally enough, in the Egyptian Sculpture Court, where Cecil B. DeMille sphinxes looked down on red-satin tablecloths and an array of canapés—smoked turkey, steak, shrimp, and caviar. Finally, the intricately draped Fountain Café became a discotheque where a breakfast of omelettes, crêpes, and coffee jogged dancers back into 1970.

True to history, the chairman of the Ball Committee was Mrs. Vincent Astor, who welcomed a cosmic cross-section of New York's present: editors and publishers, critics and collectors, actors and producers, restaurateurs and food experts, educators, musicians, doctors, designers, lawyers, financiers, and gallery owners, along with a heavily represented old guard—Vanderbilts, Whitneys, and Rockefellers. Dress ran from grandmothers' carefully preserved antique silks to Corrège evening space suits. One young woman waltzed through the 1870 ballroom splendidly topless. Meanwhile, throughout the festivities, a "quiet, non-action, dignified" picket line paraded in front of the museum's entrance, protesting the Metropolitan's expansion into the park. Black tied and ballgowned picketers carried signs wistfully asking "Is Central Park Destined to be a Memory?"

The nadir of Pop entertaining was probably reached in 1972 at a record company executive's birthday party, held at the unsuspecting St. Regis Roof, in honor of rock star Mick Jagger. A silicone-implanted nude leaped from a cake and flipflopped wildly for the edification of the mixed company. *Harper's Bazaar* reported that among the guests were a "double-dating" foursome composed of the Princess Radziwill, Truman Capote, Andy Warhol, and Candy Darling, a transvestite underground superstar. The guest of honor was presented with many tokens of love and affection, including an American Indian turquoise bracelet and an exquisite gold coffer filled with cocaine. Warhol kept several of his young apprentices scurrying back and forth with cameras, freezing on film the many highlights of the party. Senator Javits' wife Marion burst into bitter tears. "Everything we fought for all those years has just gone down the drain. I won't stand for it." Few others seemed to care, although political columnist Harriet Van Horne later voiced her concern in the *New York Post:* "In the perfumed twilight of the Roman Empire unspeakable things went on. Are we entering the same twilight?"

Physically, New York is one of the least historic cities in the world. The lower part of Manhattan has been torn up and rebuilt with such regularity that archaeologists of the future might well assume repeated and devastating invasion and destruction. Practically nothing remains of Dutch New York whatsoever. By the time that the City Landmarks Preservation Commission was established in 1965, there was little enough left to save. Scraps of history remain here and there, preserved as air spaces between later structures, overlooked, or simply allowed to decay in areas so derelict that building was unprofitable. The Stuyvesant Fish House, dating from 1803, stands on land that had belonged to Peter Stuyvesant as it did when Delacroix first opened his spectacular Vauxhall Gardens. Gracie Mansion already stood many miles out in the country. The Schermerhorn block of commercial buildings in the "Federal" or "Greek Revival" style still stands on Fulton, South, and John streets, and Captain Hall well may have hurried by them on his way for a quick lunch at the Plate House. The Salmagundi Club, on Fifth Avenue and Twelfth Street, is the last of the great mansions that once lined the Avenue from Washington Square north to Central Park. It was newly built when the Delmonicos bought the Grinell home just two blocks away and moved uptown to feed the "Avenoodles."

For a diner in search of a bit of history, there still remain, at this writing, a few places where he may sense a shock of recognition. Fraunces Tavern still stands, one of the few remaining structures of its era. It has undergone any number of vicissitudes since Black

Sam's day, including a minimum of three fires. By 1904, when it was bought by the Sons of the Revolution in New York State, the house had degenerated into a tenement with stores on the street level. Still its preservation is something of a miracle, and it is the oldest building in Manhattan, although almost entirely restored, with only a few antique beams and some brickwork from the original. On the ground floor, in a wainscoted reproduction of the eighteenth-century dining room (Fraunces' long room is reproduced above), the good citizens of Wall Street dine off aromatic baked clams and broiled venison steak. A portrait of the benign founder hangs in the hall. There are surprisingly few tourists, although upstairs one may find Baron Steuben's campcase, Lafayette's bloodstained sash, a hank of Washington's reddish hair, a tooth, a piece of the stone balcony of Federal Hall, buttons, a piece of coffin, and Martha's shoe.

Surprisingly enough, too, Delmonico's Beaver Street premises is still a restaurant, a good one at that, although it has changed hands. Josephine Delmonico sold out in 1919, but the Beaver Street restaurant was purchased and reopened in 1934 by Oscar Tucci (not to be confused with Oscar Tshirky), an imaginative Italian with an eye for preserving style. Lunches and dinners are served to a Wall Street clientele in its several ground-floor rooms, of which the main dining room (The Palm Room) and the Roman Room, a tiny Egypto-Roman bijou, radiate a pronounced charm and *fin de siècle* atmosphere. And the front door is still flanked by those Pompeian columns that impressed New York as much in their day as the Picasso curtain at the Four Seasons does today.

The financial district is further historically enhanced by Sweets, the oldest extant fish house in the city; it made its debut across the street from the Fulton Fish Market in 1845. The staircase of its doddering old building at 2 Fulton Street is jammed twice daily with expectant patrons. Patience and fortitude are required at Sweets, and generations have learned to wait for their giant helpings of simply prepared gray sole, halibut, swordfish, and delicate bay scallops. Oysters, once the focal point of Sweets' menu, were banished from the house several years ago; Peter, the chef, advanced the darkly provocative theory that "something in the ocean has been eating at them."

And, of course, Luchow's still stands, giving one the eerily pleasurable sensation of being physically transported back in time. It was reverently preserved by Jan Mitchell, a fair-haired Swede of *alt deutsch* hauteur, who bought the restaurant in 1950, and determined to restore and maintain its traditions. He revived the festivals with their German band music and special menus: The October Festival, the Venison Festival, the Goose Feast, the Bock Beer Festival, the May Wine Festival and the midsummer Forest Festival. Victor Herbert had persuaded August Luchow to engage a string ensemble to perform "tunes" from Wagner, Brahms, and Victor Herbert and Mitchell maintained the tradition, having the only orchestra in New York that still played "entrances." Helen Traubel was greeted by themes from "Lohengrin," and "I Can't Give You Anything but Love" was struck up for Frank McHugh. These reverberated from the lofty ceiling; the huge mirrors in their foot-wide gilt frames reflected "Siegfried Forging the Sword" in the Nibelungen Room, Franz Snyder's "Preparing the Feast," August Hagborgs' "The Potato Gathering," and Van Dyke's "The Earl of Wentworth and His Secretary." No concessions were made to indirect lighting, and upstairs in the Lillian Russell Room, that lady's bible, the fan she was given by the Princess Polignac, and the handkerchief presented to her by Edward VII, take their comfortable and well-cared-for repose, along with her costume for "Princess Nicotine." Most important of all, Mr. Mitchell's intelligent management perpetuated a hearty cuisine and his roast Watertown Goose with stewed apples could not fall short of August Luchow's own.

Two of the splashier turn-of-the-century hotels have managed to maintain their prestige in the upwardly mobile world of the automated New York Hilton, the Americana (the world's tallest hotel) and the Regency, the only truly luxurious one of the lot. The Plaza and St. Regis have given way only inch by inch. The Plaza's various public rooms have undergone numerous incarnations. The large room on the corner of Fifty-ninth Street and the Plaza, which was called simply the "restaurant," assumed various decors as the Edwardian Room and the Green Tulip, and the Fifty-ninth Street dining room that served as the office of Jules Bache has become, and remains, the Oak Room.

Finally, the Plaza houses New York's one functioning Palm Court, and it has a busy day. Breakfasts and salad lunches are served, and no sooner are the last leaves of lettuce carried away than the violinist and pianist tune up and a flame is put under the tea kettles and cocoa in the kitchens. This does not mean, however, that the Plaza has not plunged ahead into the future. Not only does it provide its guests with closed circuit television and a choice of two movies daily, but troubleshooting hostesses called "service coordinators," together speaking all of fifteen languages, patrol the lobby and halls where once private maids and lackies scurried obediently.

The St. Regis has retained less of its historic quality than the Plaza, but its King Cole bar harbors another notable monument, Maxfield Parrish's Old King Cole, late of the Knickerbocker Bar and the Racquet Club, and always a curiosity to the younger generation.

Not all the gastronomic landmarks of New York curried favor with the carriage trade. McSorley's Saloon, which was already considered ancient at the turn of the century, survived Prohibition, changing not a whit during ensuing decades. Harry Kirwan, the late owner's son-in-law, took over the reigns, the pot-bellied stove, and the ale pumps; McSorley's classic Bermuda onion and Liederkrantz sandwiches continued to perfume the pipe-smoke-laden air. Tradition was enforced with courteous tyranny, and no woman could enter McSorley's until, in 1971, the Women's Liberation movement took hand and "liberated" McSorley's. One of the first women who brazenly attempted to patronize the premises was attacked by a doddering but irascible habitué who poured a mug of ale over her head.

Nor are all keepers of the flame preoccupied with nineteenth-century New York. Rivaling Jan Mitchell and Harry Kirwan as super-traditionalist is the Algonquin's Ben Bodney who bought the hotel in 1946, shortly after the death of Frank Case. Years before, he had brought his bride from South Carolina to honeymoon at the home of the Round Table. She was so ecstatic that he promised to buy the hotel for her, and, with the passage of time and the amassing of a fortune in oil, he did just that.

Bodney's sons-in-law, Andrew Anspach and Sidney J. Colby, eventually assumed management of the hotel, and the family has held a unanimous position against change. When the wear and tear of the famous had taken its irrevocable toll on the lobby furnishings, they sought the advice of Oliver Smith, renowned for his evocative stage sets. Little by little and bit by piece, the furniture was smuggled out in the dead of night, reupholstered, and sneaked back as soon as possible. In the same conspiratorial manner, walls were refinished, draperies hung, and a new carpet in-

McSorley's back room, by John Sloan. *Courtesy of the New York Public Library.*

stalled. Miraculously, the last accessory turned out to be an exact replica of the hotel's original, a pattern which Mr. Smith had never previously seen or heard of. With further devotion to authenticity, the Rose Room was redecorated as dictated by an illustration of it in a 1920s' Camel cigarette advertisement.

"The day we need self-service elevators is the day I sell the Algonquin," said Mr. Bodney. The small attractive rooms of the hotel seem much as they always have, shoes are left in the hall to be shined, and the coffee is ground on the premises. As the turn-over in staff is practically nil, little has changed in the kitchen or dining rooms. The net result of all this cheerful necromancy is that the Algonquin never loses its old regulars as long as they are in this world. Helen Hayes still floats through the lobby, and Thornton Wilder and Mrs. James Thurber greet their friends. The Algonquin is still a Broadway tradition. Sir Laurence Olivier and Noel Coward, Albert Finney, Julie Christie, Sir John Gielgud and Peter O'Toole, Jean-Luc Goddard, Simone Signoret and Yves Montand, among others, would not stay anywhere else.

In 1962 the Rainbow Room was completely overhauled and returned to its devastating 1930s' estate. New wine racks stood at attention at the head of the Rainbow Room's chrome-outlined double staircase, bending with the lunging, swerving "ocean liner" curve characteristic of the thirties, where once Bea Lillie had made her hilarious entrance; otherwise the room has been restored to look exactly as it did before World War II forced it into the status of an undistinguished cocktail lounge. At the auxiliary Rainbow Grill, Duke Ellington, Peter Duchin, Jonah Jones, and Louis Armstrong could hold sway in a twinkling red-tableclothed atmosphere suggesting a homogeneous cruise ship. Nowhere else in New York could one recapture the sleekly nostalgic essence of the thirties as poignantly as at the Rainbow Room, and for this reason it should be included in any listing of the city's last living relics.

Others of Lucius Beebe's Café Society haunts have suffered various fates. El Morocco, too, came back from the dead. It had gone into eclipse after the passing of John Perona, and the desert setting quite literally presented vistas of empty wastes, but it was resuscitated to a second period of glamor when, in the style of Le Club, it became a private club. And, as if through some act of divine vengeance, a pocket park and torrential waterfall comman-deered the site of the Stork Club, with children dripping ice cream on the site of the bar and waters crashing down where Winchell once held forth.

The Pavillon and the Colony died, more or less together. The combined event was so cataclysmic that seasoned New York watchers proclaimed it the death of haute cuisine itself. Within a very short period, an extraordinary number of New York's best restaurants, all French and all of the haute cuisine caliber, had closed their doors, including Le Voisin, and the Café Chauveron, owned by the founder of the Chambord. Others, less well known but each with a devoted following, also succumbed, including Maud Chez Elle and L'Armorique.

The fact was that by 1971 the restaurant business itself had fallen off from 10 to 40 per cent. What was the reason? Teams of analysts examined the patient. Unquestionably, rising costs had a great deal to do with the problem. Restaurant labor was among the most poorly paid until 1960. Now that is all changed, the finest restaurants have been unionized (they were far too visible, as targets, to be passed over), and a waiter's wages, which stood at thirty-eight dollars in 1960 rose to sixty-seven dollars by 1972. Dishwashers earned one hundred dollars a week, and a master chef might command twenty thousand dollars, or much more. The owner of one of New York's costliest restaurants claimed that in the five years between 1966 and 1971 his payroll had risen 55 per cent, laundry 53 percent, and garbage collection 120 percent. Moreover, restaurants cannot mechanize, or cut their highly trained staff in a time of lull. Other costs had risen as well: the price of food, for example, and rent. The highrise offices that rose relentlessly in midtown Manhattan generally displaced streets full of small brownstones that made convenient and inexpensive homes for restaurants. Similar ground-floor space in the new buildings might rent for six times as much. To further compound the problem, there was a dearth of French chefs. America has never seemed to be able to train master chefs, and when, in 1970, the Department of Immigration suspended the "skilled worker" classification because of unemployment in America, they cut off a vital source of talent. "In a few years it won't be a business anymore" complained a distressed Mme. Henriette. "You won't find the chefs." And while costs rose, volume dropped. The last years of the 1960s saw a long and debilitating recession.

Each one of the deceased had claimed a different malady. When the Café Chauveron's building was torn down and replaced by a skyscraper, the rent more than tripled. "At that price you work for the landlords," declared Roger Chauveron after four decades of masterful service. Gene Cavallero, Jr. made no secret of the Colony's complaint. A curt sign on his locked door read quite clearly; "Due to Unrealistic Terms in our Union Contract, the Colony is not Open." As the Pavillon died, its final owner, Stuart Leven, wept. "There simply are not enough patrons to keep a restaurant of this stature in the style it should be kept."

In the final analysis, who or what had been responsible for the accelerated death rate? In the opinion of a professional connoisseur, Baron de Groot, the fault lay with the quality of the patrons themselves:

The New York Restaurant business is sick, and may, in fact, be dying. The reason, I believe, is that almost all the basic factors essential to the making and maintenance of a great restaurant are lacking in New York. A great restaurant needs the support and discipline of a loyal, gastronomically-experienced circle of patrons, intimately familiar with its wine list and menu, knowing exactly what to expect from everything they order, and ready to send every bottle or dish back if it is one iota less than perfect. Instead, some of our fanciest restaurants are mainly supported by tourists, unlikely to come back for another five years and unaware that Ris de Veau is anything more than a casserole of rice with veal. Even the average New York diner behaves as if he were a tourist. When he has had three meals at one restaurant, he is bored with it. He wants something new. He believes the loudest publicity. He races to the new restaurant which is carefully rumored to have "the beautiful people"—or where the gossip columns say that Jackie Onassis was seen dining—or which is alleged to be the haunt of "the literary set," or "the stars," or some other ingroup. The stress is on everything *except* the excellence of the food and service.

The publicity for Charles Masson's La Grenouille stresses that he has the most expensive flower arrangements. The Lafayette has ridden to fame on the journalistic cliché that its owner, Jean Fayet, is ruthlessly rude to his customers. The Four Seasons is known for the Picasso in its entrance hallway—The Rainbow Room, for its view across the city—and every major restaurant, even in these hard times, still used the pretentious phrase "reservations essential." Robert Benchley may have said it first: "If you can get a reservation at the place, it can't be any good."[22]

According to de Groot, when a restaurant in New York was no longer fashionable, it could no longer function regardless of its food. The Pavillon ceased to be fashionable when Henri Soulé died in 1966, and the Colony ceased to be fashionable for no other reason, perhaps, than that the *locomotive* had moved on, and when a restaurant loses its clientele, like a man without his accustomed friends, it becomes dispirited. Gael Greene's final meal at the Colony suggested a feast at a waxworks:

Today lunch in the Colony dining room is like lunch at Forest Lawn except that here the flowers are mostly plastic. A captain picks his nose. A waiter nibbles toast. Gene Cavallero, Jr. picks his teeth with a matchbook. Young couples arrive bubbling and cheerful . . . and soon grow grim and silent in the vast unoccupied spaces of Marienbad in the off-season.[17]

In fact, grave and more fundamental changes seem to be afoot. The remaining haute cuisine restaurants, and there are still many flourishing establishments, have responded to economic necessities by cutting their menu and trimming their staffs. At the "21," business is booming. "But it's luxury dining" says Robert Kriendler, "not gourmet dining. We try to make the best sauce Madère, for instance, but we are not consumed with tasting and retasting it." The question remains: is haute cuisine dining itself going out of fashion? Certainly the love of good food and interest in restaurants is not. By the late 1960s, restaurant critiques were read as avidly as the theatrical reviews of the thirties, and Craig Claiborne of the *Times* had the following once accorded George Jean Nathan. Although Baron de Groot might shudder, the result was that people went to new restaurants as they might go to new plays. In final rebellion against the prohibitive prices of restaurants and domestic help alike, New Yorkers finally, and with a passion, learned to cook themselves. They are as anxious for self improvement as ever, and cookbooks have replaced the books of etiquette, as the books of etiquette very nearly replaced the Bible. The great names in cooking were no longer Escoffier, but James Beard and Julia Child.

And what about the future? Will the younger generation, as they grow older, take any interest in haute cuisine? For their elders, the gulf between the Stork Club and the Pavillon was a small one. But between the Cheetah and the likes of Henri Soulé yawns a chasm few would try to leap, or even want to. The hushed respect, the lunging white-gloved servitors, are so out of keeping with the spirit of the younger generation that it would be hard to believe the tradition could survive.

And so, with considerable nostalgia, New York contemplates the end of yet another era, with the usual hankering for the past.

Newsweek spoke of the demise of the Pavillon in the spirit with which Oscar had lamented the passing of the ten-course meal:

What breathless extravagance to dump three tins of Beluga caviar down the sink at $150 each because he didn't think it tasted just right. What extraordinary arrogance to turn Bing Crosby away when he appeared in sports shirt, sans tie—"an insult to my restaurant." And what nostalgia is evoked by his remark to Martin Decre—his maître d' hotel for twenty-five years—as he admired the black-tied gentlemen and bejeweled ladies seated on his banquettes: "Look how beautiful our Pavillon is tonight."[23]

Finally, the most pessimistic observers saw the demise of great New York restaurants as a symptom of the city's own impending doom. For New York, they insisted, using the generation's favorite hyperbolic cliché, the apocalypse was near. The composition of the city had irrevocably changed, they said, the middle class had deserted to the suburbs, and fear and crime ruled the streets.

It was true that the New York of the sixties presented a more frightening picture than the New York of the thirties. But then the New York of 1860 had presented a more frightening picture than the New York of 1830. In retrospect, had New York ever been a comfortable city of the middle class? Had the streets ever been completely safe? Only a scholar could pinpoint the moments when they had. Granted that the Broadway area had become a sink of pornography shops, "massage parlors," perambulating prostitutes of every sex, and knife-wielding muggers, but had Water Street and the Tenderloin been so conveniently obliterated from memory that this kind of public menace seemed to be something disturbingly new?

The desegregation of schools by busing students to distant neighborhoods stirred ugly demonstrations which in turn stirred outraged rhetoric. But the fact that ten thousand racists had rioted during the Civil War, looting, lynching, and burning down an orphanage was a fact so appalling that it had been routed from the collective consciousness. Perhaps, as F.P.A. had pointed out, "Nothing is more responsible for the good old days than a bad memory."

What the pessimists failed to recognize was Nature's predilection for life cycles, her capricious habit of first giving and then taking away. Philosophically speaking, the city's older, grander restaurants should be thought of now in terms of senior members of a distinguished club, slightly in their dotage, of an age when it should not come as too much of a surprise to occasionally find one dead in his chair.

Like an old club member at the bar, New York history keeps repeating itself. Drawing parallels could become an endless game. However, one might consider just a few recurring phenomena in New York history. Many contemporary personalities mirror the past, as, for example, "the New Journalists"—Gay Talese, Willie Morris, *et al.*—and their hefty protectress, Elaine. A self styled *salonière*, Elaine might recall several figures from New York's early restaurant days—Loosely, coddling his favored clientele; Charlie Pfaff, nursing artistic egos while filling artistic bellies. Pfaff's special pets, the writers led by *Atlantic Monthly* editor William Dean Howells, were known to the public then as "the New Journalists." When the Leonard Bernsteins ingenuously undertook to introduce the Black Panther Party to the liberal rich at their Park Avenue apartment, it was New Journalist Tom Wolfe who publicly crucified them and "Radical Chic," an epithet no one had thought to append to Mabel Dodge after she invited anarchist Emma Goldman to sound off in her Fifth Avenue salon. Many others are brought to mind: Mike Todd playing Diamond Jim to Elizabeth Taylor's Lillian Russell, inviting eighteen thousand friends to a circus-like party at Madison Square Garden as, across the nation, millions watched the cake-throwing climax on television; Elsa Maxwell dying the lonely death of Ward McAllister, her funeral attended by a skimpy handful of mourners; astronaut John Glenn's returning to Earth like a Space Age Lindbergh for the heaviest tickertape reception in history.

Broadway of the twenties would return to life in the night clubs of another generation. The Follies reappearing at Billy Rose's Diamond Horseshoe, burlesque in the cages of go-go girls, the intimate revue in Julius Monk's witty ateliers. The plate houses would have their renaissance in "fast food" chains and "pretty waiter girls" in the Playboy Club's "bunnies."

With the swinging pendulum of time, New Yorkers would once more entertain at home. When caring not to cook, however, pizza, fried chicken, or a complete Chinese dinner might now be delivered to their doors rather than an eighteenth-century oyster stew. A passion for wine would once again consume them as it had the gentlemen of colonial times, and restaurateurs would rediscover the practical beauties of the limited menu. Once more musicians would begin to play in small restaurants and clubs around the city, and in the age of divorce and the brief affair, the silvery-templed *New York Times* would notice that married couples had started to make pilgrimages to the "singles bars" where first they had spoken before whisking each other home. With its free love, free Shakespeare, free concerts in a city meadow, New York was still Leonard Bernstein's "Wonderful Town."

It had taken an Irishman, Brendan Behan, to best sum up the city's eternal attraction. Upon returning to his native isle, wife Beatrice wished to be reassured that New York had not fatally seduced him. Wasn't it wonderful to be home again, she said, wasn't it wonderful not to have to live in a great, ugly, noisy . . .

"Listen Beatrice," interrupted Behan, "it's very dark."

Afterword

***

The Coming of the American Food Revolution:
1973 to the Present

Little Birds are writing   *Read, I say, not roasted—*

Interesting books   *Letterpress, when toasted,*

To be read by cooks: *Loses its good looks.*

                                 Lewis Carroll

Time to spill the truth. The first completed manuscript of this book, or what we presumed to be the completed one, originally had been contracted for publication in the late 1960s. But no. The feet of the initial publisher, like Lewis Carroll's wrinkled "Little Birds" introducing our final chapter, had regrettably turned "pale with sudden cold." The reason given was blunt and craven: The time was no longer right. New York City had entered what looked to be a prolonged season in socioeconomic hell, worsened by a relentless battering in the media. The city teetered on bankruptcy, crime was shooting to an all-time high. Even the serene groves of academe had come under assault, as militant students faced off police from the shattered windows of Columbia University. In response, tourists were giving Manhattan a wide berth. So who, the retreating publisher frowned, would want to buy a book about the historic joys of living in a city now in limbo?

A few years passed, and New York's outlook began to brighten, as did our manuscript's, when Charles Scribner's Sons, themselves a proud piece of local history, stepped in to publish an updated version in 1973. A happy ending? All things considered, yes. But viewed from 1998, at the far end of the so-called American Food Revolution and a bull market Wall Street run of unprecedented duration, a number of comments and predictions we had made in the closing passages of the 1973 manuscript today astound us with their gloom. Had we really written that "America has never been able to train master chefs"? Or worried "Will the younger generation, as they grow older, take an interest in haute cuisine"?

Little could we realize that the most accurate prophecy for the final quarter of the twentieth century lay buried in one of the nineteenth-century chapters. Specifically, a prescient letter written from New York by James Fenimore Cooper's fictitious *Traveling Bachelor* to a Parisian count: "I know of no spot of the habitable world to which the culinary scepter is so likely to be transferred, when the art shall begin to decline in your own renowned capital, as this city."

Today, in fact, many respected critics point to New York City as the restaurant capital of the world, hardly the case in 1973, when the term "American chef" still had the dissonant ring of an oxymoron, and the upper stratum of Manhattan's gastronomic deities was crumbling.

How had this unforeseeable, and hugely profitable, transfiguration come about? A number of key signposts and revolutionary banners can be identified.

First, a brief survey of the most relevant social and professional revolutions. The counter-culture, flower power, greening-of-America movements of the commotional Vietnam War era tangentially provoked several by-products of formidable staying power, namely a rehabilitated respect for the hand crafts, a vigilant concern for the freshness and purity of foods, and moralistic dedication to the "nurturing" of others. In tandem with this came the steamrolling feminist movement, which would have profound effects on restaurant, hotel, and nightlife business, both in the front and back of the house, to use trade argot.

On one hand, young women's mass defection from home kitchens, as emblematic as bra burning, would create a new restaurant clientele of swiftly multiplying proportions. Not only did fresh waves of newly liberated singles demand to be seated without the once requisite male escort, but rapidly swelling ranks of women in the workplace spelled double-income couples with scant time to cook but sufficient funds to split frequent restaurant checks. Because of this, dining out came to be considered more and more a workweek necessity, and less the special-occasion festivity it had been in the past. This served as a powerful fertilizer to New York restaurant expansion. Thick sproutings could be spotted in once barren residential stretches of the Upper East and West

Sides, along with faded precincts undergoing either "gentrification," such as SoHo and, later, TriBeca, or reclamation, such as Chelsea, Ladies' Mile, and the Flatiron district.

With gender taboos being jettisoned left and right, young men began to edge from the outdoor grill to the indoor stove without apparent loss of machismo; soon unexpected numbers of college-aged sons announced to their conventional American parents (for whom a chef still meant a temperamental foreigner with a white smokestack hat, pointy moustache, and marginal social standing) startling ambitions to build careers as professional cooks and hosts.

Expressing similar intentions came a new pro-cooking breed of self-assured feminist, now politically determined to buck male supremacy in the hieratic professional kitchen and show that they, too, could rule the roost as executive chefs. Or as restaurant operators, and forget the ladylike tearooms of an earlier, meeker sisterhood.

Thanks to the messianic promptings of Hungarian immigrant Louis Szathmary, noted Chicago chef-restaurateur and fervent American culinary archivist, the American Culinary Federation, with a membership blanketing the country, finally persuaded the U.S. Department of Labor to reclassify chefs as professionals rather than domestics. And this not until 1977. But a potent augury nevertheless.

Stylistically speaking, the culinary quake that would most alter the terrain of American professional cooking, counter to today's revisionist theories, was nouvelle cuisine, as coalesced in France, beginning in the 1960s, by a radical archipelago of chefs including Paul Bocuse, Michel Guerard, Roger Vergé, and the Troisgros brothers, most of whom owned their own restaurants. Today, nouvelle cuisine is often dismissed as a precious fad: too little undercooked food, archly composed on outsized plates like Japanese flower arrangements, at too steep a price. To the contrary, the style's seminal influence was, and is, sweepingly broad and deep.

The first and most vocal cheerleaders for the movement were the French restaurant critics, guidebook and magazine publishers Henri Gault and Christian Millau. Their introduction of a tiny red toque ideogram to identify restaurants embracing the tenets of nouvelle cuisine had chefs across France scrambling to retool their menus à la ultramodern mode.

And what were these tenets and traits? In 1976—a fateful year, by the way, for American cooking—Gault and Millau, with Gallic ardor for academically structured systems, summed up the commonalities. Above all, unnecessarily complicated cooking was to be dumped. In order to "reveal forgotten flavors," the cooking time for a litany of foods and dishes should be reduced, and steaming reintroduced when applicable. Equally crucial, the new style was designated a *cuisine à marché,* literally market cookery, reliant on the freshest available seasonal produce, a proposition that would also lead to collaborative efforts between chefs and breeders of livestock, cheese artisans, vegetable and fruit growers (shades of Delmonico's nineteenth-century truck garden in New Jersey) to secure the finest. Enter the golden age of the specialist purveyor. Next, disguising of inferior ingredients with domineering sauces was disallowed, particularly ones thickened with newly despised flour roux, along with war-horse hollandaise, espagnole, and velouté. Game was now preferred served fresh, rather than hung to an odorous high. Regional dishes deserved reexamination; the old provincial repertoires of France's rustic *terroirs* suddenly seemed a more attractive and fertile inspirational resource than stiflingly codified Parisian haute cuisine. Dietary evangelism was now permissible in the stylish kitchen, with lightness in all forms a recurrent goal. Technological breakthroughs were welcomed in the pursuit of convenience, speed, and virtuosity; the revolutionary Robot Coupe food processor, for example, soon spawned a nationwide tide of silken fish quenelles, once too laborious and time consuming for the average kitchen staff. The New World would quickly follow suit. Tediously long menus, hallmark of the petrifying grand hotel style of "continental cuisine" should not only be shortened but honor the season; this last dictum would lead, years later, to menus printed in-house by any establishment that could afford its own laser system, empowering chefs to finalize a seasonal market menu honoring that *day's* best fresh produce mere hours before meal service.

The frisky inventiveness of the pioneer nouvelle cuisiniers was applauded by Gault and Millau as something to admire unequivocally; the implicit message that it was worthy of wholesale emulation would not only be heard by a new generation of chefs across France but across the Atlantic. And though the two critics did not list it in their rundown, the new common preoccupation with the look of food, pensively arranged on individual plates as if by graphic or fashion designers, would forever alter the way chefs seeking recognition on either side of the ocean would compose their latest *specialités*—not only with a mind to flavor and texture, but to color and line. Decades earlier, in his memoirs, Escoffier wrote that cooking was always prone to fashion. Not until the incessant menu retailoring incited by nouvelle cuisine, however, did the world learn how dizzyingly correct the "father of modern cuisine" had been.

New York was already known as a design hub for its Seventh Avenue runways. By the 1980s, some of the hottest collections anywhere were parading down the tables of the city's latest slew of voguish restaurants. In 1977, a young French countess named Marina de Brantes, a relative of the president of France, had opened a tiny restaurant in the east sixties. She called it Coup de Fusil, French slang for a bandit holdup, which was the first wholly nouvelle cuisine restaurant in New York, and in America. Not only did she present a culinary style to latch on to, but the novel concept of cramming a cooking classroom and catering concern on an upper floor. If that weren't enough, she launched her own cooking program in an embryonic new medium, cable TV. In the sizzling new climate of restaurant entrepreneurship, such ideas crackled and flamed as if by spontaneous combustion, spreading the fires far.

From today's perspective, Gault and Millau's 1976 summation of nouvelle cuisine can be recognized as an international manifesto, one which coincided with a signal episode in the saga of American cuisine, the marathon observation of the nation's Bicentennial. Recalling the furor of Lafayette's triumphal return to the United States for its fiftieth birthday party, New York City and its harbor were jammed with millions of celebrants, including an armada of tall-masted ships under thirty-one flags, and a twenty-two nation fleet of pennant-festooned naval vessels. Eventually the nonstop nationwide ballyhoo and festivities stretched to a couple of years. National self-esteem, woefully dampened by Vietnam and other ails, was reignited. Awakening to newfound pride in their ethnic and regional roots, Americans took action to honor and protect them, much like the architects and preservation groups which at that time were rushing in to salvage decayed inner cityscapes and bestow landmark status on threatened cultural treasures. New York watchdog committees would rescue countless buildings, including the irreplaceable Beaux Arts Grand Central Station, its starry vaulted ceilings just now restored, and its tiered levels brightened by a galaxy of new restaurants and food ventures.

Professional American cooking, too, would benefit from this new group custodianship of traditions which, it must be added, was and remains a selective one, precisely like the post-Modernist schools of art and architecture, in which traditional or classic forms are "deconstructed," arbitrarily edited, then reintegrated with avant-garde elements to create fresh unities that appear at once familiar and unfamiliar. Prior to this, regional American cuisines, with isolated exceptions like the touristic Creole repertoire of New Orleans, were still largely the picket-fenced domain of Mom, hereditarily installed as chief cook, bottle washer, and keeper of the ancestral culinary flame. Since World War II, however, heirloom cooking had been buckling under an onslaught of convenience foods, cheaply prefabricated in the same homogenizing spirit as postwar mass housing. Now, swept along by the combined forces of aspiring native-born chefs; reanimated national and ethnic pride; a growing reverence for heirlooms (including recipes and plants), regionalism, and seasonality; and the "creative" maxims of nouvelle cuisine, American restaurant cooking would be transformed not only into an inspired new arts-and-crafts form, but a business vibrant enough for Wall Street eventually to declare it a dynamic branch of the entertainment industry.

How New York, from the time of the Bicentennial to the present, became the restaurant capital of the world, and the ultimate Broadway stage for chefs as performing stars, could, as they say, fill a whole book. Unfortunately, because of space limitations, not this one.

Apart from general observations already made, certain telling episodes in our own professional gastronomic life together in New York, over the same quarter of a century, might spotlight other key evolutionary phases in the amorphous phenomenon known today as the American Food Revolution.

First off, in 1973: the launch of this book's publication, staged after closing hours at Bloomingdale's flagship department store on 59th Street, a location considered decidedly "uptown" not so long before (as ever, New York's carriage trade vortex continued to roll northward ten or twenty blocks every generation). Widely aped as the avatar of merchandising and marketing daring—its "gourmet" aisles and counters at that time outshone others—"Bloomie's," as nicknamed by quasi-resident fans, was on the cusp of making gala parties in a retail setting the height of fashion. Their clout was autocratic: When Queen Elizabeth paid a formal call, management indeed managed to have Lexington Avenue's one-way traffic reversed, creating total chaos, so that Her Majesty might descend from the side of the limo dictated by protocol.

This was the disco age, as mad about fashion as gadding about, seemingly anywhere at any hour. Offbeat party venues were particularly popular, terrorizing fashionable caterers with the Herculean logistical demands. Had the receding specter of the Metropolitan Museum's one-hundredth birthday bash unloosed hosts' imaginations and bank deposits with an abandon even Elsa Maxwell would have applauded? In any case, our Bloomingdale's book launch was reported in the press to have been a milestone New York event, for a reason that today would hardly cause a stir: buffet tables set up and staffed by two dozen of the city's leading restaurants and hotels, at which over four hundred guests, including Mayor John Lindsay and James Beard, could sample at their ease. Barbetta brought white truffles and a string quartet in eighteenth-century costume; Luchow's dispensed bratwurst and the blare of a German oompah band in lederhosen; Hippopotamus hauled dance records and a live discoteur. In retrospect it would appear that a new template for public entertainments had been devised: serial food and drink "stations" showcasing the specialties of famous establishments. Famous *establishments*, we must underscore, not chefs, whose status as stars would take a few more years to twinkle.

A second milestone was reached in 1979, at a three-day first anniversary party for *Food & Wine* magazine, of which we were the founding editors, staged at Warner LeRoy's Tavern on the Green in Central Park. A series of lunches and dinners entitled La Jeune Gastronomie, it made both national and international wire services as being the first gastronomic event, anywhere, to present American chefs, namely Alice Waters, mother of "California cuisine," and Paul Prudhomme, big daddy of contemporary Cajun cooking, on the same footing with recognized stars of France and Italy's nouvelle kitchens. That many food journalists not only found the Americans' menus every bit as innovative and deftly produced, but possibly more so, was related with almost the same degree of incredulity as had the Gallic wine establishment when, in 1976, a California cabernet sauvignon placed first in a blind tasting of mostly French bottlings in Paris.

It should be noted that in the 1970s superior wines began to shed their image as either intimidating restaurant amenity or closed conversational topic of men's club snobs. An outpouring of oenophilic periodicals and books certainly helped the cause, but it was principally restaurateurs, once they discovered the promotional magic of "wine events," choreographed tastings, and special wine-and-food matched menus (often in the company of the featured wine's maker) who managed to transform a wine-shy dining public into a discerningly bibulous, and happier, one. In New York, probably the most influential wine-driven programs were the annual California wine barrel tastings staged in the Four Season's pool room with parade ground panache recalling the synchronized waiter drills of the grand hotel gong-beating days. For over a decade, fourteen wineries each year poured preview samplings in tandem with Chef Seppi Renggli's groundbreaking wine-matched menus, each of his dishes eerily prescient in forecasting food trends to come, including revamped American regional classics and Asian-European fusion cuisine.

The idea that chefs could be promoted like movie or rock stars would also have elicited disbelief on both sides of the ocean until Yanou Collart, a French film and celebrity publicist, took command of the international media destinies of the likes of Paul Bocuse and Roger Vergé in the '70s. Her launch of Parisian pastry titan Gaston LeNotre's restaurant in New York, complete with shoving paparazzi and an enormous dairy cow disrupting sidewalk traffic, is today remembered more vividly than the short-lived bistro itself. Collart would also be responsible for a truly epic milestone event, a three-day Monte Carlo celebration in 1990 of *New York Times* restaurant critic Craig Claibourne's seventieth birthday. This marked the first time a large delegation of America's finest chefs and restaurateurs was formally invited to join a principality-sized army of Michelin-starred European luminaries as professional confreres of equal standing. This coincided with a mounting ambition amongst young European chefs to migrate to the United States to ply their craft.

The New York restaurant scene of the Reagan-era 1980s grew more deafening by the year, thanks principally to a Wall Street boom that resonated around the world. The media declared a restaurant "Feeding Frenzy." Tourism thrived, fueled by real-estate-hungry Europeans and free-spending Japanese. The advent of the Manhattan Yuppie masses, often derided for their manic consumerism, but meanwhile possessed of bottomless "discretionary dollars," banished our earlier concern in 1973 that the new generation would show no interest in haute cuisine. To the contrary, they seemed interested in *any* cuisine, as long as it was packaged with sufficient style to conform with the contours of a fashionable trend.

A precipitous market plunge in late 1987 had doomsayers croaking that the end of good times had come to stay: An era of greed and excess, like the Gilded Age of the century before, had brought on its own demise. The '90s would be a chastened affair, they predicted, days of spaghetti and tap water. But not for long, as it transpired. Within a couple of years, New Yorkers were back at the restaurant races in even greater force than ever.

With something to offer practically any taste, New York restaurants had become essential to the factional forms of daily self-expression that the trend-fixated media now fingered as "lifestyles." Reaganite supply-side jargon helpfully defined a widespread urge "to upscale," and flush New Yorkers vied to put their mouths where their money was. An ideal supply-side day might start with "power breakfast" in a hotel dining room rich in chandeliers and influence barterers and end with grappa or ancient Armagnac and chocolate truffles at LeCirque, suavely seated by owner Sirio Maccione, late of The Colony, next to the odd ex-President, Secretary of State, or stiffly coiffed blonde whose dresses and amours were vigilantly chronicled in Suzy's vestigial society gossip column. That a restaurant was reputed to be the most expensive of its kind was no longer a drawback, particularly if it was the object of critical adulation, as in the case of Breton Chef Gilbert Lecoze and his sister Maguy, whose importation of their Paris Le Bernardin irrevocably altered the image and essence of a "fish house" with its iconoclastic recipes (tending to the rare or raw), ethereal dessert samplers, plutocratic wine list, huge marine oil paintings, extravagant flowers, and Maguy's haute couture wardrobe, also diligently scrutinized by the hovering fashion press.

Once, the discovery of a good ethnic restaurant had been cause for frugal celebration: An ethnic menu meant a cheap one. But with the advent of upscaling (a term Joe Baum would have demolished had it been applied a generation earlier to Fonda del Sol, his sublime vision of Mexico) came a rush of expensive and stagily redesigned global "dining experiences." And more continue to open than ever. On the other hand, as in the past, successive tides of immigrants continued to lap at New York's shores, discharging a mixed cargo of cuisines from India, Pakistan, the Caribbean, Central America, Southeast Asia, Russia, Korea, China, the Mideast, Poland, Afghanistan, et al. As usual, ethnic neighborhoods redolent of small mom-and-pop eateries, fragrant food shops, and open-air stalls steadily mushroomed. Post-modern chefs descended like culinary desperadoes, carting off booty to weave into their experimental cross-cultural recipes which would be lumped together as "fusion cuisines" by the '90s food press.

New York, in the post-'70s long run, has prevailed as epicenter of the transmutation of the restaurant industry into what, in food-service-speak's darkest hour, was baptized "eatertainment." A shining sea of spotless chef jackets had produced a second "Great White Way." Counting fin de siecle restaurant affinities with big-time Broadway theater biz is irresistible. As always, tracing back to Niblo and beyond, there are the vivid impresarios, New York hybrids of Hannibal and Ziegfeld. Such as Joe Baum, who, after inadvertently siring the "theme restaurant" in his Restaurant Associates halcyon days, has leapt from skyscraper pinnacle to pinnacle, from Windows on the World to the reborn Rainbow Room and back again, just this year blurring his smoke and mirrors with those of David Copperfield at a spectacular Broadway-area theme restaurant bearing the magician's name. Such as Warner LeRoy, who, when not adding literally millions of lights to the galactic glitter of Tavern on the Green, tends to a stack of smoking irons on his development company's fires. As to the introduction of international star power to gastronomic marquees, the initial visitations of Bocuse and Vergé earned them the sort of guest artiste status, at least in epicurean circles, bestowed on Jenny Lind and Fanny Kemble before the Civil War. It could be said that a coveted copper mold had been cast. From then on, New York was the stage on which out-of-town chefs, celebrated or striving, from whatever corner of the country or world, wished to perform, if only for a day. They alighted like flocks of white gulls, bright eyes fixed on the New York media banquet. The fall and spring seasons, in particular, saw their names and faces on piles of culinary playbills. Charities showcased their talents to draw full houses: the March of Dimes' Gourmet Galas staged in grand hotel ballrooms, dinner dance recipe contests at first limited to amateur celebrity cooks; Rockefeller Center food and wine station fund-raisers to feed elderly shut-ins, orchestrated by Meals-on-Wheels, the valiant organization founded by restaurant critic Gael Greene and

American food guru James Beard; Share Our Strength, and many, many more.

After Beard's death in 1985, Julia Child led a mission to buy his Greenwich Village private townhouse-cum-cooking school and turn it into a non-profit educational foundation and "little Carnegie Hall" for performing chefs. Under the baton of cooking school proprietor and Beard Foundation president Peter Kump came the premier of the national food world's answer to the Broadway "Tony" trophy, the annual James Beard Awards, telecast countrywide on the new New York-headquartered TV Food Network, itself yet another powerful local lure for exposure-seeking chefs and restaurateurs from every state in the union.

Also analogous with Broadway has been the rise in public awareness of marquee-value designer/architects such as Adam Tihany, David Rockwell, and Hugh Hardy, masters of dramatic visual atmosphere. As a revivifier of faded historic properties, Hardy has bridged the restaurant-theater gap by bringing intensified new life to the Rainbow Room, Windows on the World, the new restaurant on the Bryant Park backside of the 42nd Street library, and, in concert with the "family values" sanitization of the titillatingly lurid Times Square area, the renaissance of the glorious New Victory and New Amsterdam Theatres, the last the dilapidated realm of Flo Ziegfeld salvaged by the forces of Disney.

What else has been preserved? What has perished? Many of the old ethnic enclaves are gone or going, their inhabitants' offspring preferring to row onto the real estate mainstream. One by one, the old German restaurants and food emporia of Yorkville—the Jaeger House, the Bremen House, the Kleine Konditorei—have closed their premises, usurped by high-rise apartments. When we last looked, however, Kramer's Bakery remained to supply almond-scented stollen and the laboriously engineered baumkuchen, while the Hansel-and-Gretel-windowed Elk Candy Shop, relocated on Second Avenue near Schaller and Weber's pungent sausage-strung pork butcher shop, once again dispenses gingerbread houses and good luck marzipan pigs with golden rings in their snouts.

Little Italy continues to loop its streets with festival lights, although its boundaries have shrunk with the brisk expansion of Chinatown. Among its more endearing old haunts is the double-decker flagship Ferrara pastry shop and café, humming arias to itself since 1892. Sidling into the West Village, one finds Grand

Ticino, the ristorante that found Italo-American immortality in Hollywood's *Moonstruck*, as well as the godfather of New York expresso lairs, the richly patinated Caffe Reggio, which opened, fittingly, at the height of the Roaring Twenties. For the succor of nostalgic clientele, Little Italy, from stem to stern, is still heroically awash in oregano-doused southern Italian "red sauces," oblivious to the fashionable, i.e. northern, wild mushroom risottos, truffled game and polenta dishes served uptown.

Stacked to the eaves with new immigrants legal and illegal, Chinatown, still the most exciting place to sample authentic mainland cuisines, is literally a movable feast: Chefs here are notoriously fickle, evaporating with their kitchen warrior cleavers to work for the competition on a moment's notice; connoisseurs must regularly trade insider information to keep up with change of casts.

Certain cherished relics and hallowed halls still persist, at least in the physical sense, and can be visited on variably satisfactory field trips into the city's past. Zagat's 1998 New York City Restaurant Survey, compiled from readers' reports, lists no less than eighty-six "Old New York" establishments, from grandiose to grungy, dating from 1716 to 1945. Fraunces Tavern, frequently reconstructed in the wake of everything from early fires to 1975 terrorist bombs, still retains elements of the original setup in which Black Sam served Washington and his pantheon of national forebears. Keen's Chop House, or Steakhouse as it is now designated, still exhibits its vast collection of long-stemmed clay pipes, the sort that nineteenth-century regulars used to puff on after cutting into the house's famous mutton chops. McSorley's Old Ale House since 1854 has been a sudsy retreat for dedicated schooner hoisters.

Just this year, one of Delmonico's grander incarnations was refurbished by a Milanese restaurant entrepreneur. And the cacophonous, white-tiled Oyster Bar in Grand Central Station, recently ravaged by fire, has once more opened its doors for the sampling of their staggering lists of net-to-jet global seafood and excellent wines. The legendary backroom bar at the "21" club, under new ownership, has just curved its way into the front lounge, spreading cocktail joy practically to the street; the citywide appetite for more worldly and varied bar food is being stoked by postmodern artistry in the low-ceilinged kitchen that once serviced Jack and Charlie's speakeasy crowd.

To please traditionalists, the dwindling ancestral signature dishes of the city's old houses on today's menus are often set apart in preservationist boxes. Either that, or starred in asterisks to indicate their longevity. "21"'s legendary hamburger may have undergone more facelifts than an aging Hollywood idol, but their mild chicken hash, given a fresh presentation in ornate silver chafing dishes (shades of Delmonico's lobster Wenburg-turned-Newberg epoch) still soothes the faithful. Necks crane, as ever, to scan the indelibly painted blackboard menus at P. J. Clarke's and Billy's almost floor-to-ceiling lineup, a '40s time capsule starting with fresh fruit cocktail and French onion soup and progressing through broiled bay scallops or flounder, lobster tails, boneless shell steak, linguine with marinara sauce, and nutmeggy rice pudding. The Sunday Night special chalked in the window is "baked lamb" with mint jelly, mashed potatoes, and green peas. To complete the timewarp picture, Portuguese rosé is featured front and center on the short wine list illustrated with labels in plastic sleeves.

Barbetta's baroque menu card, like an art historian's monograph, states the precise year of each dish's admittance to the collection. And the Rainbow Room continues to pack them in, not only with live dance bands and the famously revolving dance floor, but dazzling retreads of retro pleasures like bird-bath-sized deco cocktails and cumulous Baked Alaskas, along with the latest in contemporary couture dishes, whenever possible composed of fresh, devotedly gathered New York State products, in tune with the chef-invented New American Regionalism times.

The city itself had undergone a regional reidentification of its own in many people's minds: Whereas in the past the Upper East and West Sides had symbolized a geographic great divide between conservative and progressive attitudes, 14th Street now became the line of demarcation between the old Uptown Establishment in business suits and Downtown New Bohemia, armies of the after-midnight funky "club scene" cultishly dressed in black.

Modern "Downtown" dining would be elevated to star-winning stature by two sets of American pioneers, Chanterelle chef David Waltuck and his wife Karen and Montrachet restaurateur Drew Nieporent, who within a few years would demonstrate Joe Baum-like visionary impresario talents, including juggling a number of his own differently themed operations at a time. Nieporent's first chef, David Bouley, would also ascend to the critical firmaments, striking out to open his own restaurants and, hopefully this year, an ambitious complex corralling retail foods, research facilities, cooking classes, and a couple of dining options on a single block on Duane Street.

Once established, the new age of the celebrity chef led inexorably to the chef as competitive entrepreneur, ambitious to build proprietary mini-empires of diversified food ventures. At the head of the New York pack today are the Americans Larry Forgione, Charley Palmer, Matthew Kenney, and Bobby Flay; French born Daniel Boulud and Jean-Georges Vongerichten; and the northeastern Italian Lidia Bastianich. Similarly, the marauding growth of national and regional hotel and fast-food chains, on an obviously reduced scale, suggested a profitable business format for restaurateurs and their increasingly eager investors to explore. The result has been a new phenomenon, the "independent restaurant group" exemplified by Nieporent, Danny Meyer, Tony May, Michael Weinstein, Steve Hanson, the Santos family, and Alan Stillman, whose Manhattan principality of scintillating steak, seafood, and contemporary Americana houses now rules satellite fiefdoms as far afield as Chicago and Miami. Not uncommonly, the group restaurateur came armed with a business background, MBA, or accountancy diploma, as well as a shrewd appreciation for high powered—and high priced—marketing and public relations consultants. New York was now not only the financial capital of the world, but its restaurant business hub. Indeed, so many ambitious restaurants continued to tumble over each other to open that in 1997, guidebook publisher and critic Peter Meltzer was obliged to take Mrs. Astor's cue and retitle his annual overview *Passport to New York, the 400 Restaurants that Matter Most*.

Gross profits jumped. Eatertainment, indeed: If E.B. White had complained about the lengthening lines at Schrafft's because too many New Yorkers were going out to eat, what would he have said of having to make a reservation months in advance, as if for a Broadway hit, when a place was known to be "hot"?

And what would Baron de Groot make of modern New Yorkers' lascivious pawing of the new, their lust to embrace the latest anointed trend? Ironically, few of them try to distinguish between the truly new and replays, often unintentional, of the old. The French presence, for example; even the elderly think it all began with Henri Soulé after the 1939 World's Fair, when history states that antecedents of the young Gallic eminences of the 1990s, say Jean-Georges Vongerichten or Daniel Boulud and pastry chef François Payard, whose recently opened pâtisserie-glacerie bistro is cheered as the first of its kind in New York, are in fact Delacroix and Corré, transplants of the French Revolution. That the four-star replacement of moribund Continental hotel cuisine by inspired European chefs at the Waldorf-Astoria, Essex House, and St. Regis echoes the glory days of the city's nineteenth-century seminal grand hotels. That the current steak house renaissance, promising the finest available cuts, traces back to butcher Henry Astor and his Boston Post Road forays. That trendy champagne bars simulate the days of the debonair Champagne Charlies, or fine wines-by-glass programs the Germanic weinstube beloved by the vanished Academy of Music crowd. That the extravagance of the Bradley-Martin ball has become almost a commonplace, with half-million dollar weddings not unknown to the mirrored walls of the Plaza Hotel. Or that calorie-counting "spa menus" still resonate with Dr. Graham's breed of watchdog barking, i.e. "There's death in the pot!" and "Pies kill!"

Certainly there is much that is new to celebrate. The Italianization of New York, for example, from Madison Avenue boutiques to the countless menus, on whatever restaurant rung, offering everything from omnipresent pizza to the most esoteric regional *cucina*. The extent to which Italy has made incursions on local die-hard Anglo-Saxon menus can be savored on a December 30, 1997 Fraunces Tavern Restaurant specimen, offering (along with a "priced-fixed colonial repast" of Mrs. Glasse's 1789 recipe for New England clam chowder, baked chicken à la Washington with root vegetables and savory crust, and Hannah Davis' apple crunch, "original family recipe") a choice of melon and prosciutto, handmade mozzarella and tomatoes with fresh basil pesto vinaigrette, penne with shrimp, marinara sauce and parmigiana, smoked chicken ravioli, and chicken marsala. Special thanks should go to Naples-born group restaurateur Tony May, whose "Italian Fortnight" staged at the Rainbow Room in 1971, featuring imported guest chefs, foods, wines, and musicians, set into motion an unbroken stream of similar New York epicurean events promoting the kitchens and cellars of too many nations and American states to count.

Also new is New York's coming of age as an education mecca for the study of culinary, gastronomic, and nutritional matters, on professional and post-graduate levels. The French Culinary Institute, the most prestigious, with a faculty headed by Jacques Pépin, Alain Saillac, André Soltner, and Jacque Torres, four of the most prominent French chefs in the country, coincidentally resides at the corner of Broadway and Grand, epicenter of the city's French quarter in the second half of the last century. And only a few years ago, New York University announced the formation of a department of Nutrition and Food Studies not only for undergraduates, but for scholars pursuing master's and doctoral degrees, that promises to set a new curriculum standard for the times. And, as mentioned, the James Beard Foundation on West 12th Street has promoted itself to the position of supreme arbiter in bestowing awards on the American restaurant universe.

What is dead and gone forever? Much of the old Greenwich Village scene. A lot of Little Italy. Almost all of Yorkville. The Tenderloin, as little mourned as spittoons. A forlorn directory of shuttered institutions, from Luchow's to Schrafft's. But, looking at it from the bright side, the death that has New Yorkers dancing on a mass grave is that of the captive audience. Now, wherever we seem to prowl around the city—corporate dining room, art museum, department store, let alone a restaurant, hotel dining room, tea parlor, café, bistro, trattoria, or bar—one finds it increasingly difficult to find a boring or ineptly cooked meal. James Fenimore Cooper's *Traveling Bachelor* had hit the nail squarely on the head.

Bibliography

———————⚬⚬⚬———————

The Age of Elegance, by Arthur Bryant. Harper Brothers, New York, 1950.

The Age of Paranoia, by the Editors of Rolling Stone. Pocket Books (Div. of Simon and Schuster, Inc.), New York, 1972.

The Algonquin Wits, edited by Robert E. Drennan. The Citadel Press, New York, 1968.

American Cookery, by Amelia Simmons. Simeon Butler, Northampton, Hartford, 1798.

The American Heritage Cookbook. American Heritage Publishing Company, New York, 1964.

The American Hotel, by Jefferson Williamson. Alfred Knopf, New York, 1930.

American Notes, by Charles Dickens. Chapman and Hall, London, 1850.

The American Scholar Reader, edited by Hiram Haydn and Betsy Saunders. Atheneum Publishers, New York, 1960.

The Americans, a Social History of the United States, 1587–1914, by J. C. Furnas. G. P. Putnam's Sons, New York, 1969.

Around the World in New York, by Konrad Bercovic. The Century Company, New York, 1924.

The Battery, by Rodman Gilder. Houghton Mifflin Co., Boston, 1936.

Be My Guest, by Conrad N. Hilton. Prentice-Hall, Inc., Englewood Cliffs, N.J., 1957.

The Beautiful People, by Marilyn Bender. Coward-McCann, Inc., New York, 1967.

Belshazzar Court, or Village Life in New York City, by Simeon Strunsky. Henry Holt and Co., New York, 1914.

The Big Spenders, by Lucius Beebe. Doubleday, New York, 1966.

Bite, by Gael Green. W. W. Norton & Co., Inc., New York, 1971.

The Bon Vivant's Companion, or How to Mix Drinks. Alfred A. Knopf, Inc., New York, 1927–28.

Book of Recipes, Daughters of the American Revolution. Ellen Harden Walworth Chapter, New York, 1925.

The Bowery, by Edward Ringwood Hewitt and Mary Ashley Hewitt. Putnam, New York, 1897.

The Bowery, Saturday Night, by W. O. Stoddard.

Brendan Behan's New York, by Brendan Behan. Bernard Geis Assoc., New York, 1964.

Brownstone Fronts and Saratoga Trunks, by Henry Collins Brown. Dutton, New York, 1935.

Carroll's New York City Directory. Carroll and Co., New York, 1859.

The Century Cookbook, by Mary Ronalds. The Century Co., New York, 1899.

Charles Dickens in America, edited by W. G. Wilkins. Charles Scribner's Sons, New York, 1911.

Chinatown, U.S.A., by Calvin Lee. Doubleday and Co., New York, 1965.

The Colony, by Iles Brody. Greenberg, New York, 1945.

Crusades and Crinolines, by Ishbel Ross. Harper and Row, New York, 1963.

Cue's New York, by Emory Lewis. Duell, Sloan and Pearce, New York, 1963.

The Delectable Past, by Esther B. Aresty. Simon and Schuster, New York, 1964.

Delmonico's, a Story of Old New York, by Henry Collins Brown. Valentines's Manual, New York, 1928.

The Diary of Philip Hone, 1828–1851, edited by Bayard Tuckerman. Dodd, Mead and Co., New York, 1910.

Dining in New York with Rector, by George Rector. Prentice-Hall, Inc., New York, 1939.

Dining, Wining and Dancing in New York, by Scudder Middleton. Dodd Publishing Co., New York, 1938.

Domestic Manners of the Americans, by Frances Trollope, edited by Donald Smalley. Alfred A. Knopf, New York, 1949.

Do Not Disturb, by Frank Case. Frederick A. Stokes Co., New York, 1940.

The Epicurean, by Charles Ranhofer, New York, 1903.

Fanny Kemble, A Pasionate Victorian, by Margaret Armstrong. Macmillan, New York, 1938.

Feeding the Lions, an Algonquin Cookbook, by Frank Case. The Greystone Press, New York, 1942.

50 Soups, by Thomas J. Murrey. White, Stokes and Allen, New York, 1884.

The Flavor of the Past, compiled and edited by Leland D. Baldwin. The American Book Company, New York, 1969.

The Ford Treasury of Favorite Recipes, by Nancy Kennedy. Golden Press, New York, 1959.

The Gangs of New York, by Herbert Asbury. Alfred A. Knopf, New York, 1927.

Good-bye, Union Square, by Albert Halper. Quadrangle Books, Chicago, 1970.

Greenwich Village, by Anna Alice Chapin. Dodd, Mead and Co., New York, 1917.

Harlem: Negro Metropolis, by Claude McKay. E. P. Dutton and Co., New York, 1940.

Here is New York, by E. B. White. Harper and Brothers, Publishers, New York, 1949.

Historic Hotels of the World, by Robert B. Ludy. David McKay, Philadelphia, 1927.

A History of Old New York Life and the House of Delmonico's, by Leopold Rimmer. Private Printing, 1898.

How to Do It, or The Lively Art of Entertaining, by Elsa Maxwell. Little, Brown and Co., Boston, 1957.

An Illustrated History of French Cuisine, by Christian Guy, translated by Elisabeth Abbott. Bramhill House, New York, 1962.

Incredible New York, by Lloyd Morris. Random House, New York, 1951.

The Iron Gate. Published by the Jack Kreindler Memorial Fund, Inc., New York, 1940.

The Jazz World, by Don Ceruli, Burt Korall and Mort Nasatir. Ballantine Books, Inc., New York, 1960.

Journal, by Frances Anne Kemble (Mrs. Pierce Butler). Two vols., Philadelphia, 1835.

Letters from New York, by L. Maria Child. Second series, C. S. Francis and Co., Boston, 1845.

Lewis Bridge, A Broadway Idyl, by Mary Eliza Tucker. M. Dolady, New York, 1867.

Lights and Shadows of New York Life, by James D. McCabe. The National Publishing Co., Philadelphia, 1879.

The Liveliest Art, by Arthur Knight. The Macmillan Co., New York, 1957.

Luchow's German Cookbook, by Jan Mitchell. Doubleday and Co., Garden City, New York, 1952.

The Lucius Beebe Reader, edited by Charles Clegg and Duncan Emrich. Doubleday and Co., Garden City, N.Y., 1967.

McSorley's Wonderful Saloon, by Joseph Mitchell. Grosset and Dunlap, New York.

Memories of Colonial Benjamin Talmadge, Continental Light Dragoons, 1776–83. Gillis Press, New York, 1904.

Meyer Berger's New York, by Meyer Berger. Random House, New York, 1953.

Mirror for Gotham, by Bayard Still. New York University Press, New York, 1956.

The Museum, by Leo Lerman. The Viking Press, New York, 1969.

New Travels in the U.S.A. Performed in 1788, by J.P. Brissot de Warville. Berry and Rogers, New York, 1792.

New York, by Paul Morande, translated from the French by Hamish Miles. H. Holt and Co., New York, 1930.

New York, a Serendipiter's Journey, by Gay Talese. Harper and Row Publishers, New York, 1961.

New York by Gas Light, by G. G. Foster. Rewitt and Davenport, 1850.

New York City and State in the Olden Time, by John F. Watson. Henry F. Anners, Philadelphia, 1846.

New York Cookbook, by Maria Lo Pinto. A. A. Wyn, Inc., New York, 1952.

New York in Slices by an Experienced Carver, by George G. Foster. W. F. Burgess, New York, 1848.

New York, New York, by Stuart Hawkins. Wilfred Funk, Inc., New York, 1957.

The New York of Hocquet Caritat and His Associates, 1797–1817, by George Gates Raddin, Jr. The Dover Advance Press, 1953.

New York, Places and Pleasures, by Kate Simon. World Publishing Co., Cleveland and New York, 1964.

New York, Places and Pleasures, by Kate Simon. Harper and Row Publishers, New York, 1971.

New York Society on Parade, by Ralph Pulitzer. Harper and Bros., New York, 1910.

Night Clubs, by Jimmy Durante and Jack Kofoed. Alfred A. Knopf, New York, 1931.

Nightlife, Vanity Fair's Intimate Guide to New York After Dark, by Charles G. Shaw. John Day Co., New York, 1931.

The Night Side of New York, by Members of the New York Press. Excelsior Publishing House, New York, 1866.

No Cover Charge, by Robert Sylvester. The Dial Press, New York, 1956.

Now I'll Tell, by Carolyn Rothstein. The Vanguard Press, New York, 1934.

Old Bowery Days, by Alvin F. Harlow. D. Appleton and Co., New York, 1931.

Old New York: Or Reminiscences of the Past Sixty Years, by John W. Frances, M.D., LL.D. W. J. Widdleton, Publisher, New York, 1866.

Old Taverns of New York, by W. Harrison Byles. Frank Allaben Genealogical Company, 1915.

The Old-Time Saloon, by George Ade. Ray Lord and Richard R. Smith, Inc., New York, 1931.

The Old World and the New, by Frances Trollope.

Once Upon a City, by Grace M. Meyer. Macmillan, New York, 1958.

Only Yesterday, by Frederick Lewis Allen. Harper and Brothers, New York, 1931.

Oscar Wilde Discovers America, by Lloyd Lewis and Henry Justin Smith. Harcourt, Brace and Co., New York, 1836.

Palaces of the People, by Arthur White. Taplinger Publishing Co., New York, 1970.

Peacock Alley, by James Remington McCarthy. Harper Bros., New York, 1931.

Peacocks On Parade, by Albert Stevens. Crocket Sears Publishing Co., Inc., New York, 1931.

The Physiology of New York Boarding-Houses, by Thomas Butler Gunn. Mason Brothers, New York, 1857.

Physiologie du Gout, by Anthelme Brillat-Savarin. Paris, 1842.

The Plaza, 1907–67, by Eve Brown. Duell, Sloan and Pearce, New York, 1967.

Poems, by Philip Freneau. New York, 1786.

The Polite Americans, by Gerald Carson. Macmillan and Company, New York, 1966.

The Private Affairs of George Washington, by Stephen Decatur, Jr. Houghton Mifflin, Boston, 1953.

Prohibition: Era of Excess, by Andrew Sinclair. Little, Brown and Co., Boston, 1962.

The Pump House Gang, by Tom Wolfe. Farrar, Straus and Giroux, New York, 1968.

Quaint Customs of Former New Yorkers, by Katherine Lydig Brady, University Press, Cambridge, 1915.

Queen of the Plaza, a Biography of Adah Isaacs Menken, by Paul Lewis. Funk and Wagnalls, Co., New York.

The W. Johnson Quinn Hotel Collection, in the Library of the New York Historical Society. The primary source material in the vast Quinn collection is organized chronologically.

R.S.V.P., Elsa Maxwell's Own Story, by Elsa Maxwell. Little, Brown & Co., Boston, 1954.

The Rag-Bag, a Collection of Ephemera, by N. Parker Willis. Charles Scribner, New York, 1855.

The Real New York, by Helen Worden. Bobbs-Merrill, Indianapolis, 1932.

Rector's Naughty 90s0's Cookbook, by Alexander Kirkland. Doubleday and Co., Inc., New York, 1949.

Reminiscences of an Octogenarian of the City of New York, by Charles H. Haswell. Harper and Brothers, Publishers, New York, 1896.

The Restaurants of New York, by George S. Chappell. Greenberg, Publisher, Inc., New York, 1925.

Retrospections of America, 1797–1811, by John Bernard, edited by Mrs. Bayle Bernard, Harper Brothers, N.Y., 1887.

The Romantic Rebels, by Emily Hahn, Houghton Mifflin, Boston, 1967.

Salmagundi, by Washington Irving. Two vols., D. Longworth, New York, 1808.

Sardi's, the Story of a Famous Restaurant, by Vincent Sardi, Sr. and Richard Gehman. Henry Holt and Co., New York, 1925.

Since Yesterday, by Frederick Lewis Allen. Harper and Brothers, New York, 1940.

Sketches of America, by Henry B. Fearon. London, 1819.

The Story of Louis Sherry, by Edward Hungerford. William Edwin Rudge, New York, 1929.

The Story of New York, by Susan E. Lyman. Crown Publishers, New York, 1964.

The Story of Sylvie and Bruno, by Lewis Carroll. Macmillan and Co., Ltd., London, 1922.

The Streets of Old New York, by J. Ernest Brierly. Hastings House, New York, 1953.

Third Avenue, New York, by John McNulty. Little, Brown and Co., Boston, 1946.

The Thirties, a Time to Remember, edited by Don Congdon. Simon and Schuster, New York, 1962.

Thomas Jefferson's Cook Book, By Marie Kimball. Garett and Massie, Richmond, 1949.

Those Were the Good Old Days, by Edgar R. Jones. Simon and Schuster, New York, 1959.

A Tour in the United States, by Archibald Prentice. Seventh ed., London, 1850.

The Traveling Bachelor; or, Notions of the Americans, by James Fenimore Cooper. W. A. Townsend and Co., New York, 1859.

Travels Through Canada and the United States of North America, in the Years 1806, 1807, and 1808, by John Lambert. Two vols., London, 1814.

Welcome to Our City, by Julian Street. John Lane Company, New York, 1912.

Where to Dine in Thirty-Nine, by Diana Ashley. Crown Publishers, New York, 1939.

Who Killed Society, by Cleveland Amory. Harper and Brothers, New York, 1960.

The Women of New York, by George Ellington. New York Book Co., New York, 1870.

A. Woollcott, by Samuel Hopkins Adams. Reynal and Hitchcock, New York, 1945.

The Year of the Century: 1876, by Dee Brown. Charles Scribner's Sons, New York, 1966.

References

1. Arthur Bryant, *The Age of Elegance,* Harper Brothers, New York, 1950.
2. Cleveland Amory, *Who Killed Society?,* Harper Brothers, New York, 1960.
3. Elsa Maxwell, *R.S.V.P., Elsa Maxwell's Own Story,* Little, Brown and Company, Boston, 1954.
4. F. Scott and Zelda Fitzgerald, "A Millionaire's Girl," first published in the *Saturday Evening Post,* May 17, 1930. Reprinted by permission of Harold Ober Associates Incorporated. Copyright 1930 The Curtis Publishing Company. Copyright renewed 1957 by Frances Scott Fitzgerald Lanahan.
5. Ring Lardner, *Say It with Oil,* Doubleday and Company. Copyright 1923 by G.H. Doran Co.
6. Quoted in *The Jazz World,* edited by Dom Cerulli, Burt Dorall, and Mort Nasatir, Ballantine Books, New York, 1960.
7. *The Glass of Fashion,* copyright 1954 by Cecil Beaton. Reprinted by permission of Doubleday and Company, Inc.
8. *The Lucius Beebe Reader,* edited by Charles Clegg and Duncan Emrich, Doubleday and Company, Garden City, New York, 1967.
9. Marshall Sterns, *The Story of Jazz,* Oxford University Press, Inc., New York, 1956.
10. Albert Halper, *Goodby, Union Square,* Chicago, Quadrangle Books, 1970.
11. From the book *Be My Guest* by Conrad Hilton. © 1957 by Prentice-Hall, Inc., Englewood Cliffs, N.J. and used with their permission.
12. John McNulty, *Third Avenue, New York,* Little, Brown and Company, 1946.
13. E. B. White, *Here Is New York,* Harper Brothers, 1949.
14. From *Brendan Behan's New York,* copyright © 1964 by Brendan Behan with Paul Hogarth. By permission of Bernard Geis Associates, publishers.
15. Gay Talese. First Published in *Esquire* magazine.
16. Kate Simon, *New York Places and Pleasures,* 3rd edition, Meridian Books, World Publishers.

17. Gael Green, *Bite*, W. W. Norton and Company, New York, 1972.
18. James Beard, *American Cookery*, Little, Brown and Co., 1972.
19. Reprinted by permission of *Esquire* magazine, © 1972 by Esquire, Inc.
20. Marylin Bender, *The Beautiful People*, Coward McCann Geoghegan, Inc., New York, 1967.
21. Barbara Rose, first printed in *New York* magazine, May 31, 1971.
22. First printed in *Vintage* magazine, June 1972.
23. First printed in *Newsweek*, October 9, 1972.

Index